AF323680

Tributes to Yuan-Cheng Fung on His 90th Birthday

Biomechanics: From Molecules to Man

TRIBUTES TO YUAN-CHENG FUNG ON HIS 90TH BIRTHDAY

Biomechanics: From Molecules to Man

Editors

Shu Chien

Peter C-Y Chen

Geert W Schmid-Schönbein

Pin Tong

University of California, San Diego, USA

Savio L-Y Woo

University of Pittsburgh, USA

NEW JERSEY · LONDON · SINGAPORE · BEIJING · SHANGHAI · HONG KONG · TAIPEI · CHENNAI

Published by

World Scientific Publishing Co. Pte. Ltd.

5 Toh Tuck Link, Singapore 596224

USA office: 27 Warren Street, Suite 401-402, Hackensack, NJ 07601

UK office: 57 Shelton Street, Covent Garden, London WC2H 9HE

British Library Cataloguing-in-Publication Data
A catalogue record for this book is available from the British Library.

TRIBUTES TO YUAN-CHENG FUNG ON HIS 90TH BIRTHDAY
Biomechanics: From Molecules to Man

ISBN-13 978-981-4289-87-0
ISBN-10 981-4289-87-6

Printed in Singapore by Mainland Press Pte Ltd

CONTENTS

PREFACE

Professor Yuen-Cheng (Bert) Fung had his 90th Birthday, on September 14, 2008 according to the Chinese calendar. His family and friends gathered in San Diego for a joyful celebration of this marvelous occasion. The program started with a luncheon in the Engineering Courtyard in the University of California, San Diego, during which the participants from far and near were able to get together for a happy reunion, particularly with Dr. and Mrs. Fung. The *International Symposium on Genomic Biomechanics: Frontier of the 21st Century* was held, most appropriately, in the elegant Y.C. Fung Auditorium of the Powell-Focht Bioengineering Hall. Following the welcome remarks by Geert Schmid-Schönbein and Pin Tong, there were excellent presentations on topics related to biomechanics, the field created by Dr. Fung. A group picture was taken during the intermission. Following this most successful symposium, everyone gathered in the Jasmine Seafood Restaurant in San Diego for a wonderful banquet that was delicious and enjoyable, with Savio Woo giving a heart-felt speech on Dr. Fung, his respected mentor and cherished friend; everyone could resonate with that.

We wish to thank Dr. K.K. Phua of the World Scientific Publishing Co., Ltd. (WSPC) in Singapore for making the wonderful suggestion of publishing a book to commemorate this joyful event. WSPC has published several of Dr. Fung's books, including the two volumes of Dr. Fung's *Selected Works on Biomechanics & Aeroelasticity* (close to 2000 pages). We wish to express our appreciation to all our colleagues who have contributed to this book, which will be a most fitting gift to Dr. Fung on September 15, 2009, his 90th Birthday, according to the Western Calendar. With this book, all of Dr. Fung's friends and students wish to send their warmest wishes to him and Mrs. Fung for a Wonderful 90th Birthday. We all look forward to the celebration of their 100th Birthday!

Chapter 1

PHYSICAL MECHANISMS OF SOFT TISSUES RHEOLOGICAL PROPERTIES

YORAM LANIR

Faculty of Biomedical Engineering, Technion, Haifa 32000, Israel

Soft tissues rheology is determined by their internal structure and by their constituents' properties and mutual interactions. Specifics of these relationships are analyzed in terms of four constitutive properties: 1) The tissues' non-linear stress-strain relationship is consistent with their collagen fibers non-uniform undulation and gradual straightening with stretch. Response anisotropy is attributed to the fibers non-uniform orientation distribution. 2) The fibers gradual recruitment is also consistent with the tissues' viscoelastic non-linearity. It is shown that under protocols where no fibers buckle (e.g., stress relaxation and creep tests) the fibers recruitment process is compatible with the quasi-linear viscoelastic theory. 3) Preconditioning adaptation of tissues to its loading is an essential response feature, induced by the preconditioning properties of the fibers. The latter are both strain and time dependent. Excellent fit to data of multiple uniaxial (tendon) and biaxial (skin) data is obtained only if preconditioning is incorporated into the constitutive formulation. 4) Residual stress in unloaded state stems from three levels of interactions between the tissues' constituents (the micro, meso and macro levels, respectively), which must all be relieved if a true stress-free reference is desired. In summary, modeling based on structural consideration provides mechanistic insights and facilitates reliable constitutive formulation.

1. Introduction

Y.C. Fung established the foundations for studies of soft tissues mechanics based on the concepts of quasi-linear viscoelasticity, preconditioning and residual stress. In parallel, attempts have been made by Fung and others to model these response features and establish their validity. Here we analyze the mechanistic origins of these rheological characteristic based on the tissue structure, the constituents' properties and their mutual interactions, and show how these considerations facilitate reliable representation of the tissue constitutive properties.

Y. Lanir

2. Analysis and Results

2.1. *Tissues Nonlinear Stress-Strain Relationship*

The tissues uni-axial stress-strain response is convexly non-linear – the tissue becomes stiffer with increasing strain. Fung [1] proposed an exponential type uniaxial stress-strain relationship which was later generalized to multi-dimensions [2] and applied by Fung and others to the aorta [3], arteries [4], the lung tissue [5], the skin [6] and to other tissues.

Fibers recruitment in the tendon: A first clue to the possible origin of tissues' non-linearity was provided by Viidik [7, 8]. He observed in tendon that while at rest all collagen fibers are undulated, upon stretch there is a process of gradual fibers recruitment during the non-linear "toe" region. This is followed by a linear response when all fibers are straight. A similar process was observed under biaxial stretch of the mesentery [9]. Hence the stress-strain non-linearity may stem from gradual recruitment of fibers: with increasing stretch, more fibers become active (stretched), thus increasing the tissue stiffness. Based on these experimental observations, a mathematical framework was developed for incorporating the intrinsic properties of the fibers and their waviness distribution into constitutive laws [10-13].

For "elastic" fibers, the strain-energy function w_f of an individual fiber is a function of its uniaxial strain e_f. Based on Vidiik and others it is assumed that the long and thin fibers possess only axial tensile stiffness and their compressive and bending stiffness are negligible. Hence the fiber transmits only axial force and only when stretched. Its second Piola-Kirchoff axial stress is given by $s_f = \partial w_f(e_f)/\partial e_f$, with $s_f(e_f \leq 0) = 0$. Since the uniaxial response of the tendon is linear when all fibers are stretched, then this suggests that the intrinsic fiber stress-strain law $s_f(e_f)$ is also linear, or very close to it. If the fiber is wavy at rest, then the true fiber's strain is related to the global finite Lagrangian strain e by [13]:

$$e_f(e,e_s) = (e - e_s)/(1 + 2 \cdot e_s) \tag{1}$$

where e_s is the fiber straightening strain. The uniaxial strain-energy function of the entire population of parallel and non-uniformly undulared collagen fiber-bundle W_b is equal to the sum w_f of all fibers:

$$W_b(e) = \int_0^e \hat{D}(x) \cdot w_f[e_f(e,x)] \cdot dx, \quad w_f(e_f \leq 0) = 0 \tag{2}$$

where $\hat{D}(x)$ is the waviness density distribution such that the fraction of fibers becoming straight between the strain levels x and $x+dx$ equals $\hat{D}(x)\cdot dx$. Physically, the straightening strain x represents the fiber's stress-free gage-length. From the theory of hyper-elasticity, the second Piola-Kirchoff uniaxial stress of the entire fiber bundle is given by:

$$S_b^e(e) = \partial W_b(e)/\partial e = \int_0^e \hat{D}(x)\cdot \frac{\partial w_f(e_f)}{\partial e_f}\cdot \frac{\partial e_f}{\partial e}\cdot dx \quad w_f(e_f \le 0) = 0 . \quad (3)$$

Using Eq. 1 and substituting the stress s_f for $\partial w_f(e_f)/\partial e_f$ in Eq. 3 one gets:

$$S_b^e(e) = \int_0^e D(x)\cdot s_f(e_f)\cdot dx \quad s_f(e_f \le 0) = 0 \quad (4)$$

where $D(x) = \hat{D}(x)/(1+2x)$ is the modified waviness distribution function. Eq. 4 is the uniaxial elastic stress-strain law for the gradually recruiting fiber bundle. Importantly, although the stress-strain law $s_f(e_f)$ of the individual collagen fiber is linear, the response of the fiber bundle is convexly non-linear due to the gradual recruitment of fibers.

Generalization to 3D tissues: Unlike the tendon, most tissues consist of multi-dimensional networks of different types of fibers (e.g., collagen, elastin). Extension of the above structural approach to the general three dimensional case is straightforward by incorporating into the model the fiber orientation distribution [11, 13]. Tissue anisotropy is induced by non-uniform orientation distribution of its fibers.

Models based on these concepts have been developed for the tendon [14], the aortic valves [15, 16], the pericardium [17], the skin [11, 18, 19] the myocardium [20, 21] and blood vessels [22, 23].

Other structural models: A different class of structure-based models considers the collagen fiber as a rod whose non-linear uniaxial stress-strain response stems from its bending (or bending and twisting) during its gradual flattening with stretch, usually following Euler's elastica theory. The shapes considered were planar zigzag [24, 25], planar wavy [26] and helical [27, 28]. In reality however, a rod cannot reliably represent the collagen fiber since the latter is a multi hierarchy aggregate containing axial subunits (collagen molecule, micro-fibril, sub-fibril, and fibrils) interconnected laterally by various cross-links being all distinctly different in nature from the collagen molecules. In addition, unlike the rod, subunits as well as the whole fiber respond differently to tension versus compression and probably buckle under negligible small compressive load.

4   *Y. Lanir*

2.2. *Tissues Quasi-Linear Viscoelasticity*

Soft tissues manifest viscoelstic characteristics [2]. Fung observed that under many circumstances, the tissue response can be separated between an immediate ("elastic", i.e., time-independent) non-linear stress-strain relationship, and a strain-independent function of time. The resulting quasi-linear viscoelastic (QLV) theory [1] carries a significant advantage over non-linear viscoelastic theories by being compatible with the mathematical machinery of linear visco-elasticity. QLV is currently the most widely used viscoelastic representation for soft tissues. Fung [1] formulated the QLV law in two alternative forms,

$$S(e,t) = \int_0^t G(t-\tau) \cdot \dot{S}^e\left[e(\tau)\right] \cdot d\tau, \quad S(e,t) = \int_0^t G(t-\tau) \cdot \frac{dS^e}{de} \cdot \dot{e}(\tau) \cdot d\tau \quad (5)$$

where $G(t)$ is the reduced (normalized) relaxation function and the dot designates time derivative. Eq. 5a relates the current stress $S(e,t)$ to the history of the instantaneous ("elastic") response $S^e[e(t)]$, while the alternative (and equivalent) form (Eq. 5b) relates the current stress to the strain history $e(t)$.

    Viscoelasticity and fibers recruitment: Assuming that the fibers are linear viscoelastic, then the viscoelastic stress in the fiber is given by the Boltzman hereditary integral (in analogy to Eq. 5a), as follows:

$$s_f(e_f,t) = \int_0^t G(t-\tau) \cdot \dot{s}_f^e[e_f(\tau)] \cdot d\tau \quad (6)$$

where $s_f^e(e_f)$ is the immediate ("elastic") response of the fiber. The viscoelastic stress in the fiber bundle is obtained by summing the contributions of all fibers as in Eq. 4:

$$S(e,t) = \int_0^e D(x) \cdot s_f[e_f(x,t),t] \cdot H(s_f) \cdot dx \quad (7)$$

where H is the unit step function which guaranties that the stress in buckled fibers will be taken as zero. Eqs. 6 and 7 prescribe the viscoelastic response of the tendon fiber bundle.

    It was recently shown [29], theoretically, numerically and experimentally, that in fully preconditioned tendon (see below), under protocols at which no fiber buckles, the system of equations 6, 7 is equivalent to the QLV theory. If no fiber buckles then two consequences result: first, the fiber viscoelastic stress $s_f(e_f,t)$ in fibers that were stretched during the protocol is always positive, so that $H \equiv 1$. Eq. 6 can then be substituted into Eq. 7 to yield:

$$S(e,t) = \int_0^e D(x) \cdot \int_0^t G(t-\tau) \cdot \dot{s}_f^e[e_f(\tau)] \cdot d\tau \cdot dx \qquad (8)$$

In addition, since no fiber buckles then the argument functions are single-single valued so that the order of integration in Eq. 8 can be interchanged, resulting in:

$$S(e,t) = \int_0^t G(t-\tau) \cdot \int_0^e D(x) \cdot \dot{s}_f^e[e_f(\tau)] \cdot dx \cdot d\tau = \int_0^t G(t-\tau) \cdot \dot{S}^e[e_f(\tau)] \cdot d\tau \qquad (9)$$

The second equality (derived by using Eq. 4) is identical to the expression for the QLV stress (Eq. 5a). If on the other hand fibers do buckle, then their stress vanishes for all $S_f \leq 0$ and their stress-strain relationship is flat and thus no longer single-single valued.

Hence, the tissue nonlinear viscoelastic features stem from its fibers gradual recruitment. Furthermore, micro-structural considerations show that the tendon viscoelastic response is reliably represented by the QLV theory if no fibers buckle during the stretch protocol. Importantly, this condition is met during both stress relaxation and creep tests.

2.3. *Tissues Preconditioning*

In addition to viscoelasticity, experimental observations commonly manifest another time-dependent response feature: under repeated cycles of stretch, there is a decay and shift to the right of the stress-strain response. This preconditioning adaptation process is at times only partially reversible, and requires rest periods which are orders of magnitude longer than characteristic viscoelastic relaxation times. The concept of preconditioning was introduced to tissue biomechanics by Fung [30] who proposed that constitutive formulation should be based on stable response obtained from fully adapted (pre-conditioned) tissue samples. Previous attempts to model tissue preconditioning were phenomenological (not relating to the tissue structure). Preconditioning was represented by the strain softening (Mullins) effect in arrested left ventricle [31] and in the small intestine [32]. Rubin and Bodner [33] incorporated preconditioning in a phenomenological elasto-viscoplastic model for the skin, consisting of a composite of elastic and dissipative components. Sverdlik and Lanir [14] proposed that tissue preconditioning under stretch results from, and can thus be reliably modeled, based on the preconditioning response of its fibers.

Preconditioning of tissues' fibers: Experimental observations revealed that the physical processes of preconditioning in collagen and elastin fibers are mutually different. In collagen [34, 35], the fibers' stress-free length (gage-length) was observed to increase during preconditioning as a function of stretch and time, but the slope of the fibers' stress-strain relationship seems to be preserved [14]. In the recruitment model (Eq. 3), this process implies a gradual increase of the straightening strain x, which increases during preconditioning as a function of stretch and time. It was found [14] that two processes are required to adequately account for the tendon's collagen preconditioning: a rate ("viscous") process and a plastic one. Good agreement with data under multiple stretch protocols to different strain levels was obtained assuming a linear dependence of the viscous process on the fiber strain as follows:

$$\frac{dx}{dt} = \begin{cases} p_{1c}(e_f - p_{2c}), & e_f > p_{2c} \\ 0, & e_f \le p_{2c} \end{cases} \qquad \left(e_f = (e-x)/(1+2x))\right) \qquad (10)$$

with initial condition $x|_{t=0} = x_0$, where x_0 is the distributed reference waviness. Here $p_{1c}(\ge 0)$ is a positive rate constant, and $p_{2c}(\ge 0)$ is the threshold below which there is no viscous preconditioning. The second process is a perfect plastic one above a strain threshold [14], so that:

$$\frac{dx}{dt} = \frac{de_f(t)}{dt} \qquad e_f > p_{3c} > 0; \qquad \frac{de_f(t)}{dt} > 0 \qquad (11)$$

$$\frac{dx}{dt} = 0 \qquad \text{otherwise}$$

where $p_{3c}(\ge 0)$ is the threshold.

In elastin fibers the extent of preconditioning-induced stress decay is smaller than in collagen, and the gage-length seems to be preserved [19]. There is however a decay of the fibers stiffness. This type of preconditioning adaptation is termed "strain softening" or Mullins effect [36]. Mullins attributed the magnitude of stiffness reduction in rubbers and polymers solely to the highest level of previously imposed strain. In contrast, our observations revealed that elastin strain softening is also time-dependent. Good fit to uniaxial stretch data of the skin [37] at the elastin dominated low strain levels was obtained [19] when the elastin stiffness was assumed to decay with time from its reference level in a first order process depending linearly on the strain.

The structure-based mechanistic approach to modeling preconditioning in tissues was applied for the tendon [14], and for the skin under both uniaxial [19] and biaxial [38] deformations. It was shown that preconditioning is an essential response feature and that a reliable representation of responses under multiple uniaxial and biaxial (Fig. 1) tests is obtained only if preconditioning is incorporated into the constitutive equations.

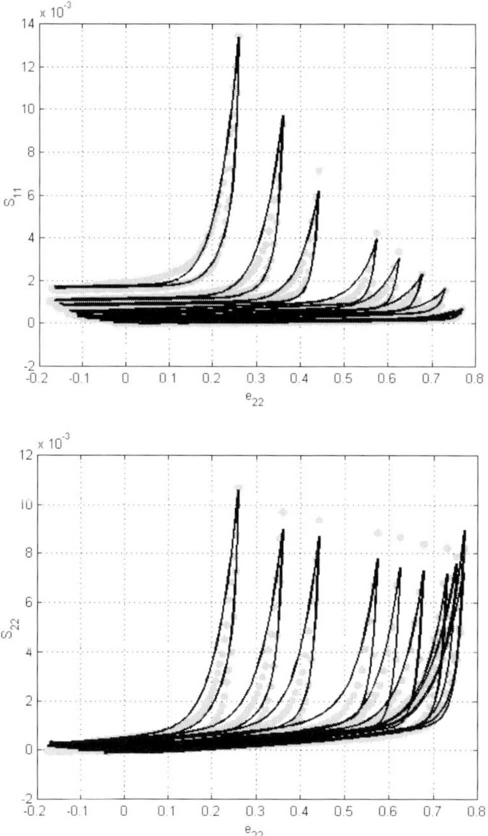

Figure 1. Comparison between Measured Skin Biaxial Response (curcles) and Model |Predictions (dark line) for Stress Components S_{11} (upper panel) and S_{22} (lower panel) to Multiple Level Constant Rate Stretch in the 2-Direction.

2.4. *Tissues Residual Stress*

Chuong and Fung [39] found that arteries, rather then being stress-free in the unloaded state (i.e., free vessel segment, zero pressure) are internally loaded by residual stress (RS). When a vessel ring is cut longitudinally, it springs open to a presumably truly stress-free configuration with an opening angle (OA) whose magnitude is a measure of the RS. Similarly, OA was later found also in myocardial left ventricle rings [40]. Importantly, Fung and coworkers [39-42] later showed that RS significantly affects the tissue's stress and strain distributions, and that it carries substantial functional benefit by reducing stress concentration near the internal surface in these organs, thereby reducing their energy consumption.

The importance of knowing the level of RS and the related stress-free configuration is two-fold. First, the stress-free configuration of the organ is an essential reference needed to evaluate the true stress and strain which is exerted on the tissue's cells, thereby determining their biological signaling response and the ensuing tissue biological remodeling. Second, often RS are associated with increased stiffness of the tissue. This stiffening, is likely to modify the tissue function.

Mechanisms of RS: The mechanistic origin of tissues' residual stress is yet unclear. In a recent report [43] it was proposed that there is a hierarchy of different RS producing mechanisms. The micro level (tissue interstitium) RS is induced by local interactions between the tissue constituents (fibers, cells, ground substance matrix). The second meso-level RS results from internal interactions induced by non-homogeneities in the tissue micro-structure or constituents composition. This meso RS is determined by several factors including the local micro RS, the mechanical interactions between composite tissue elements (e.g., between the media and adventitia in the arterial wall [44]), and by in-homogeneity in the tissue mechanical properties [45]. The third, macro (organ) RS arises from kinematical constraints on the tissues structures which produce additional internal loading (e.g., forces and bending moments required to close the wall into an intact vessel).

The implication of this hierarchy of RS producing mechanisms is that a true stress-free state can only be achieved if all RS mechanisms are neutralized. Relieving just one source of RS such as a radial cut through the arterial wall is insufficient and may leads to erroneous estimate of the state of stress and strain in the tissue.

The micro-level RS: While the macro- and meso-levels RS are fairly well studied and defined [39, 41, 42, 45-52], the mechanism underlying the micro-level one are still unclear. Yet its existence has been experimentally verified in both the left ventricle [53] and in arteries [44, 54]. A recent analysis based on the tissues' multi-constituents structure and on the mixture theory [43], suggests that micro-level RS can stem either a) from contact stresses between the tissue's solid constituents resulting from their incompatible growth and remodeling [55, 56], or b) from the mechanical interaction between the solid constituents and the extra-cellular swelling-induced fluid pressure. This interaction can be understood from the following mixture theory equation relating the total stress T^{tot} to the stress T_i^s of the solid constituent i and to the pressure of the extra-cellular ground substance P, and from the thermodynamic equilibrium equation between P and the osmotic pressure π^{osm} :

$$T^{tot} = \sum_i \phi_i \cdot T_i^s - P \qquad P = \pi^{osm} \qquad (12)$$

where ϕ_i is the volume fraction of the *i-th* solid constituent. Eq. 12 shows that in the un-loaded state ($T^{tot} = 0$), internal residual stress can exist as a result of mutual interaction between the solid constituents, or between them and the fluid osmotic pressure. The latter results from the osmotic Donnan effect of the negatively charged glycosaminoglycan (GAG) side chains in the large proteoglycans macromolecules (PG, primarily decorin and versican), which are immersed in the fluid-like matrix (the ground substance).

Analysis of previous experimental observations suggested [43] that under swelling levels similar to the *in vivo* ones, osmotic-induced tissue swelling is a major contributor to the micro-level RS. This conclusion is strongly supported by a recent study [44] on the aorta, in which it was possible to separate between the effects of swelling of smooth muscle cells on one hand, versus that of the extra-cellular space (which contains charged PGs) on the other. The results suggest that the osmotic charge effect in the extra-cellular matrix was the predominant underlying mechanism of the observed RS.

3. Discussion and Conclusions

The analysis presented here shows that tissues rheological response features can well be accounted for by their constituents' structure and properties, and their mutual interaction. Specifically, non-linearity and anisotropy result respectively from the fibers distributed undulation and from their non-uniform orientation distribution. The fibers gradual recruitment with stretch is also responsible for the tendon non-linear viscoelasticity, and the analysis shows that this micro-structural consideration is consistent with the QLV theory, but only if no fibers buckle during the test protocol. Tissue preconditioning adaptation to the loading cycle is attributed to the preconditioning response of its fibers. Experimental observations suggest that the physical processes of preconditioning in collagen and elastin fibers are mutually different, but both are strain and time dependent. Finally, residual stress in unloaded tissues is attributed to interactions between constituents at three levels of organization, from the micro tissue space, via the meso tissue elements to the macro organ level.

The analysis presented here provides a unifying micro-mechanistic basis to various rheological response features of tissues. The merit of such a unifying outlook is that it can be readily generalized to other more complex structured tissues since all soft tissues are composed of similar constituents. In addition, results obtained thus far indicate that the structure-based approach provides for a reliable representation of the tissues properties.

References

1. Y.C. Fung, *Biomechanics - Its Foundations and Objectives*, ed Y.C. Fung, N. Perrone, M. and M. Anliker, (Prentice-Hall, Englewood Cliffs, NJ, 1972), p. 181.
2. Y.C. Fung, *Biorheology,* **10**, 139 (1973).
3. T.T. Tanaka and Y.C. Fung YC, *J. Biomech.* **7**, 357 (1974).
4. Y.C. Fung, K. Fronek, P. and Patitucci, *Am. J. Physiol.* **237**, H620 (1979).
5. Y.C. Fung, *Circ. Res.* **37**, 481 (1975).
6. P. Tong and Y.C. Fung, *J. Biomech.* **9**, 649 (1976).
7. A. Viidik, *Z. Anat. Entwicklungsgesch* **136**, 204 (1972).
8. A. Viidik, *Int. Rev. Connect. Tissue Res.* **6**, 127 (1973).
9. B.M. Chu, W.G. Frasher, and H . Wayland, *Ann Biomed Eng* **1**, 182 (1972).
10. Y. Lanir, *J. Bioeng.* **2**, 119 (1978).
11. Y. Lanir, *J. Biomech.* **12**, 423 (1979).
12. Y. Lanir, *J. Biomech. Eng.* **102**, 332 (1980).
13. Y. Lanir, *J. Biomech.* **16**, 1 (1983).
14. A. Sverdlik and Y. Lanir, *J. Biomech. Eng.* **124**, 78 (2002).
15. K.L. Billiar and M.S. Sacks, *J. Biomech. Eng.* **122**, 327 (2000).
16. M.S. Sacks, *J. Biomech. Eng.* **125**, 280 (2003).
17. P.A. Shoemaker, D. Schneider, M.C. Lee, and Y.C. Fung, *J. Biomech.* **19**, 695 (1986).
18. S.M. Belkoff and R.C. Haut, *J. Biomech.* **24**, 711, (1991).
19. O. Lokshin and Y. Lanir, *J. Biomech. Eng.* **131**, 031009 (2009).
20. A. Horowitz, Y. Lanir, F.C Yin, M. Perl, I. Sheinman and R.K. Strumpf, *J. Biomech. Eng.* **110**, 200 (1988).
21. E. Nevo and Y. Lanir, *J. Biomech. Eng.* **111**, 342 (1989).

22. T.C. Gasser, R.W. Ogden, and G.A. Holzapfel, *J. R. Soc. Interface* **3**, 15 (2006).
23. M.A. Zulliger, P. Fridez, K. Hayashi, and N. Stergiopulos, *J. Biomech.* **37**, 989 (2004).
24. C.P. Buckley, D.W. Lloyd, and M. Konopasek, *Proc. Roy. Soc. Lond.* **A372**, 33 (1980).
25. J. Diamant, A. Keller, E. Baer, M. Litt, and R.G. Arridge, *Proc. R. Soc. Lond. B. Biol. Sci.* **180**, 293 (1972).
26. M. Comninou and I.V. Yannas, *J. Biomech.* **9**, 427 (1976).
27. D.E. Beskos and J.T. Jenkins, *J. Applied Mechanics* **42**, 755 (1975).
28. D.C .Stouffer, D.L. Butler, and D. Hosny, *J. Biomech. Eng.* **107**, 158 (1985).
29. E. Raz, and Y. Lanir, *J Biomechanical Eng. (Accepted* 2009*)*.
30. Y.C. Fung, *Biomechanics - Mechanical Properties of Living Tissues* (Springer- Verlag, New York, 1981).
31. J.L. Emery, J.H. Omens, and A.D. McCulloch, *J. Biomech. Eng.* **119**, 6 (1997).
32. H. Gregersen, J.L. Emery, and A.D. McCulloch, *Ann. Biomed. Eng.* **26**, 850 (1998).
33. M.B. Rubin, S.R. Bodner, and N.S. Binur, *J. Biomech. Eng.* **120**, 686 (1998).
34. Y. Lanir, E.L. Salant, and A. Foux, *Biorheology* **25**, 591 (1988).
35. M Abrahams, *Med. Biol. Eng.* **5**, 433 (1967).
36. L. Mullins, *Rubber Chem. Technol.* **42**, 339 (1969).
37. H. Eshel and Y. Lanir, *Ann. Biomed. Eng.* **29**, 164 (2001).
38. O. Lokshin and Y. Lanir, *Biomaterials* **30**, 3118 (2009).
39. C.J. Chuong and Y.C. Fung, *J. Biomech. Eng.* **108**, 189 (1986).
40. J.H. Omens and Y.C. Fung, *Circ. Res.* **66**, 37 (1990)
41. Y.C. Fung, *Ann. Biomed. Eng.* **19**, 237 (1991).
42. Y.C. Fung and S.Q. Liu, *Am. J. Physiol. Heart Circ. Physiol.* **262**, H544 (1992).
43. Y. Lanir, *J. Biomechanical Eng.* **131**, 044506 (2009).
44. E. U. Azeloglu, M.B. Albro, V.A. Thimmappa, G.A. Atcshian and K.D. Costa, *Am. J. Physiol. Heart Circ. Physiol.* **294**, H1197 (2008).
45. L.A .Taber and J.D. Humphrey, *J. Biomech. Eng.* **123**, 528 (2001).
46. R.N. Vaishnav and J. Vossoughi, *J. Biomech.* **20**, 235 (1987).
47. T. Matsumoto, K. Hayashi, and K. Ide, *J. Biomech.* **28**, 1207 (1995).
48. X. Lu, A. Pandit, and G.S. Kassab, *Am. J. Physiol. Heart Circ. Physiol.* **287**, H1663 (2004).
49. S.Q. Liu and Y.C. Fung, *Diabetes* **41**, 136 (1992)
50. G.A. Holzapfel, G. Sommer, M. Auer, P. Regitnig, and R.W. Ogden, *Ann. Biomed. Eng.* **35**, 530 (2007).
51. H.C. Han and Y.C. Fung, *J. Biomech.* **24**, 307 (1991).

52. S.E. Greenwald, J.E, Moore Jr., A. Rachev, T.P. Kane, and J.J. Meister, *Biomech. Eng.* **119**, 438 (1997).
53. Y. Lanir, G. Hayam, M. Abovsky, A.Y. Zlotnick, G. Uretzky, E. Nevo, and S.A. Ben-Haim, *Am. J. Physiol.* **270**, H1736 (1996).
54. X. Guo, Y. Lanir, and G.S. Kassab, *Am. J. Physiol. Heart Circ. Physiol.* **293**, H2328 (2007).
55. E.K. Rodriguez, A. Hoger, and A.D. McCulloch, *J. Biomech.* **27**, 455 (1994).
56. R. Skalak, S. Zargaryan, R.K. Jain, P.A. Netti, and A. Hoger, *J. Math Biol.* **34**, 889 (1996).

Chapter 2

BIOMECHANICS OF AN ISOLATED SINGLE STRESS FIBER[*]

MASAAKI SATO

Department of Biomedical Engineering, Tohoku University
Sendai, Miyagi 980-8579, Japan

SHINJI DEGUCHI

Department of Biomedical Engineering, Tohoku University
Sendai, Miyagi 980-8579, Japan

Knowledge of mechanical properties of stress fiber (SF), a bundle of actin filaments, is crucial for understanding its role in mechanotransduction in adherent cells. Here, we characterized tensile properties of single SFs by *in vitro* manipulation. SFs were isolated from cultured vascular smooth muscle cells with a combination of low ionic-strength extraction and detergent extraction and were stretched until breaking. The breaking force of the SFs for stretching was, on average, 377 nN, which was greater than actin filaments, 600 pN. The Young's modulus was estimated as 1.45 MPa, which was three orders of magnitude lower than actin filaments. Strain-induced hardening, a common mechanical behavior of living adherent cells, was observed in a physiological strain range of the force-strain curves. Estimated force level of physiological tension in single SFs was the same order of magnitude with that of the substrate traction force of adherent cells required for maintenance of cell integrity. These results suggest that SFs are a principal subcellular component in bearing intracellular stresses.

1. Introduction

It has been proposed that mechanical forces transmitted via cytoskeleton in adherent vascular cells are important for activation of proteins localized at focal adhesions, plasma membrane, intercellular junctions, and so on, which induces the downstream signaling related to gene expression and protein synthesis [1-4]. Hayakawa et al. [5] directly elucidated that mechanical forces were transmitted from cell surface to focal adhesions through stress fibers (SFs: bundles of actin filaments).

---

[*] This work is supported by the Grant-in-Aid for Scientific Research (Scientific Research A #17200030 and Specially Promoted Research #20001007) by the Ministry of Education, Culture, Sports, Science and Technology, Japan (MEXT), and the Mitsubishi Foundation.

Knowledge of the mechanical properties of SFs is then crucial for understanding intracellular stress levels and force transmission [6]. Although mechanical properties of actin filaments had been extensively investigated at single protein levels [7, 8], the only substantive measurement of the higher order structure has been made on biopolymer gels containing synthesized actin filaments and α -actinin, which is localized along SFs in living cells to cross-link actin filaments. These studies only reported spatially averaged macro-scale properties of the gels. Thus far, no description of the mechanical properties of single SFs has appeared. The lack of detailed information about the mechanical properties of SFs seems largely due to the fact that SFs are unstable in the cytoplasm. However, Katoh et al. [9] showed that SFs are isolable from the cells, suggesting that SFs are a mechanically stable organelle in vitro. This approach overcomes this difficulty and enables us to experimentally evaluate mechanical properties of SFs.

We performed tensile tests of single SFs isolated from cultured smooth muscle cells (SMCs) by *in vitro* manipulation with a pair of cantilevers [6]. Force-strain relation, stretching stiffness, and breaking force were evaluated.

2. Materials and Methods

2.1. *Cell culture*

SMCs were isolated from the bovine thoracic aortic media by an explant technique. Cell culture was done in Dulbecco's modified Eagle medium (Invitrogen, The Netherlands) containing 10% heat-inactivated fetal bovine serum (JRH Biosciences, USA), 1% penicillin, and 1% streptomycin in an incubator at 37°C and 5% CO_2/95% air. Cells between passages 5–10 were seeded in a 35-mm-diameter glass base culture dish (Asahi Techno Glass, Japan) for experiments.

2.2. *Isolation of stress fiber*

SFs were isolated from SMCs according to the reported technique [9]. Briefly, cells were washed with PBS (hereafter containing 1 µg/ml leupeptin (Wako, Japan) and 1 µg/ml pepstatin (Wako) and cooled to 4°C before use) and treated with an ice-cold low-ionic-strength extraction solution consisting of 2.5 mM triethanolamine (Wako), 1 µg/ml leupeptin, and 1 µg/ml pepstatin in distilled water. Extracted cells were then treated with PBS containing 0.05% NP-40 (Wako, pH 7.2) for 5 min and 0.05% Triton X-100 (Wako, pH 7.2) for 5 min. Extracted SFs were then washed gently with a cytoskeleton stabilizing buffer (10 mM imidazole (Wako), 100 mM KCl, and 2 mM EGTA (Wako), pH 7.2) for

15 min at 4°C to thoroughly remove Triton X-100. Next, rhodamine-conjugated phalliodin (Molecular Probes, USA) was added at 1 nM to stain F-actin, a major component of SFs, for 5 min at 4°C. SFs were scraped off from the dish using a rubber scraper, put through an injection needle of 23-gauge twice, suspended in the stabilizing buffer with oxygen-removal reagents (2.3 mg/ml glucose, 0.018mg/ml catalase, 0.1mg/ml glucose oxidase) in a 35-mm-diameter suspension culture dish.

2.3. *Tensile test of stress fiber*

A carbon fiber (Toray, Japan), with approximately 7 μm in diameter, was attached to the tip of a rigid glass rod with epoxy resin (Figure 1). The carbon fiber–glass rod, referred to hereafter as cantilever, was used for tensile tests of the isolated SFs. Tensile tests were carried out at room temperature (20°C) on an inverted microscope (IX-71, Olympus, Japan) fixed on a vibration-free table. Immediately prior to tests, the tips of both cantilevers were thinly coated with epoxy resin (Araldite, Vantico, Japan). Under illumination from a halogen and a mercury light, the cantilevers were positioned on a targeted single SF and moved toward ends of the SF by using two hydraulic micromanipulators equipped on both sides of the microscope. Here, the one cantilever was placed vertically to the specimen, and the other in parallel. Great care was taken not to apply any tension to the specimen before tests.

Tensile tests were initiated while controlling the position of the parallel cantilever with a piezo-electric actuator (AE0505D16, NEC Tokin, Japan) connected to the base glass rod to stretch the specimen at 0.02 s^{-1} strain rate with a program written in LabVIEW programming language (National Instruments, USA). Leverage was applied at the base part of the parallel cantilever to increase its maximum extension. The output characteristics of piezo-electric actuator were calibrated before each experiment to compensate for its nonlinear behavior. The deflection of cantilever and the displacement of specimen were calculated by using NIH image software from the images taken at 5 s intervals through a digital CCD camera (C4742-95, Hamamatsu, Japan), yielding force-displacement relation.

3. Results and Discussion

Images of a stress fiber during tensile testing are shown in Figure 1. During stretching, the deflectable cantilever was gradually bent with increases in stress transfer. A typical example of the force-strain (defined as the ratio of the displacement to the original length, i.e. the zero-stress length) curve in cyclic loading/unloading processes at strain rate of 0.05/s was shown in Figure 2.

Maximum strain was fixed at 1.0. SFs showed perceptible residual strain and hysteresis in the cycles, indicating viscous or plastic behavior. We also obtained force-strain curves examined up to the breaking points. The summarized data showed initial length of the specimen was 14.0 ± 9.8 µm (mean $\pm$ SD, n = 20). Six specimens were found to reach breakage by stretching because the torn SFs were left on both the cantilevers when the deflectable cantilever returned to the original position, whereas we could not distinguish the other specimens were torn nor just detached from either cantilever. The breaking force was 377 ± 214 nN (mean $\pm$ SD, n = 6) and the breaking strain 1.99 ± 0.63. It was assumed that the zero-stress length was passed when the slope of the force-strain curves began to change. The maximum strain reached 2.75. In most cases, SFs showed high extensibility together with nonlinear increases in stiffness.

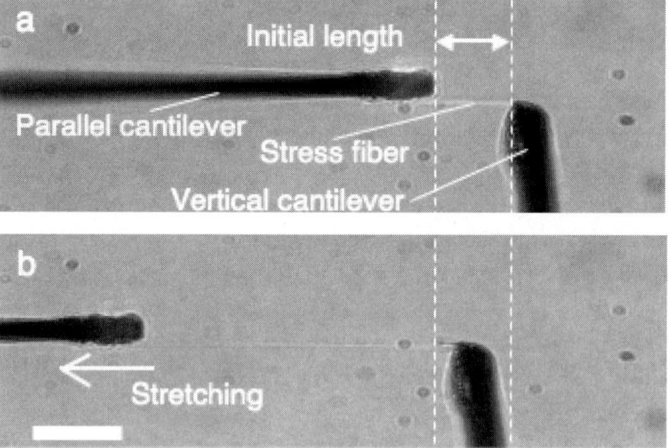

Figure 1. Sequential images of a single SF during tensile test. (a) Before stretching. SFs were fluorescently visualized. Cantilevers were viewed under transmission light illumination. (b) During stretching. Horizontal arrow, direction of the left cantilever displacement.

SFs were capable of large deformation as much as they achieve a maximum length that is 3.75-fold of the zero-stress length. There is no report of the breaking strain of actin filaments though their physiological strain reaches 0.2–0.42% of the initial length. The breaking force of SFs was, on average, 377 nN, which was larger than that of untwisted actin filaments, 600 pN [10]. In living cells, tensile load existing in SFs would be ~10 nN according to the measurements of substrate traction force at focal adhesion sites [11], suggesting that single actin filament solely is not sufficient for bearing the physiological loading and it would be bundled up in parallel to be strengthened. Variations in the composition and the amounts of the subfilaments in single SFs are probably responsible for the data dispersion of the force-strain curves since diameters of each specimen were not evaluated in the present study.

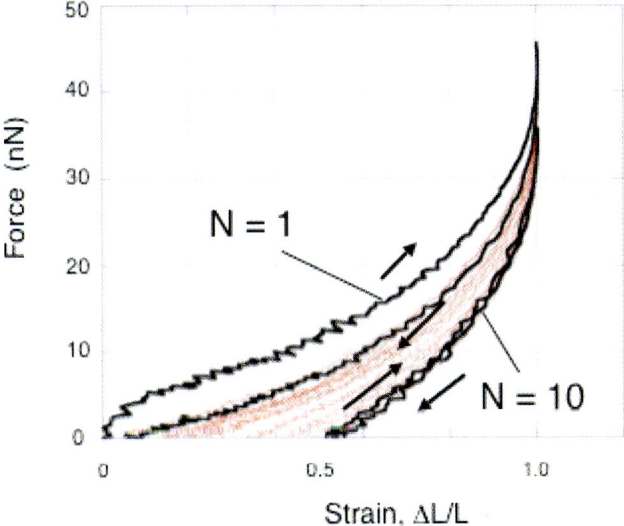

Figure 2. Relationship between force and strain of the isolated SFs in cyclic loading/ unloading processes. N: repetition number.

When it is assumed that SFs were homogeneous, and the cross-section was a circle with a 0.1 μm radius based on electron microscopy, the relationship between nominal stress and strain of SFs can be then obtained. Initial slope of stress-strain curve at strain = 0, which is equal to the Young's modulus, was determined to be 1.45 MPa. The Young's modulus is comparable to the previous results obtained from atomic force microscopy [12], in which localized increases in elasticity of more than 100 kPa along SFs were detected. The Young's modulus was at least three orders of magnitude lower than that of synthesized F-actin, 2.6 GPa reported by Gittes *et al.* [8] or 1.8 GPa by Kojima *et al.* [7] estimated under the similar analytical assumptions of homogeneity.

The elastic modulus of SFs at the breaking point (strain = 1.99) was also evaluated with another assumption that SFs were incompressible. The modulus is then obtained as 104 MPa, which is approximately 70-fold of the Young's modulus at the zero-stress state. Thus, elastic modulus of SFs increases nonlinearly with increasing strain as if it approaches the order of actin filaments or thick filaments (1.0 GPa [13]).

The high extensibility and strain hardening of single SFs should be associated with the molecular architecture. SFs consist of actin filaments cross-linked by α-actinin, and myosin II enters the spacing between actin-filaments [9]. Such loose packing may allow SFs to exhibit strain-dependent mechanical behavior. As strain increases, the initially buckled subfilaments might be stretched and entangled in each other until their intrinsic rigidities gradually appear as suggested by biopolymer gel studies. Otherwise, deformation rate-dependent α-actinin cross-link rearrangement or sliding between actin filaments and α-actinin may be responsible for the extensibility of SFs. Experimental evaluation of such viscoelastic properties (the effect of temperature or strain rate on SF's stiffness), diameter changes, and Poisson's ratio is left as a future study.

祝　卒寿

馮 元楨 先生

Photos with Professor and Mrs. Y.-C. Fung. (a) At UCSD, September 2008. (b) At UCSD, July 1984 with Prof. Nerem, Maya Sato (daughter) and Yoshika Sato (wife).

References

1. P. F. Davies, *Phys. Rev.* **75**, 519 (1995).
2. D. E. Ingber, *Annu. Rev. Physiol.* **59**, 575 (1997).
3. N. Wang and Z. Suo, *Biochem. Biophys. Res. Comm.* **328**, 1133 (2005).
4. A. J. Maniotis, C. S. Chen and D. E. Ingber, *Proc. Natl. Acad. Sci., USA* **94**, 849 (1997).

5. K. Hayakawa, H. Tatsumi and M. Sokabe, *J. Cell Sci.* **121**, 496 (2008).
6. S. Deguchi, T. Ohashi and M. Sato, *J. Biomech.* **39**, 2603 (2006).
7. H. Kojima, A. Ishijima and T. Yanagida, *Proc. Natl. Acad. Sci., USA* **91**, 12962 (1994).
8. F. Gittes, B. Mickey, J. Nettleton and J. Howard, *J. Cell. Biol.* **120**, 923 (1993).
9. K. Katoh, Y. Kano, M. Masuda, H. Onishi and K. Fujiwara, *Mol. Biol. Cell* **9**, 1919 (1998).
10. Y. Tsuda, H. Yasutake, A. Ishijima and T. Yanagida, *Proc. Natl. Acad. Sci., USA* **93**, 12937 (1996).
11. J. L. Tan, J. Tien, D. M. Pirone, D. S. Gray, K. Bhadriraju and C. S. Chen, *Proc. Natl. Acad. Sci., USA* **100**, 1484 (2003).
12. G. T. Charras and M. A. Horton, *Biophys. J.* **83**, 858 (2002).
13. D. Dunaway, M. Fauver and G. Pollack, *Biophys. J.* **82**, 3128 (2002).

Chapter 3

THE ORIGIN OF PRE-STRESS IN BIOLOGICAL TISSUES — A MECHANO-ELECTROCHEMICAL MODEL: A TRIBUTE TO PROFESSOR Y.C. FUNG

LEO Q. WAN, X. EDWARD GUO, VAN C. MOW[†]

Department of Biomedical Engineering, Columbia University
New York, NY 10032, U.S.A.

The existence of residual stresses within biological tissues was demonstrated more than two decades ago by Professor Y. C. Fung (1984) when blood vessels sprang open with a longitudinal cut. This simple experiment has been repeated in many load-bearing tissues including articular cartilage. It is believed that these residual stresses and strains play important physiologic roles in functionally reducing the stress *in situ*. In the present study, the *in vitro* swelling and curling behaviors of thin strips of cartilage were analyzed with a model using the triphasic mixture theory with a collagen-proteoglycan solid matrix composed of a three-layered laminate with each layer possessing a distinct set of orthotropic properties. A cone-wise linear elastic matrix has been incorporated to account for the well-known tension-compression nonlinearity of the tissue. It has already been shown that this theory can account for the curvatures found in published experimental results. The results suggest that for a charged hydrated soft tissue, such as articular cartilage, the balance of proteoglycan swelling and the collagen restraining within the solid matrix is the origin of the *in situ* residual stress, and that the layered collagen ultrastructure, *e.g.*, relatively dense and with high stiffness at the articular surface, play key roles in determining curling behaviors of such tissues.

1. Introduction

More than two decades ago, Professor Yuan-Cheng Fung demonstrated the existence of pre-stresses (*i.e.*, residual stresses) within biological tissues with a simple experiment that blood vessels sprang open with a longitudinal cut [1]. Residual stresses and strains play important roles in the biomechanical function of the load-bearing tissues such as vein, cardiac ventricle, and cartilage [2]. An adult articular cartilage sample will curl and warp toward its articular surface after removal from the underlying subchondral bone. The curvature is found to vary with the saline concentration in the external bathing solution [3]. This curling behavior is an evidence of inhomogeneous residual stresses and strains inside the solid matrix of articular cartilage, and is believed to be contributed by

---

[†] Corresponding author. Tel.: (212) 854-8462; Fax:(212) 854-8725; E-mail: vcm1@columbia.edu

21

both swelling properties and layered inhomogeneous ultrastructure of articular cartilage [4].

Compositionally, tissues such as blood vessels and articular cartilage are a mixture composed of solid matrix and interstitial fluid. The solid matrix itself consists mainly of collagen fibrils (or if larger in diameter, the collagen exists as a woven fibers) and proteoglycans [5]. While the former bio-macromolecule is a slender, long, relatively stiff and neutral molecule, the latter is a globular, soft gel like substance that is highly negatively charged; these charges mainly derive from the presence of carboxyl (COO^-) and sulfate groups (SO_3^-) along with the chondroitin and keratan sulfate chains. Collagen and proteoglycans are the main load bearing structural macromolecules within cartilage and other connective tissues (ligament, tendon, meniscus, intervertebral discs, *etc*). These fixed charges in such porous-permeable interstitial space require a high concentration of water, counter-ions (Na^+) for electro-neutrality to be maintained and also introduce an imbalance of total mobile ion concentration between the fluid compartment inside the tissue and the bathing fluid outside the tissue. The physico-chemical colligative property of this difference in total ion concentration is the Donnan osmotic pressure, which causes the tissue to swell (dimensionally and by weight) [5, 6]. The extent of tissue swelling is restrained by the surrounding stiff collagen network surrounding the proteoglycan molecules. The balance of this swelling pressure and the collagen restraining stresses developed within the solid matrix is the origin of residual stress.

The layered inhomogeneties of chemical contents and macromolecular organization *in situ* cause variations of osmotic pressure and solid matrix stresses throughout the depth of the tissue, and thus giving rise to the often-observed curling behaviors of articular cartilage. Many light and electron microscopy studies have been reported on the variation of collagen fibrils (or fiber) architectural organization through the depth of articular cartilage [5]. Collagen fibrils are aligned tangentially to the surface in the superficial tangential zone (SZ), randomly in the middle zone (MZ), and vertically in the deep zone (DZ) (Figure 1). These micro-structural features of collagen seem to correlate well with the relative amounts of the substances found within the matrix, such as water and collagen contents and fixed charge density (FCD); in other words, composition also varies layer-wise throughout the tissue depth from the surface to the deep layer attached to the bone. As a result, mechanical properties and swelling effects of the tissue also show the depth dependency, leading to the curling behavior of articular cartilage.

While the intuitive picture for the cartilage curling behavior has been known for many years [6], and quite a few models have been developed to explain the swelling behavior (as measured by weight change or dimensional change; [6, 7]), to our knowledge, only two previous attempts have been made to model the cartilage curling behavior. Setton *et al* modeled the curling behavior by assuming a linear increase of negative FCD from articular surface through the depth of articular cartilage [4]. However, their results predicted a highest curvature when the external concentration is around 0.3M, which is contrary with their earlier later experimental data [3]. More recently, in 2002, Olsen and Oloyede modeled cartilage to be transversely isotropic with a finite element method and assumed an *ad hoc* swelling pressure distribution across thickness and its change with external ion concentration [8]. The focus of our current study is the development of a quantitative model based on our triphasic constitutive law [9] to describe the swelling and curling behaviors of articular cartilage; this model also utilizes the known layer-wise variation of collagen (stiffness variation) and proteoglycan (charge density variation) throughout the depth. Thus, the specific aims of our study are: 1) to develop a multi-layer triphasic orthotropic model based on the structure and composition of articular cartilage; 2) to compare the predicted deformation on the curling behavior of thin strips of cartilage specimens with previous experimental results; and 3) to analyze the contributions of these various intrinsic physical parameters on the curling behavior of articular cartilage.

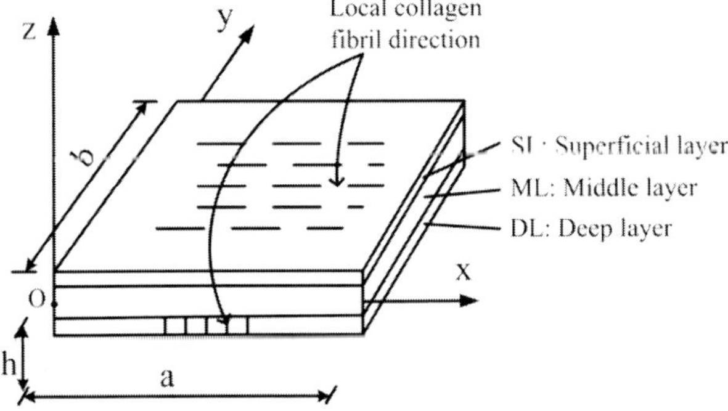

Figure 1. The layer structure of articular cartilage strip with its dimensions and the predominant orientation of collagen fibrils.

2. Methods

At the hypertonic state (asymptotically $c^* \to \infty$), we assume the swelling effects associated with the FCD are negligible and the entire tissue (length a × width b × thickness h) is assumed to be flat initially as shown in Fig. 2(A). All physical parameters after swelling are determined relative to those at this hypertonic reference state (HRS). The material properties and chemical parameters are assumed to vary with depth (*i.e.*, between the three layers), but to be homogenous within each layer. If these three layers were separated from each other, and allowed to swell independently, then each layer would experience a different deformation as shown in Fig. 2(B) and remain straight [7]. However, if the layers were joined together before swelling, then the in-plane strains (in x-y plane) at the interface would have to match, which would produce curling (Fig. 2(C)).

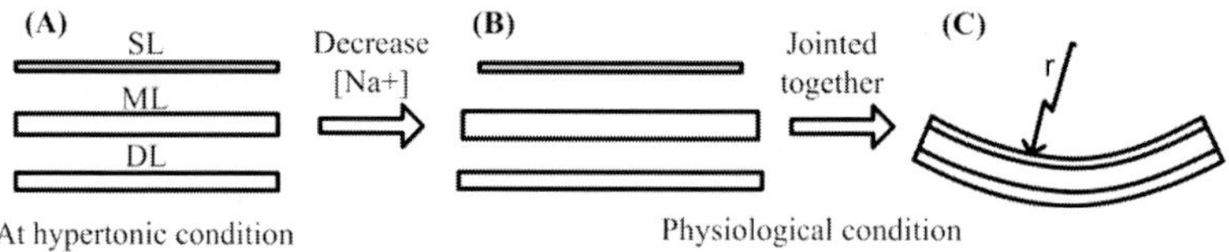

Figure 2. Schematic explanations for the origin of curling behavior of articular cartilage. Three layers (A) have different swelling potentials (B), which lead to curling behavior of cartilage strips (C).

The triphasic theory [9] is used to model the swelling behavior (Fig.2:A→B) of each layer. The total stress (σ) consists of elastic stresses inside the solid matrix (σ_s) and the osmotic pressure (p) as shown below:

$$\sigma = -p\,\boldsymbol{I} + \boldsymbol{\sigma}_s \quad \text{with} \quad p = \pi - \Pi\,e$$

where $\pi = \phi RT(c^k - 2c^*)$ and $\Pi = \phi RT\,(c_r^F)^2/(\phi_r^w c^k)$ [10, 11]. Here ϕ is the osmotic coefficient, R is the universal gas constant, T is absolute temperature, c^k is the total ion concentration, e is the dilatation, and ϕ_r^w and c_r^F are the porosity and FCD at HRS, respectively.

Table 1. Intrinsic parametric values for articular cartilage, based on previous literatures [5,13].

Position	Thickness (mm)	Porosity	FCD (mEq/ml)	λ (MPa)	μ (MPa)	λ_{+1} (MPa)	λ_{-1} (MPa)
STZ	0.3	0.7	0.12	0.2	0.4	10	3.0
MZ	1.2	0.8	0.25	0.2	0.4	MZ=Isotropic	—
DZ	0.5	0.6	0.20	0.1	0.15	3.0	0.9

An orthotropic constitutive equation with tension-compression nonlinearity is used to describe the mechanical property of the solid matrix for each layer based on the collagen fibril microstructure [7, 12]. Since the cartilage strip is assumed to be very thin, *i.e.*, $h << a$ and $h << b$, the problem can be solved based on a classical lamination theory and the usual assumption of pure bending (Fig. 2:B→C). To compare with Setton *et al*'s previous experimental results [3], the typical parameters adopted for patellar cartilage are as the base case and shown in Table 1 [5, 13], in which λ_{+1} and λ_{-1}, the elastic moduli for tension and compression, respectively, account for the tension-compression nonlinearity of tissue solid matrix [14]. The elastic modulus was set to be 3MPa for the SL in the direction perpendicular to the split lines [13].

3. Results

3.1. *Variation of Curvature and Stretch with Saline Concentration*

Figures 3 and 4 represent variations of surface stretch ($\Lambda = 1 + \varepsilon$) and curvature (K) as a function of the external ion concentrationsin the directions both parallel and perpendicular to the predominant collagen fibril directions [13]. It can be seen that the tissue at the bottom of the sample is under tension (*i.e.*, convex side with stretch $\Lambda > 1$), while at the top (concave side) the strain will be very close to zero, or even slightly negative (*i.e.*, in compression with stretch $\Lambda \leq 1$). The curvature increases with the decrease of external saline concentration, and is always larger in the collagen fibril direction (x)

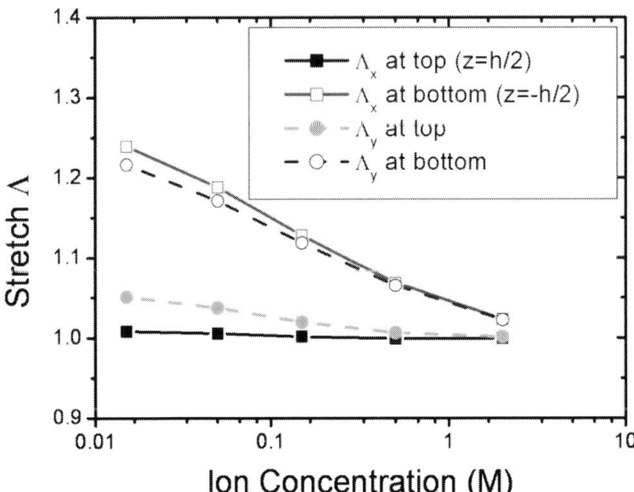

Figure 3. Variation of stretch (*Λ=dx/dX*) of cartilage strip in different external ion concentrations (0.015M, 0.05M, 0.15M, 0.5 M, and 2M). The stretch is defined as the ratio of the length after deformation (*dx*) and the original length (*dX*).

                                     L. Q. Wan et al.

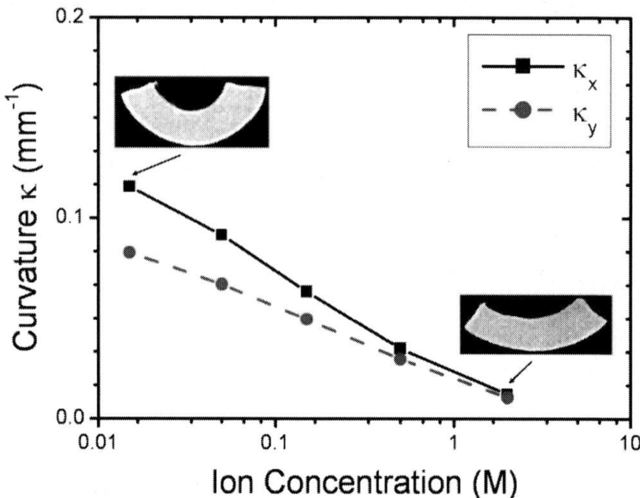

Figure 4. Variation of curvature of cartilage strip in different ion concentrations for the base case parameters listed in Table 1.

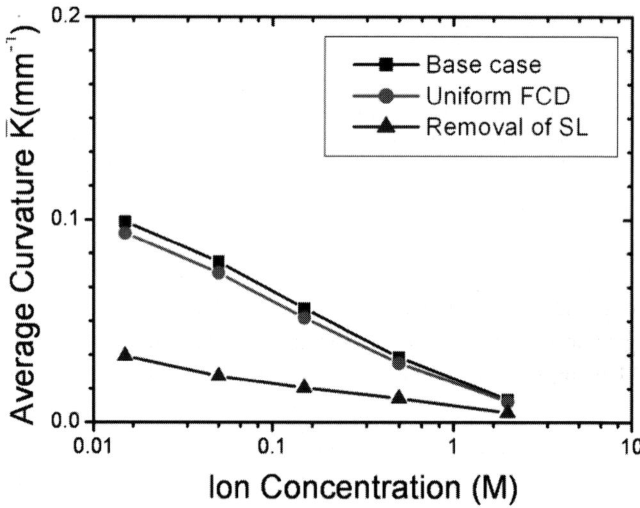

Figure 5. Variation of average curvature $\bar{\kappa}\ (=(\kappa_x + \kappa_y)/2)$ with external ion concentration for the case with uniform fixed charge density (FCD) distribution and the case with the absence of superficial layer (SL) (also shown the results for the base case).

3.2. *Roles of FCD Inhomogeneity and High Stiffness of Superficial Layer*

There has always been the question of whether collagen and/or proteoglycan stratification predominate the curling behavior. To answer this question, two cases were modeled. In the first case, we assumed that the FCD to be uniform, and has a value of 0.218mEq/ml, which is the average value for the base case (Table 1). It is found that there was almost no predicted change both in x and y directions in the curling response of the tissue (Fig.5). In the second case, we assumed the surface to be absent. The average curvature $\bar{\kappa}$ in physiological saline solution (0.15M), which is defined as $(\kappa_x + \kappa_y)/2$, will decrease from 0.055 mm^{-1} to 0.03 mm^{-1}(Fig. 5).

4. Discussion

The objective of this study was to develop a mechano-electrical chemical model to describe the swelling and curling behaviors of thin cartilage strips under osmotic loading. For simplicity, the full thickness cartilage strips have been modeled as a thin laminate composed of three fibril-enforced layers with different chemical contents and mechanical properties. Based on the classic lamination theory [15], simple mathematic solutions have been obtained for the variation of curvatures of the freely swelling cartilage strips in different external saline solutions.

The predicted stretch and curling are comparable to experimental data reported previously by Setton *et al* [3]. Just as shown in their study, the in-plane strains (ε_x and ε_y) at the articular surface is predicted to be very close to zero for the direction parallel to the split lines (the predominant collagen fiber direction at the surface) as well as for the direction perpendicular to the split lines. The experiment results also show that the strain changes about 0.07-0.11 in the deep zone of the strip when the ion concentration varies from 2M to 0.15M, which implies that the overall curvatures changes about 0.035 - 0.055mm^{-1} for a 2mm thick sample. Our values for stretch change (~0.1) and curvature change (~0.05 mm^{-1}) are in these ranges. Considering that the typical material parameters adopted in this study are based on experimental data of chemical contents and mechanical properties in previous literatures, we believe that our model can capture the actual mechanism of swelling and curling behavior of articular cartilage and therefore be used to predict mechano-electrochemical events happening inside the tissue during osmotic loading.

The validation of our model with previous published experimental data gives us confidence that the predicted *in situ* residual stresses and strains should be very close to the actual values one could obtain experimentally. The in-plane strains (ε_x and ε_y) are continuous and linear, which is a direct consequence from the pure bending assumption as described before. It agrees with the experimental finding that the lateral edge is almost always perpendicular to the surface and bottom of articular strip as shown in the figure 1 of Setton *et al*'s paper [3]. This is also the reason that the shear stress inside the tissue was ignored and the classical lamination theory was chosen instead of the first-order approximation of the shear-deformed laminated plated theory [15]. However, note part of the residual stresses could have already been released when the cartilage strips are removed from the subcondronal bones. Therefore, the residual stresses could be larger in the intact cartilage, especially in the superficial zone.

Our numerical results show that there is no significant change in curvature when the variation of FCD is replaced by a uniform distribution of average FCD throughout articular cartilage while the curvature decreases to about one half when the stiffer superficial zone (see Fig. 5) is removed. In previous reported experiments, although the maximum curvature found near the sample mid-length had no significant changes after removal of the superficial layer, the overall curvature change at physiological condition ($c^* = 0.15$M) relative to the hypertonic condition ($c^* = 2$M) did decrease after the SL removal [3]. Physically, this significant curvature change is a result of the substantial lateral stiffness of the superficial layer, which essentially prevents horizontal strains at the top of the sample and thus forcing the neutral plane (where the strain ε_x or ε_y is zero) to be situated close to the upper surface. The combination of high tensile stiffness and the location of the superficial layer near the articular surface thus appears to play a key role in the curling behavior of articular cartilage and controlling the *in situ* stress; this may indeed by the physiological protective mechanism for exposing normal articular cartilage to high stresses, even though the loading stress is rather high.

The technique of removing articular cartilage from the subchondral bone and measuring the extent of curling (κ) can become a potential method to evaluate the mechanical function of articular cartilage. With osteoarthritis and its progression, both compositional and structure changes will occur which include the fibrillation of the superficial zone of articular cartilage, the decrease of the PG concentration and the imbibitions of water. The collagen fibrillation largely decreases the tensile modulus of the superficial layer. As a result, the cartilage sample will produce less curling after removal from the subchondral bone since the gradient of the mechanical property with depth is significantly reduced. From the cartilage curling behavior, it is possible to calculate the change in mechanical property with cartilage degeneration based on our mechanical model. This technique may provide a quick way to examine the quality of articular cartilage. Compared to conventional mechanical tests, such as compression or tensile tests, the observation of swelling-induced curling behavior is technically more feasible for the cartilage in small animal joints, such as rabbits and guinea pig.

Acknowledgments

This study is supported by Stanley Dicker and Shelly Ping Liu endowments.

References

1. Y. C. Fung, *Biomechanics in China, Japan and USA*, eds Y. C. Fung, and E. Fukada, (Science Press, 1984) p. 1.
2. Y. C. Fung, *Biomechanics: Motion, Flow, Stress and Growth*, (Springer-Verlag, New York, NY, 1990)
3. L. A. Setton, H. Tohyama, and V. C. Mow, *J. Biomech. Eng.* **120**, 355 (1998).
4. L. A. Setton, W. Y. Gu, V. C. Mow, and W. M. Lai, *Mechanics of Porous Media*, A. P. S. Selvadurai, ed., Kluwer (Acad Press, 1995) pp. 229.
5. V. C. Mow, W. Y. Gu, and F. H. Chen, , *Basic orthopaedic biomechanics and mechano-biology*, eds. V. C. Mow, and R. Huiskes, (Lippincott Williams and Wilkins, Philadelphia, 2005) p. 181.
6. A. Maroudas, *Adult Articular Cartilage*, ed. M. A. R. Freeman, (Pitman Medical, Kent, UK, 1979) pp. 215.
7. E. R. Myers, W. M. Lai, and V. C. Mow, *J. Biomech. Eng.* **106**, 151 (1984)
8. S. Olsen, and A. Oloyede, *Comput. Methods Biomech. Biomed. Engin.* **5(6)** 377 (2002).
9. W. M. Lai, J. S. Hou, and V. C. Mow, *Journal of Biomech. Eng.* **113(3)**, 245 (1991).
10. G. A. Ateshian, N. O. Chahine, I. M. Basalo, and C. T. Hung, *J. of Biomech.* **37(3)** 391 (2004).
11. L. Q. Wan, C. Miller, X. E. Guo, and V. C. Mow, 2004, *Mechanics & Chemistry of Biosystems* **1(1)**, 81 (2004).
12. S. Akizuki, V. C. Mow, F. Muller, J. C. Pita, D. S. Howell, and D. H. Manicourt, *J. Orthop. Res.* **4(4)**, 379 (1986)
13. V. Roth, and V. C. Mow, *J. Bone Joint. Surg.* **62(7)**, 1102 (1980).
14. M. A. Soltz, and G. A. Ateshian, *J. Biomech Eng*, **122(6),** 576 (2000).
15. R. M. Jones, *Mechanics of composite material*, (Scripta Book Company, Washington, D.C., USA, 1975).

Chapter 4

HOW BLOOD FLOW SHAPES NEOINTIMA[*]

SHU Q. LIU

Biomedical Engineering Department, Northwestern University
Evanston, IL 60208, U.S.A.

Y. C. FUNG

Department of Bioengineering, University of California at San Diego
La Jolla, CA 92093-0412, U.S.A.

Blood flow and associated fluid shear stress play a role in the regulation of vascular morphogenesis during embryonic development and adaptive alterations in response to physiological and pathological stimuli. Various flow patterns, including laminar and vortex flow, in association with different levels of fluid shear stress are present in the vascular system. As fluid shear stress negatively regulates the mitogenic activities of vascular cells and molecular transport processes in the circulator system, nonuniform shear stress exerts profound effects on the formation of thrombus and neointima. In this study, we showed, by using an experimental model of polymeric cylinder implantation to the rat vena cava, that fluid shear stress controlled the development of thrombus and neointima. The degree of thrombus/neointima development was inversely correlated to the shear stress level, resulting in the formation of thrombus/neointima with a shear stress-dependent geometry and size. Furthermore, shear stress gradients controlled smooth muscle cell migration from the host-vessel media to the thrombus/neointima via the mediation of cell density gradients established under the influence of nonuniform shear stress. This paper addresses how nonuniform fluid shear stress controls the formation of thrombus/neointima and smooth muscle cell migration.

1. Introduction

The vascular system is subject to fluid shear stress, a blood flow-associated frictional force component at the endothelial surface. Blood flow, as an environmental factor, contributes to the regulation of vascular morphogenesis during embryonic development and adaptive alterations in response to physiological and pathological stimuli [1-15]. Various blood flow patterns, including laminar and vortex flow, are present in the vascular system. In arteries with a sufficiently large Reynolds number, vortex blood flow can develop at divergent bifurcations, resulting in zero-shear stress stagnations at the flow separation and reattachment points as well as reduced fluid shear stress in

---

[*] This work is supported by NSF and AHA.

31

                              S. Q. Liu & Y. C. Fung

regions between the stagnation points [16-19]. As zero and reduced fluid shear stress activates mitogenic signaling pathways, leading to vascular smooth muscle cell proliferation and migration [20-22], vortex blood flow is often associated with an increase in the intimal thickness, resulting in the formation of neointima. Under pathological conditions, such as endothelial cell injury and activation in response to biological and chemical stimuli, vortex flow-induced mitogenic activities can facilitate thrombogenesis and atherogenesis [22-24]. Thus, it is important to understand the role of blood flow in mediating vascular morphogenesis and pathogenesis. In this report, we demonstrate how blood flow and fluid shear stress control the formation of thrombus and neointima and how vascular smooth muscle cells proliferate and migrate in response to nonuniform shear stress.

2. Role of Fluid Shear Stress in Controlling Thrombus and Neointima Formation

The role of fluid shear stress in mediating thrombus and neointima formation was tested by using an in vivo experimental model of polymeric cylinder implantation into the rat vena cava, establishing a blood flow field with known shear stress distribution on the implanted cylinder. This model allows investigating the influence of various levels of fluid shear stress on thrombus and neointima formation on the implanted cylinder. To create such a model, a polypropylene cylinder of 0.3 mm diameter coated with polyethylene glycol was implanted at the center of the rat vena cava (about 3 mm in diameter) in a direction perpendicular to blood flow (Fig. 1). The vena cava was chosen in this study to avoid the influence of flow pulsatility, which complicates the assessment and analysis of fluid shear stress.

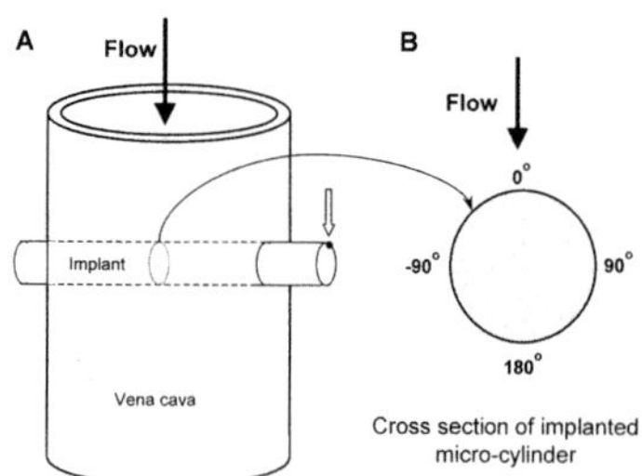

Figure 1. Schematics showing a polymeric cylinder implanted in the rat vena cava (A) and the cross section of the cylinder (B). The angles 0 and 180 deg indicate the leading and the trailing stagnations, respectively. The diameter of the cylinder was about 350 μm. Note that the marker indicated by the open arrow was the leading stagnation of the implanted cylinder. From Liu and Goldman. IEEE Transactions on Biomedical Engineering 48:474, 2001 with permission.

The polymeric cylinder implanted in the vena cava was subject to laminar blood flow in the leading region (from the 0° stagnation to flow separations at 109 and −109 deg) and subject to vortex blood flow in the trailing region (from 109 deg to the trailing stagnation at 180 deg) with the flow pattern symmetrical for the positive and negative sides (Fig. 2 and 3). Fluid shear stress is zero at the flow stagnations and separations. The distribution of blood shear stress on the implanted cylinder from 0 to 109 deg was estimated and analyzed on the basis of the boundary layer theory [25]. The rate of blood flow (Q) in the vena cava was measured with a Transonic flow meter and used for computing the distribution of blood flow velocity (*u*) along the axis of the implanted cylinder using the following equation [26]:

$$u(y) = [2Q/(\pi r^4)](r^2 - y^2) \tag{1}$$

where *y* is distance along the radius of the vena cava and *r* is the radius of the vena cava.

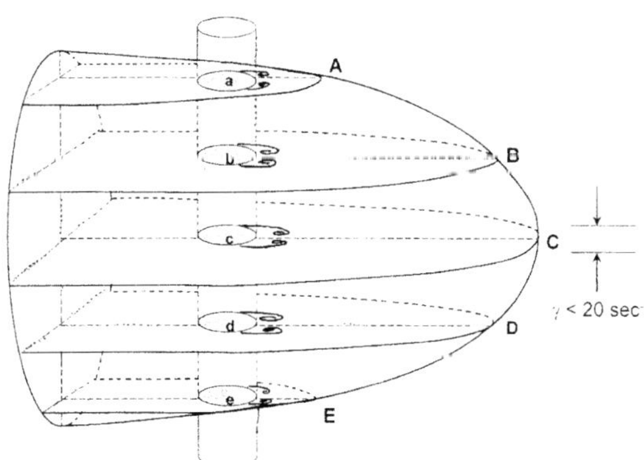

Figure 2. A schematic showing the implanted cylinder in the rat vena cava and the paraboloidal flow profile in the vena cava. Several locations, labeled with a, b, c, d, and e, were selected along the cylinder axis for analyzing flow patterns and shear stress distribution. Flow velocities estimated at A, B, C, D, and E were used to assess the distribution of shear stress along the circumference of the implanted cylinder at a, b, c, d, and e, respectively. γ is shear strain rate, which was less than 20 s⁻¹ in the core flow region about 7% of the vena cava diameter. From Liu and Goldman, IEEE Transactions on Biomedical Engineering 48:474, 2001 with permission.

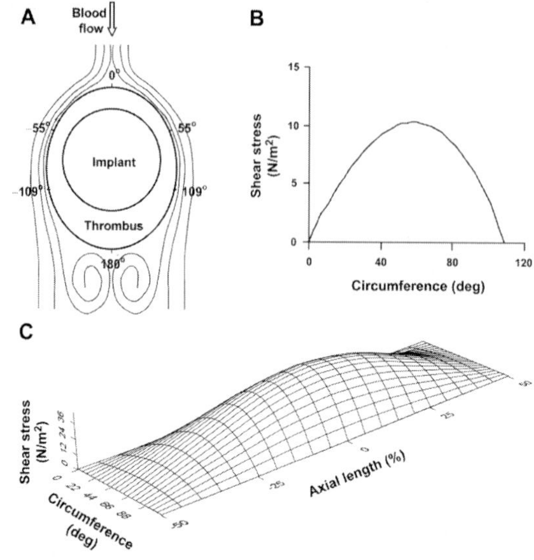

Figure 3. Flow pattern and fluid shear stress distribution on the cylinder implanted into the rat vena cava. (A) Schematic of a transverse section of the implanted cylinder. Flow streamlines are shown around the encapsulating thrombus of the cylinder. (B) Distribution of fluid shear stress along the circumference of the implanted cylinder (parallel to blood flow) from 0 to 109 deg at a selected axial location with a Reynolds number about 10. (C) Three-dimensional distribution of fluid shear stress on a thrombus-encapsulated cylinder from 0 to 109 deg at day 5 after cylinder implantation. The axial locations of 50% and –50% were the intersections of the implanted cylinder with the vena cava wall. From Liu et al., American Journal of Physiology 285:H1071, 2003 with permission.

Following the implantation of the polymeric cylinder, a thrombus layer forms on the cylinder within hours. Smooth muscle cells started to migrate from the host vena cava into the encapsulating thrombus of the implanted cylinder at about day 5 after cylinder implantation, resulting in the formation of neointima. The thrombus/neointima layer was taken into account for the analysis of shear stress distribution on the implanted cylinder. To assess the dimensions and geometry of the thrombus/neointima layer, the cylinder implant was collected at specified times, fixed in 4% formaldehyde in PBS for 20 min, and cut into serial transverse cryo-sections of 10 μm in thickness. A number of 20 equally spaced sections along the axis of the implanted cylinder were collected for measurements and analyses. For each section selected at an specified axial location of the implanted cylinder, the thickness of the encapsulating thrombus was measured at 20 equally spaced locations within the region from 0 to 109° in the circumferential direction of the implanted cylinder (parallel to fluid shear stress) and used for the calculation of the radius and circumferential surface distance (x) of the thrombus-encapsulated cylinder. The viscosity of blood (v) was measured as described previously [16]. The magnitude of fluid shear stress (τ) along the circumference of each selected transverse section was calculated from $0°$ to $109°$ using the following equation [25]:

$$\tau = \{[(1/2)\rho u^2]/[ur/v]^{1/2}\}[6.973(x/r) - 2.732(x/r)^3 + 0.292(x/r)^5 - 0.0183(x/r)^7 + 0.000043(x/r)^9 - 0.000115(x/r)^{11} + \ldots] \tag{2}$$

where ρ is blood density and u is maximal blood flow velocity at a given axial location on the implanted cylinder (Fig. 2). Reconstruction of the circumferential shear stress profiles at selected axial locations provides a distribution of fluid shear stress on the surface of the implanted cylinder (Fig. 3C). Differentiation of fluid shear stress with respect to the circumferential distance of the implanted cylinder provides shear stress gradients. Fluid shear stress in the trailing vortex flow region from 109 to 180 deg was not estimated because of the limitation of the analytic approach.

Following the implantation surgery, the implanted cylinder in the vena cava was rapidly encapsulated during the early period with a thrombus layer of non-uniform thickness. The rate of thrombus formation was dependent on the level of fluid shear stress. At a given axial location on the implanted cylinder, the maximal thickness was found at the trailing and leading stagnation points, whereas the minimal thickness was found at the maximal shear stress point (Fig. 4 and 5). It appeared that fluid shear stress shaped the thrombus into a structure with a streamlined cross-sectional profile. On the entire implanted cylinder, the maximal thrombus thickness was found near the vessel wall where the blood flow velocity in the vena cava and fluid shear stress on the implanted cylinder were zero, whereas the minimal thrombus thickness was found at the center of the implanted cylinder where the maximal flow velocity in the vena cava and maximal fluid shear stress on the implanted cylinder were found. Overall, fluid shear stress was inversely correlated with the thrombus thickness (Fig. 6). Following thrombus formation on the implanted cylinder, smooth muscle cells migrated from the vena cava media to the encapsulating thrombus along the axis of the implanted cylinder, resulting in the formation of neointima. As for the thrombus structure, fluid shear stress negatively controlled the growth of neointima. Thus, the shear stress effect resulted in the formation of a hyperboloidal thrombus/neointima structure on the implanted cylinder (Fig. 4 and 5).

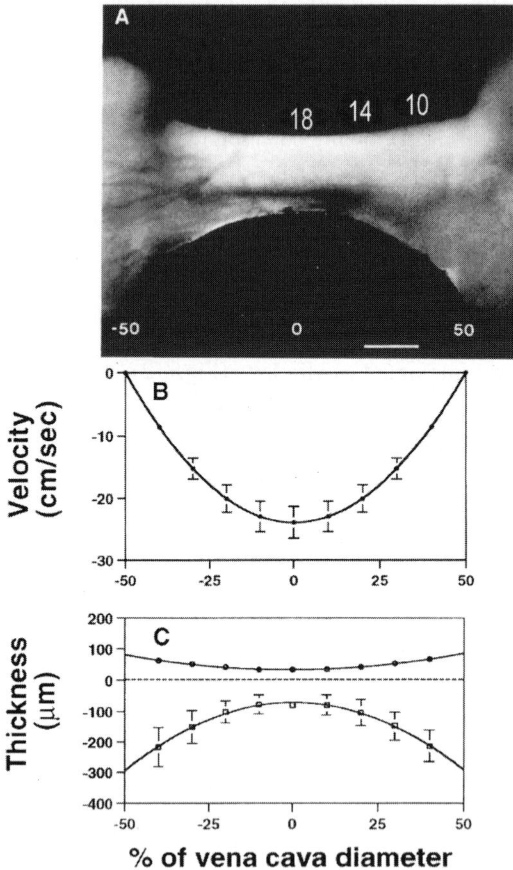

Figure 4. Influence of blood flow on thrombus formation on an implanted cylinder. (A) Photograph of hyperboloidal thrombus/neointima on an implanted cylinder at day 10 after cylinder implantation. The points at –50% and 50% indicate the intersections of the cylinder with the vena cava wall, and 0% is the mid-point of the cylinder. The numbers 10, 14, and 18 are Reynolds numbers at corresponding locations. Blood flow direction: top to bottom. Scale: 350 µm. (B) Velocity distribution in the vena cava. (C) Thickness of the encapsulating thrombus/neointima of the implanted cylinder axis at the leading and trailing stagnations (see Fig. 1 for these locations). The positive and negative signs on the vertical axis indicate opposite directions of growth of thrombus/neointima with respect to the cylinder. From Liu et al., Journal of Biomechanical Engineering 124:30, 2002 with permission.

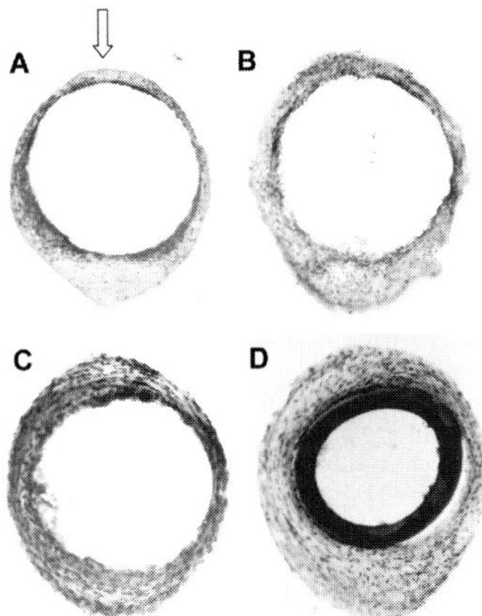

Figure 5. Micrographs showing the cross-section of the encapsulating thrombus/neointima of the implanted cylinder collected at the axial location with Reynolds number about 14 at day 1, 5, 10 and 30 (panel A, B, C, and D, respectively). Arrow: blood flow direction. Scale: 100 µm. From Liu et al., Journal of Biomechanical Engineering 124:30, 2002 with permission.

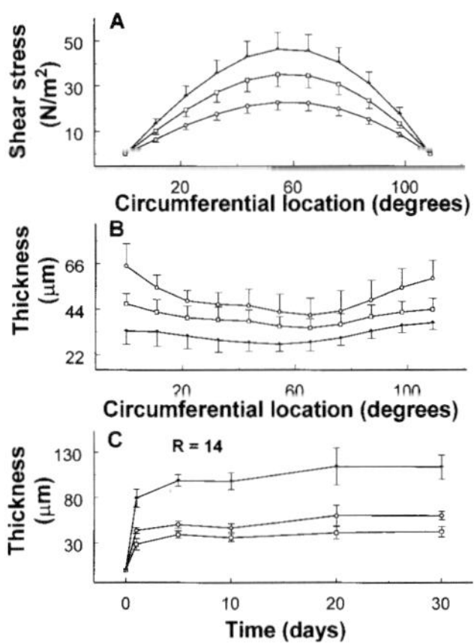

Figure 6. Influence of fluid shear stress on the thrombus/neointima thickness on the implanted cylinder. (A) Distribution of blood shear stress on the cylinder from 0 to 109 deg at three axial locations: the mid-point of the micro-cylinder (Reynolds number about 18, triangles), 15% of the cylinder axial length from the mid-point (Reynolds number about 14, squares), and 30% of the cylinder axial length from the mid-point (Reynolds number about 10, circles). See Fig. 4A for these locations. Data were reduced from 10-day specimens. (B) Thickness of the the encapsulating thrombus/neointima of the implanted cylinder from 0 to 109 deg at the three axial locations indicated in panel A at day 10 after cylinder implantation. (C) Changes in the thickness of the encapsulating thrombus/neointima measured at 3 circumferential locations 0 deg (circles), 56 deg (squares), and 180 deg (triangles) (see Fig. 3A for these circumferential locations) at an axial location with Reynolds number about 14 (see Fig. 4A for the axial location) from day 1 to 30 after cylinder implantation. From Liu et al., Journal of Biomechanical Engineering 124:30, 2002 with permission.

3. Influence of Fluid Shear Stress on Smooth Muscle Cell Proliferation and Migration

During the course of smooth muscle cell migration in the encapsulating thrombus/neointima of the implanted cylinder, fluid shear stress influenced the proliferation of these cells. As shown in a BrdU incorporation test, the density of BrdU-positive cells in regions near flow stagnations and separations was higher than that in shear stress regions (Fig. 7). Blood shear stress was inversely correlated with the density of BrdU-positive cells. These observations suggest that fluid shear stress negatively regulates cell proliferation. This effect resulted in the formation of cell density gradients. It appeared that the cell density gradient mediated the effect of fluid shear stress on smooth muscle cell migration. As fluid shear stress negatively regulates cell proliferation, the non-uniform distribution of fluid shear stress resulted in the formation of cell density gradients on the implanted cylinder. The cell density was inversely correlated with fluid shear stress (Fig. 3C and 8A). The cell density gradient drove the migration of smooth muscle cells. There were two major cell density gradients on the implanted cylinder: axial and circumferential gradients (perpendicular and parallel to fluid shear stress, respectively) (Fig. 8B). In the axial direction of the implanted cylinder, the cell density in the injured vessel wall was higher than that in the thrombus of the cylinder. This cell density gradient caused migration of vascular cells, including smooth muscle cells, along the axis of the implanted cylinder, resulting in smooth muscle cell alignment in parallel to the cylinder axis. In the circumferential direction of the implanted cylinder, the cell density at the leading and trailing zero-shear stress stagnations was higher than that in shear stress regions. This cell density gradient, established due to the inhibitory effect of fluid shear stress on cell proliferation, resulted in smooth muscle cell migration along the circumferential direction of the implanted cylinder (Fig. 9). Overall, the direction and pattern of smooth muscle cell migration were controlled by the relative levels of the axial and circumferential cell density gradients. In high shear stress regions, as the average circumferential gradient of cell density was larger than that in the axial direction of the implant, smooth muscle cells migrated primarily in the circumferential direction of the implanted cylinder. In low shear stress regions near the leading and trailing stagnations, the axial cell density gradient was dominant, and smooth muscle cells migrated primarily along the axis of the implanted cylinder (Fig. 9). These observations suggest that non-uniform fluid shear stress controls the pattern of smooth muscle cell migration via the mediation of the cell density gradient.

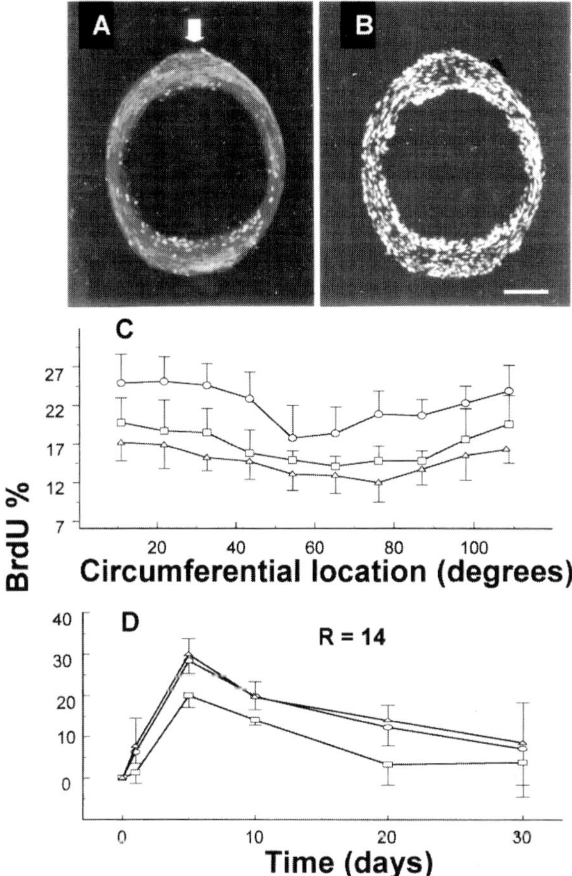

Figure 7. Influence of fluid shear stress on the density of BrdU-positive cells in the encapsulating thrombus/neointima of the implanted cylinder. (A and B) Fluorescence micrographs showing BrdU-positive and Hoechst 33258-labeled cell nuclei, respectively, in the encapsulating thrombus/neointima at day 10 following cylinder implantation. Arrow: direction of blood flow. Scale: 100 μm. (C) Distribution of BrdU-positive cells in the encapsulating thrombus/neointima of the cylinder from 0 deg to 109 deg (see Fig. 2A for these locations) at three axial locations with Reynolds numbers 10 (circles), 14 (squares), and 18 (triangles) (see Fig. 4A for these axial locations). Data were reduced from 10-day specimens. (D) Changes in the relative density of BrdU-positive cells in the encapsulating thrombus/neointima measured at 3 circumferential locations 0 deg (circles), 56 deg (squares), and 180 deg (triangles) (see Fig. 3A for these circumferential locations) at a selected axial location with Reynolds number about 14 (see Fig. 4A for the axial location) from day 1 to 30 after cylinder implantation. From Liu et al., Journal of Biomechanical Engineering 124:30, 2002 with permission.

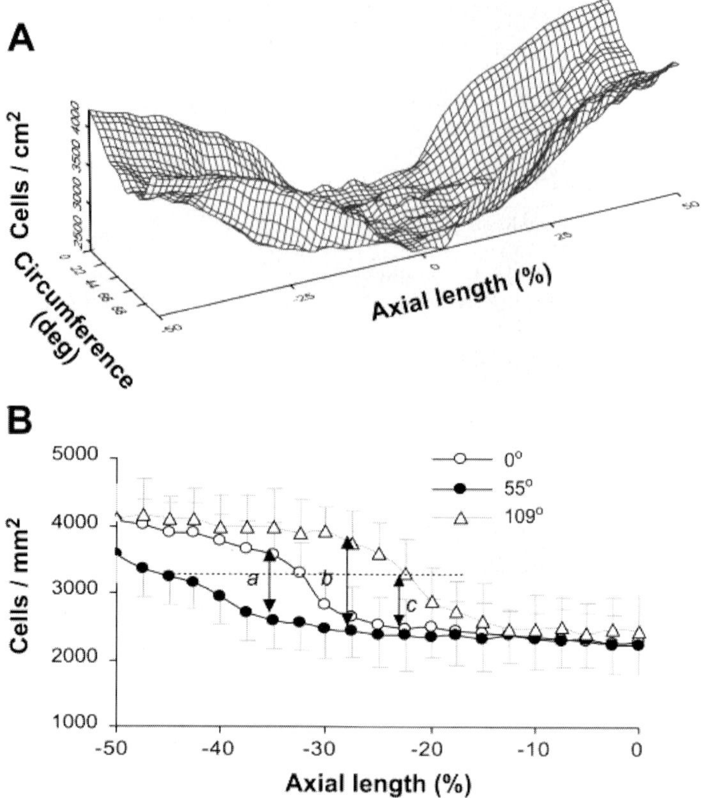

Figure 8. Distribution of cell density on a polymeric cylinder implanted into the vena cava. (A) Three-dimensional distribution of cell density from a 5-day specimen. (B) Average cell density at selected circumferential locations, including 0, 55, and 109 deg, along the axis of the implanted cylinder (perpendicular to blood flow) at day 5. Means and SD are presented (n = 5 at each data point). *a* and *b*, Circumferential cell density gradients from 0 to 55° and from 109 to 55°, respectively; *c*, axial cell density gradient at 55°. From Liu et al., American Journal of Physiology 285:H1071, 2003 with permission.

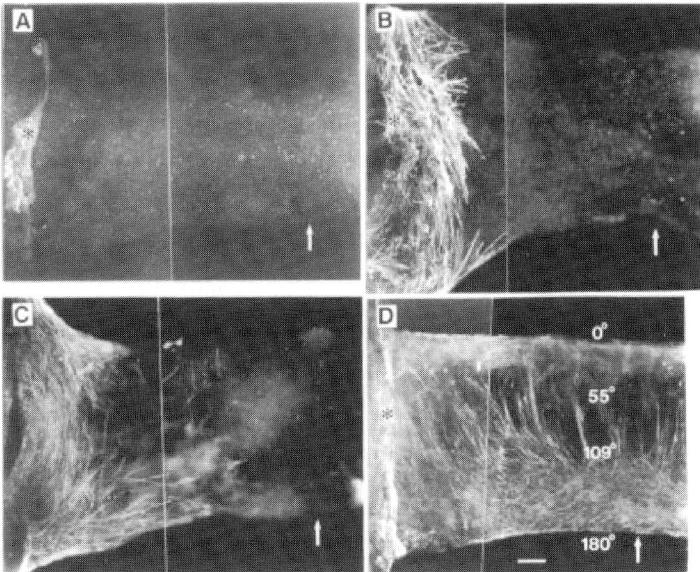

Figure 9. Fluorescence micrographs showing SMC α actin-positive cells on the implanted cylinder at (A) day 1, (B) day 3, (C) day 5, and (D) day 10. The cells were labeled with an anti SMC α actin antibody. The arrow in each panel indicates the mid point of the implanted cylinder. The degrees 0, 55, 109, and 180 in panel D indicate the leading stagnation, maximal shear stress location, flow separation, and trailing stagnation, respectively. Scale: 100 μm. From Liu and Goldman. IEEE Transactions on Biomedical Engineering 48:474, 2001 with permission.

4.　Role of PDGF Signaling Pathways in Mediating Smooth Muscle Cell Proliferation and Migration in Response to Non-uniform Fluid Shear Stress

PDGF-BB and PDGF-β receptor are expressed in vascular cells, and their activities are regulated by fluid shear stress. As fluid shear stress mediates the proliferation and migration of vascular cells, these molecules are possibly involved in mediating vascular cell proliferation and migration in response to non-uniform fluid shear stress. To test this possibility, we analyzed the relative expression of PDGF-BB as well as the relative phosphorylation of PDGF-β receptor tyrosine kinase and a downstream non-receptor tyrosine kinase, Src, in relation to the distribution of fluid shear stress and cell density in the encapsulating thrombus/neointima of the implanted cylinder. As shown in Fig. 10, PDGF-BB was upregulated in primarily endothelial cells and smooth muscle cells of the thrombus/neointima in regions near the zero-shear stagnations, but not in shear stress regions. A gradient of PDGF-BB expression was established from the stagnation to the maximal shear region in the direction of fluid shear stress (Fig. 10). A similar pattern was observed for the expression of PDGF-β receptor and the phosphorylation of Src on Y418 (Fig. 10). An inverse relationship was found between the level of fluid shear stress and the relative activity of Src (Fig. 3B and 10).

To test the role of PDGF-related molecules in the mediation of nonuniform shear stress-induced formation of cell density gradients and the pattern of smooth muscle cell migration, we applied pharmacological inhibitors AG-1296 and pp1 for PDGF-β receptor tyrosine kinase and Src, respectively, to rats with cylinder implantation. Immunoblotting analyses demonstrated that AG-1296 significantly suppressed the relative phosphorylation of PDGF-β receptor tyrosine kinase as well as Src, and pp1 inhibited that of Src at day 5 when maximal SMC migration was observed. Immunohistochemical analyses showed that AG-1296 and pp1 both significantly reduced the relative phosphorylation of Src on Y418 in regions near the zero-shear stagnations. The administration of these inhibitors resulted in a shear stress-independent, relatively uniform distribution of phosphorylated Src on Y418 in the shear stress direction of the implanted cylinder (Fig. 10). As a result, a treatment with AG-1296 or pp1 significantly suppressed smooth muscle cell migration in regions near the zero-shear stagnations, whereas such treatment did not significantly influence SMC migration in shear stress regions. The influence of these inhibitors resulted in a relatively uniform pattern of smooth muscle cell migration, thus reducing the circumferential gradient of cell density on the implanted cylinder (Fig. 11). As a result, fewer SMCs were migrating in the circumferential direction of the cylinder, and the population of circumferentially aligned smooth muscle cells was significantly reduced compared with the control model without a protein kinase inhibitor (Fig. 9 and 11). These observations suggest that fluid shear stress may suppress the activity of PDGF-β receptor and Src, and these factors may mediate nonuniform shear stress-induced smooth muscle cell proliferation and migration in neointima.

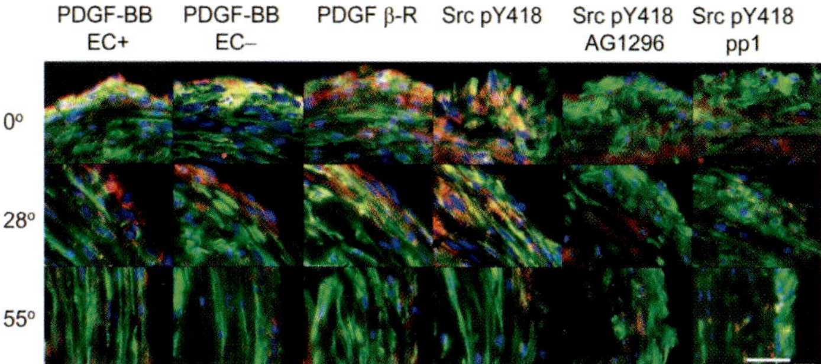

Figure 10. Fluorescence micrographs showing antibody-labeled platelet-derived growth factor (PDGF)-BB, PDGF-β receptor (PDGF-βR), and phosphorylated Src (Src pY418) (all in red) at 3 circumferential locations 0 (zero shear stress stagnation), 28, and 55 deg (maximal shear stress), as indicated in Fig. 3A, at an axial location of the implanted cylinder with a Reynolds number of about 10. The green color represents SMC α-actin, the blue color is for cell nuclei, and the scale bar represents 10 μm (for all images). EC+, with endothelial cells; EC–, without endothelial cells. From Liu et al., American Journal of Physiology 285:H1081, 2003 with permission.

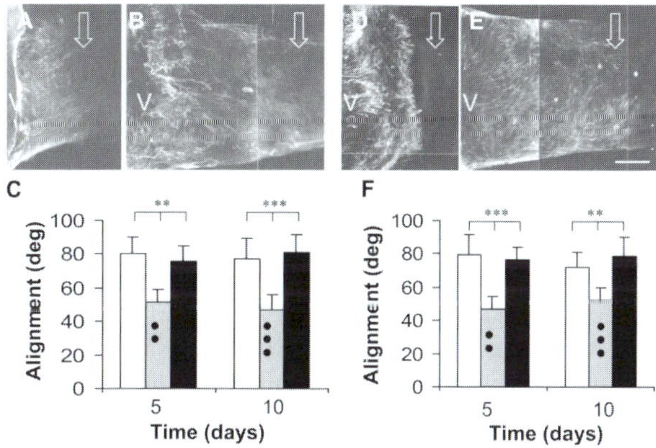

Figure 11. Analyses of SMC migration in the encapsulating thrombus/neointima of the implanted cylinder in the presence of AG-1296 and pp1. (A and B) Fluorescence micrographs (en face preparations) showing SMC alignments on the cylinder with AG-1296 at days 5 and 10, respectively. Note that smooth muscle cells migrated at a relatively constant speed over the entire cylinder. (C) Average smooth muscle cell alignment with AG-1296 at days 5 and 10 (n= 6). (D and E) Fluorescence micrographs (en face preparations) showing smooth muscle cell migartion on the implanted cylinder with pp1 at days 5 and 10, respectively. (F) Average smooth muscle cell alignment with pp1 at days 5 and 10 (n = 6). For panel A, B, D, and E, the arrow indicates the direction of blood flow, V is the vena cava wall, and the scale bar represents 100 μm. For panel C and F, the white, gray, and black bars represent sampling locations at 0–5, 53–58, and 105–110 deg, respectively, on the implanted cylinder. **$P < 0.01$ and ***$P < 0.001$, respectively, for comparisons between the three different locations at each time (by ANOVA). From Liu et al., American Journal of Physiology 285:H1081, 2003 with permission.

5. Concluding Remarks

Fluid shear stress negatively influenced cell attachment to and cell proliferation on vascular implants, resulting the formation of thrombus and neointima with a shear stress-dependent geometry. As fluid shear stress suppressed the mitogenic activities of vascular cells, the presence of nonuniform fluid shear stress resulted in the development of cell density gradients, which in turn mediated the pattern of smooth muscle cell migration and proliferation in thrombus and neointima. These observations demonstrate the significance of nonuniform fluid shear stress in the regulation of thrombus and neointima formation.

The authors, Dr. YC Fung (left) and SQ Liu, in the 1994 Annual Meeting of the American Society of Mechanical Engineers in Chicago.

References

1. C. G. Caro, *Arterioscler Thromb Vasc Biol.* **29**, 158 (2009).
2. B. C. Berk, J. I. Abe, W. Min, J. Surapisitchat and C. Yan, *Ann N Y Acad Sci.* **947**, 93 (2001).
3. S. Chien, S. Li and Y. J. Shyy, *Hypertension.* **31**, 162 (1998).
4. P. F. Davies, J. A. Spaan and R. Krams, *Ann Biomed Eng.* **33**, 1714 (2005).
5. Y. C. Fung YC and S. Q. Liu, *J Biomech Eng.* **115**, 1–12 (1993).
6. D. P. Giddens, C. K. Zarins and S. Glagov, *J Biomech Eng.* **115**, 588 (1993).
7. M. A. Gimbrone Jr, J. N. Topper, T. Nagel, K. R. Anderson, G. Garcia-Cardeña, *Ann N Y Acad Sci.* **902**, 230 (2000).
8. S. Q. Liu, M. Yen and Y. C. Fung, *Proc Natl Acad Sci USA* **91**, 8782 (1994).
9. S. Q. Liu SQ, *Crit Rev Biomed Eng* **27**, 75 (1999).
10. L. V. McIntire, *Ann Biomed Eng.* **22**, 2 (1994).
11. R. M. Nerem, *J Biomech Eng.* **114**, 274 (1992).

12. J. M. Tarbell, S. Weinbaum and R. D. Kamm, *Ann Biomed Eng.* **33**, 1719 (2005).
13. S. Weinbaum, J. M. Tarbell and E. R. Damiano, *Annu Rev Biomed Eng.* **9**, 121 (2007).
14. C. R. White and J. A. Frangos, *Philos Trans R Soc Lond B Biol Sci.* **362**, 1459 (2007).
15. D. M. Wootton and D. N. Ku, *Annu Rev Biomed Eng.* **1**, 299 (1999).
16. S. Q. Liu and J. Goldman, *IEEE Trans Biomed Eng.* **48**, 474 (2001).
17. S.Q. Liu, L. Zhong and J. Goldman, *J Biomech Eng.* **124**, 30 (2002).
18. S. Q. Liu, D. Tang, C. Tieche and P. Alkema P, *Am J Physiol.* **285**, H1071 (2003).
19. S. Q. Liu, C. Tieche, D. Tang and P. Alkema, *Am J Physiol* **285**, H1081 (2003).
20. S. Akimoto, M. Mitsumata, T. Sasaguri and Y. Yoshida, *Circ Res.* **86**, 185 (2000).
21. J. Goldman, L. Zhong, S. Q. Liu, *Am J Physiol.* **292**, H928 (2007).
22. S. Q. Liu, *Arterioscler Thromb Vasc Biol,* **19**, 2630 (1999).
23. B. L. Harry, J. M. Sanders, R. E. Feaver, M. Lansey, T. L. Deem, A. Zarbock, A. C. Bruce, A. W. Pryor, B. D. Gelfand, B. R. lackman, M. A. Schwartz and K. Ley, *Arterioscler Thromb Vasc Biol.* **28**, 2003 (2008).
24. S. Q. Liu, *Atherosclerosis* **140**, 365 (1998).
25. H. Schlichting, *Boundary-Layer Theory* (McGraw-Hill, New York, 1968) p. 149..
26. Y. C. Fung, *Biomechanics: Motion, Flow Stress, and Growth* (Springer, New York, 1990) p. 499.

Chapter 5

ILLUMINATING A PATH: THE ROLE OF BIOMECHANICS IN UNDERSTANDING ADAPTIVE REMODELING IN THE MICROCIRCULATION

THOMAS C. SKALAK

Office of the Vice President for Research and the Biomedical Engineering Department
University of Virginia, Box 400896
Charlottesville, VA 22904, U.S.A.

During the early 1980s, the capillary blood vessel networks of the human body were just being understood as dynamic entities, in which the process of angiogenesis could produce dramatic new growth to accomplish wound repair, fracture healing, and help heal the injured heart, as well as produce dangerous growth of tumors by providing them with nutrients. Other vessels, such as larger arterioles and venules, were viewed as much more passive, although there was historical evidence that they could be just as dynamic, adjusting to both biochemical stimuli as well as mechanical stresses. Dr. Fung illuminated the path for many in this field to define the quantitative ways in which physical forces affect blood vessel remodeling, both in the large vessels as well as in the microcirculation. In microcirculation, the complex network architecture and low Reynolds number flow regime combine to produce a system in which knowledge of both individual vessel constitutive properties and the vessel connectivity to other parts of the network are essential to understanding adaptive remodeling. Dr. Fung was the first to realize this and shed light on the problem in the pulmonary circulation. Others then followed the path with analyses of various tissue microcirculations, expanding our understanding of organ capacity for growth and change, and opening the way to today's 21st century designs for engineered tissues with synthetic microcirculatory networks.

1. Let There Be Light – UCSD, Dr. Fung, and Biomechanics

The motto of the University of California, San Diego is "Fiat lux", which means "Let there be light". It might be said of Dr. Fung's unmatched life in bioengineering that he dedicated himself to dispelling darkness or vague thinking about various tissue and organ systems, and replaced the darkness with light and understanding. The union of UCSD, Dr. Fung, and the young field of biomechanics produced a rate of growth for all three entities that was truly dazzling. James Thurber said "There are two kinds of light – the glow that illuminates, and the glare that obscures." Clearly, the glow of Dr. Fung's vision has been of the first type, illuminating the world of biomechanics for more than half a century since 1957, when he first became involved with biology and medicine. That was the year I was born, so my career in biomechanics has indeed been fully illuminated by the path Dr. Fung created. I cannot adequately express the gratitude I feel for his role model during the years I worked with or near his influence. Dr. Fung is a role model of the "paradoxical combination of professional will and personal humility" characteristic of great leaders [1]. His reverberating laughter, heard in the Basic Science Building of UCSD during my time there, was the surest sign that all was well and that ideas were being created daily. People there were expected to work hard, share ideas freely, and immerse themselves in the ferment of growing the world of biomechanics. For this experience, I thank Dr. Fung and hope that his example will serve the growth of new fields as we seek to solve new problems for the good of society.

2. First Meeting with Dr. Fung

Upon arriving at UCSD in the fall of 1979, I first met with Dr. Fung at his unassuming office in the middle of the fifth floor of the basic Science Building, which contained all the faculty labs at that time, including those of my advisor Dr. Geert Schmid-Schonbein, Dr. Marcos Intaglietta, and Dr. Benjamin Zweifach, the great microcirculationist whose lab would be part of my training grounds. Dr. Savio Woo's orthopedic biomechanics lab was just across the street. Dr. Fung offered me a friendly welcome and suggested that I pay attention to my initial year of studies and select a dissertation subject that would add fundamental knowledge to the biomechanics of circulation. That was a tall order for a new PhD student, but typical of Dr. Fung's method – which was to take in new acquaintances of all ages and talent levels, and share his ideas and vision freely.

3. A Great Mentor

As a classroom teacher, Dr. Fung was both inspiring and demanding. I remember the very first day of a graduate class at UCSD on Biomechanics in 1979. He asked the class to derive various equations for him on the board, including the equations of motion of continuum mechanics, and much of the rest of Chapter 10 of his famous small book on the subject, "A First Course in Continuum Mechanics" [2]. Of course most of us were much too nervous to write any such thing for such a great man, and the result was that he sent us home immediately with the admonition to be prepared to formulate new problems with field equations and boundary conditions at any time – for this was the essence of his research approach, and the thing that he most wished us to learn and practice. Needless to say, our copies of "A First Course" were soon heavily tattered, and Chapter 10 became one of our most highly consulted references. Subsequent lectures were a pleasure, filled with his vision for the methods to formulate new problems involving highly deformable materials with irregular geometries, the essence of biomechanics. The notes he used in that class became the basis for his now classic text "Biomechanics: Mechanical Properties of Living Tissues", first published in 1981 [3].

My PhD advisor, Geert Schmid-Schonbein, had produced his own dissertation work under the guidance of Dr. Fung, so I was fortunate to have two visionary biomechanicists on my own dissertation committee. Geert suggested that I pursue a quantitative reconstruction of the capillary, arteriolar, and venous networks in skeletal muscle, measure the mechanical properties of each class of vessel, and combine the information in a network analysis to understand the pressure-flow relationships in this important tissue. This was a great period of work for me, as Geert was always the optimist and always a source of ideas when the work hit any impasse. Near the end of the work, when we had nearly completed the initial whole muscle computations, I went to visit Dr. Fung to obtain his final advice for the preparation of my thesis document. After seeing all the network data and computational analyses, and the individual vessel mechanics data, he had one comment. He said "I suggest that you focus on providing a complete set of biomechanical properties for the small arterioles, venules, and capillaries for this tissue. This is unique and novel information, and others will be able to build upon it to study many problems in muscle." True to his vision, this data has indeed been used many times since then, to compute capacity to adapt to hemorrhagic shock, to estimate the arteriolar response to Type 2 diabetes, to estimate the flow effects of small venule blockage by adherent leukocytes or platelet aggregation, and most recently to estimate how bone marrow-derived stem cells might influence venular remodeling.

That work led my own lab at the University of Virginia to pursue the natural consequences of microvessels that were subjected to stresses outside the normal range, and thus to develop models of adaptive remodeling in the microcirculation. This research on microvessel network adaptation led to publications by my own PhD student Richard Price [5] and Shayn Peirce (Peirce et al., 2004) on the subjects of circumferential stresses in arteriolar growth and cellular level adaptation of whole networks, respectively. This work and that of others is summarized in Skalak (2005). I attribute much of this progress in understanding to Dr. Fung's lesson to try our best to choose important subjects and to illuminate them with proper use of biomechanics, geometry, and biology. I hope that Rich, Shayn, W. Lee Murfee, and many others who have followed will keep that lesson alive. Certainly, we have always made a point to recognize the degree to which we are standing on the shoulders of others – in this case, the

                                    T. C. Skalak

broad shoulders of Dr. Fung, Dr. Ben Zweifach, and Dr. Geert Schmid-Schonbein, among others who illuminated this path for us.

When I co-edited a book on Microvascular Mechanics with J.S. Lee at the University of Virginia in 1989, Dr. Fung again provided a role model for others in writing a concluding chapter for the book, tracing the origins of his pulmonary flow theory and connecting the methodology to that of many other chapters in the book [4]. His goal was to show how a detailed treatment of the micromechanics could be linked to and explain many important macroscale organ behaviors – this approach would later become popularly termed "multiscale modeling" and is the basis for many federal agency program emphases today.

4. Memories

One of my most distinct memories of Dr. Fung's influence on the global practice of biomechanics is from 1984, when the First World Congress of Biomechanics was held in La Jolla on the UCSD campus. I gave my first major talk to a large professional audience there, with Dr. Fung in the front row. It was remarkable that essentially the entire world of biomechanics had assembled, and each speaker had a personal story to tell of Dr. Fung's influence on their work. I presented data on the viscoelastic properties of microvessels. Dr. Fung, as always, was able to motivate my later work with a simple word of encouragement, saying "These material properties are fundamental. Keep it up."

I have particularly fond memories of holiday parties with the Fungs and the Sobins – these large faculty-student-staff gatherings were like family parties. At the Del Mar residence of Sid and Venus Sobin, we passed many a happy hour, talking of both biomechanics and the various creative processes involved in making models and simulations. When Venus, a visual artist, immortalized Dr. Fung's strain energy expression using free-form brush strokes in a painting, it symbolized for us the inherent collaborative creativity of these great thinkers and bioengineers, supported by the artistic frontier of UCSD's larger community.

At one point in my graduate career, I sustained a fractured cheekbone in a cycling accident. Dr. Fung took the personal time and attention to make sure that I received the orthopedic surgery treatment to get it repaired correctly, even calling on the surgeon, a personal friend of his, to make sure what was happening at each stage. Luckily, no osseointegration fixtures were needed, just a little bit of leverage to produce the natural geometry and some natural healing. Probably some passive micromotion and bone marrow-derived cells were involved, but none of that was yet known to medicine or bioengineering. His help was like that of a father for a son, and I will always be grateful for that.

5. Influences

In my own courses on biomechanics, I have followed extensively the key advice offered by Dr. Fung in his early book "A First Course in Continuum Mechanics" [1]. There, he focused on simplicity and a physical sense for mechanics. Some of the initial sentences of that book will always be with me, and I suspect with many students of mechanics. For example, on page one he states that "Our objective is to learn how to formulate problems and how to reduce vague questions and ideas to precise mathematical statements." His examples are: "An airplane is flying above us. How much strain are the wings subjected to? ... Ahead you see the Golden Gate Bridge. How does one design such a cable? ... How do we breathe?" These kinds of big questions were liberating for me and for many students who have used the textbooks produced by this superb author. And, Dr. Fung practiced what he taught. He illustrated this one day in class for us. He told us he had recently completed a cross-country airline flight and had flown through a very dramatic thunderstorm. At one point, he looked out his window, and estimated that the wingtips were deflecting up and down by at least six feet (2 meters). He calmly did a mental calculation and informed the alarmed passenger next to him that everything was fine ... at least until the wingtips reached about 8 feet of deflection!

In advising my own past students, some now leading academic labs of their own and some working in private industry, I followed the example of Dr. Fung to share ideas freely, and challenge even the new arrival in graduate school to dream, to define a significant vision, and to work for its solution. This has led to rewarding collegial relationships and long friendships that I treasure.

As a past-President of the Biomedical Engineering Society (BMES), I can say that Dr. Fung's influence was very large in the formation and the pervasive influence of this organization. First, many of the officers of BMES have had training or inspiration via Dr. Fung. During my term, BMES became the lead society for ABET in the field of bioengineering. This is a significant legacy of Dr. Fung's ideas, because indeed almost all ABET-accredited programs today depend on the fundamentals of biomechanics as a cornerstone of the undergraduate student's education. So, much of the fundamental work on constitutive properties and tissue mechanics done by Dr. Fung and associates is now a required part of education for the biomedical engineer. They, in turn, bring this knowledge out to industry roles and teaching roles, where it makes a positive difference for human health and dignity.

Dr. Fung was a founding member of the American Institute of Medical and Biological Engineering (AIMBE). This organization strives to lead education and advocacy efforts in Washington, D.C. and beyond, to achieve societal recognition of the past achievements of bioengineers, so that the future can produce the many innovations needed to improve human health. As the current President-elect of AIMBE, one goal is to live up to the original aspirations of the founders, including Dr. Fung, and solidify the national recognition for the economic benefits derived from bioengineering innovations and the works of social good performed by bioengineers.

In raising my own children, I will endeavor to pass on to them a sense of the joy about life and the essential nature of change itself that Dr. Fung taught through his example. It is this realization that produces the intense desire to enjoy life and work, family and friends, and urgency to make some small contribution to others, perhaps even a contribution that will stand the test of time as an example to others, even as knowledge itself and the world evolve. The theme of remodeling is so intrinsic to the cycles of life and our environment on earth that one can see how an intellect so broad as Dr. Fung's was attracted to explain the dynamic adaptive phenomena he observed in the lung and later in arteries.

To Dr. Fung, from those of us who have worked with you and known you – thank you. We regret nothing, and are honored to have glimpsed you. You have illuminated our path. Fiat lux.

References

1. J. Collins, *Good to Great,* (Harper Business, 2001).
2. Y.C. Fung, *A First Course in Continuum Mechanics,* (Prentice Hall, 1977).
3. Y.C. Fung, *Biomechanics: Mechanical Properties of Living Tissues"* (Springer-Verlag, 1981)
4. Y.C. Fung, "Microvascular Mechanics", eds. J.S. Lee and T.C. Skalak, (Springer-Verlag, 1989), p. 191.
5. R.J. Price and T.C. Skalak, *Microcirculation* **2**, 41 (1995).

Chapter 6

COMPUTATIONAL SIMULATIONS OF THE BUCKLING OF OVAL AND TAPERED ARTERIES[*]

AVIONE NORTHCUTT

Biomedical Engineering Program, University of Texas at San Antonio
One UTSA Circle San Antonio, TX 78249, U.S.A.

PARAG DATIR

Dept of Mechanical Engineering, University of Texas at San Antonio
One UTSA Circle San Antonio, TX 78249, U.S.A.

HAI-CHAO HAN

Dept of Mechanical Engineering/Biomedical Engineering Program, University of Texas
at San Antonio, One UTSA Circle San Antonio, TX 78249, U.S.A.

It has been shown that cylindrical arteries buckle under hypertensive pressures or reduced axial tension and become tortuous. Arteries often demonstrate geometric variations such as elliptic cross sections and tapering along the longitudinal axis but the effects of these variations on the buckling behavior of the arterial walls have not been investigated. The objective of this study was to determine the buckling pressure and deformation patterns of elliptic and tapered arteries. Models of cylindrical arteries with circular and elliptic cross sections, and circular tapered arteries were generated with the walls modeled as a homogenous isotropic nonlinear material. The buckling analysis of these arteries under lumen pressure was implemented using finite elemental analysis. Our results demonstrated that arteries buckle into tortuous shapes under internal pressure. Oval initial cross sections increased the critical pressure and arteries tend to buckle in the short axis direction. In contrast, tapering reduced the critical buckling pressure and led to skewing of the deformation profile towards the distal end of the artery. The post-buckling behavior of the arteries showed kinking towards the distal end. We concluded that oval arteries are more stable than the circular arteries while tapered arteries are more prone to buckling and kinking with a more tortuous shape in the distal portion. These results will help us better understand the development of vessel tortuosity, such as aorta tortuosity in the abdominal region.

1. Introduction

Arteries are under large lumen pressures and longitudinal tensions *in vivo* [1, 2]. Longitudinal tension is important in maintaining the mechanical stability of long artery and vein segments since reduced longitudinal tension is associated with

---

[*] This work was supported by a CAREER Award (#0644646) from the NSF.

53

aging or vascular surgery and has been shown to lead to tortuous arteries [2-4]. We have recently demonstrated that arteries can buckle under hypertensive lumen pressures or reduced axial tension [5-7] and become tortuous.

Arteries often demonstrate geometric variations such as oval cross sections and tapering along the longitudinal axis [1, 8-12]. These variations may affect the buckling behavior of the arterial walls and need to be investigated. Previous studies have demonstrated that tapering significantly affects the blood flow and shear stress in arteries [13-17] and stents [18] and thus was linked to the development of atherosclerosis [19]. However, the effects of tapering on the buckling behavior of the arterial walls are still unknown.

The objective of this study was to determine the buckling pressure and deformation patterns of elliptic and tapered arteries using finite element analysis of model arteries.

2. Methods

The buckling analysis was performed for model arteries with circular and elliptic cross-sections and arteries with tapering under lumen pressure and axial stretch using finite elemental analysis.

2.1. *Artery Models*

Models of cylindrical arteries with circular and elliptic cross sections, or with tapering along the longitudinal axis, were generated using the commercial software SolidWorks®. The lumen diameter, wall thickness, and the length of the circular cylindrical vessel were assumed to be 20 mm, 2 mm, and 254 mm, respectively, to represent a typical human aorta [20, 21]. Cross-sectional minor to major axis ratios at levels of 1/5, 2/5, 1/3, 1/2, and 3/4 were created for the elliptic models to examine the effect of the oval shape (Figure 1). Specifically, the major diameter and wall thickness remained unchanged at 20 mm and 2 mm, respectively, but the minor diameter was reduced to 4 mm, 8 mm, 6.7 mm, 10 mm, and 15 mm accordingly to simulate different levels of an ellipse.

The tapered artery was modeled as an idealized cylindrical vessel with the same length, same lumen diameter and wall thickness (254 mm, 20 mm, and 2 mm, respectively) at the proximal end but with smaller lumen diameters and wall thickness at the distal end. The tapering level was described by:

$$Tan\ \theta = \frac{R_{proximal} - R_{distal}}{L} \tag{1}$$

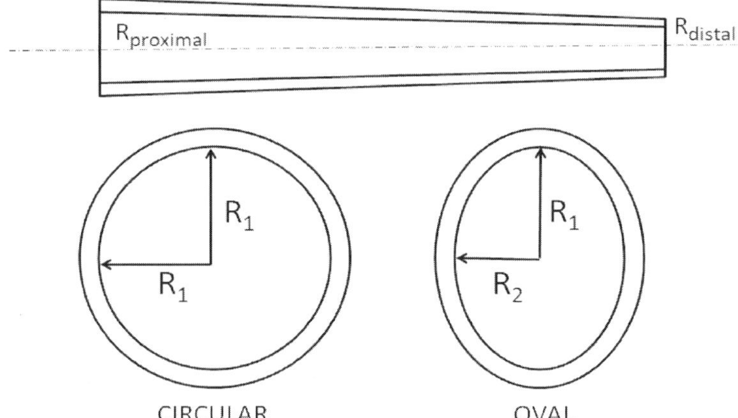

Figure 1. Schematics of models of arteries with circular, elliptic cross-sections, and an aneurysm. The oval shape of the cross section is defined by the major to minor axis ratio R_1/R_2.

where L, $R_{proximal}$, and R_{distal} are the axial length and the lumen radius at the proximal and distal ends, respectively (Figure 1). A lumen diameter of 16 mm, 10mm, and 6 mm was used for the distal end to simulate a taper angle of 0.45°, 1.13°, and 1.58°. The distal wall thickness was determined by maintaining the same wall thickness to radius ratio, i.e. 1.6, 1, and 0.6 mm, accordingly.

The three dimensional artery models were meshed in ABAQUS by using two layers of hybrid, three-dimensional tetrahedral elements (C3D4H).

2.2. *Material Model*

The arterial wall was modeled as a homogenous, incompressible, isotropic nonlinear material with a Mooney-Rivlin type strain energy function given in the form of:

$$w = C_1(I_1 - 3) + C_2(I_1 - 3)^2 + C_3(I_1 - 3)^3 \tag{2}$$

where C_1, C_2, and C_3 are material constants. We used the material constants from Raghavan et al. [22] that described the average of axial and circumferential material constants and were $C_1 = 3.757$ kPa, $C_2 = -0.9931$ kPa, and $C_3 = 1.409$ kPa.

For comparison, the critical pressure of the cylindrical arteries were also obtained using the buckling equations established by Han with the Fung-type strain energy function [6].

2.3. *Loads and Boundary Conditions*

A uniform internal pressure was applied to the lumen of the arterial models and the external pressure was set at zero. Both ends of the arteries were assumed to be fix-supported with no lateral displacement or rotation but were allowed to expand radially. These end conditions simulated the expansion of arteries under pressure *in vivo* and minimized the possible edge effects at the ends. Since arteries are under significant longitudinal strain *in vivo* [1, 2]. Axial stretch ratios of 1.0 (control), 1.3, 1.5, and 1.7 were applied to the artery models by applying designated axial displacements to all nodes comprising the distal end of the arteries to achieve the given stretch ratios.

2.4. *Finite Element Analysis*

All artery models were analyzed under lumen pressure and given boundary conditions using ABAQUS. Two types of analysis were performed to determine the buckling mode and critical pressure. A linear perturbation buckling analysis was performed using the Lanczos solver [23]. The results provided eigenvalues and buckling modes; the eigenvalues were then used to find the critical pressure.

The second was a Static General Analysis under stretch. A small initial bend of 5 degree along the central axis of the artery was created to facilitate the buckling analysis. The critical pressure of the arteries was determined by applying a range of pressure values to the arteries and obtaining the maximum deflection profile to find the pressure at which the deflection starts to increase from baseline (i.e. when the arteries buckle or bend). This approach also yielded the deflection of the arteries post-buckling.

3. **Results**

We have simulated the buckling behavior of arteries with circular and various oval cross sections and with tapering using finite element analysis. Distributions of the maximum and minimum Von Misses stresses of the arteries as well as displacements were obtained. We have also determined the effect of stretch ratio on the critical pressure of the arteries and the results are summarized below.

3.1. *Buckling of Circular Cylindrical Arteries*

Arteries buckled when the lumen pressure exceeded the critical pressures. When the pressure was below the critical pressure, arteries only expanded radially. The diameters increased as the pressure increased. When the lumen pressure reached a critical pressure, arteries started to buckle and the deflection started to increase

(Figure 2). The deflection continued to increase nonlinearly with increasing pressure post-buckling. The buckling led to local stress concentrations at the peaks and valleys (Figure 3). The higher-order buckling modes occurred at higher critical pressures and led to higher maximum stresses in the arterial wall.

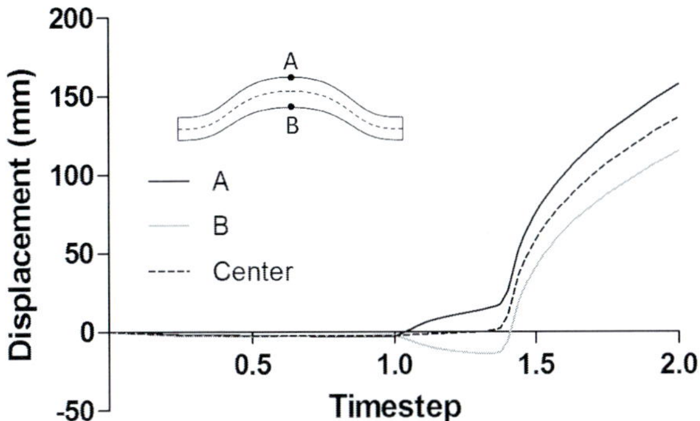

Figure 2. Peak displacements of top and bottom edges (points A & B) as well as the center of a cylindrical artery plotted with the loading step in an FEA simulation. It is seen that the diameter first decreased under axial stretch (timestep 0-1) and then gradually increased when an increasing pressure is applied (timestep 1-2) since the displacement of points A & B moved evenly in the opposite directions. The displacements of points A & B started to move in the same direction at a critical pressure (timestep 1.4). Accordingly, the center (dotline) started to deflect in the direction while the diameter continued to increase under the increasing pressure.

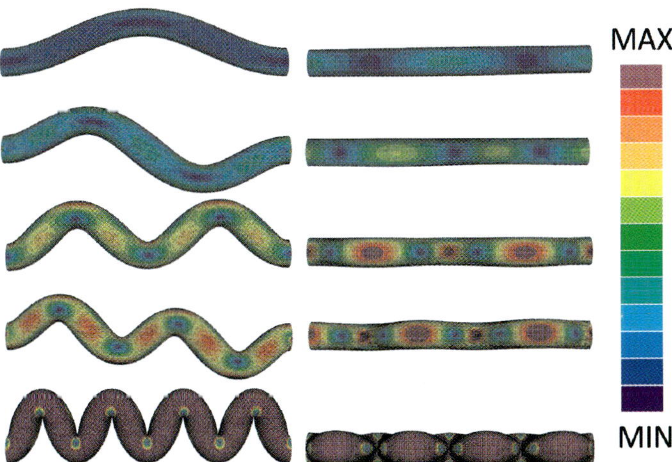

Figure 3. Buckling modes and von Mises stress (color coded) distributions of a circular cylindrical artery. The first four buckling modes and a higher order 7th buckling mode were summarized. The left images show a side view and the right show a top view.

Furthermore, the axial stretch ratio had a significant effect on the critical pressure of arteries (Figure 4). The lower the stretch ratio the easier it was to make the artery to buckle. At stretch ratios of 1.3, 1.5, and 1.7 the arteries buckled at 0.7, 1.6, and 3 kPa, respectively. These results match well with the theoretical results obtained using the buckling equation proposed by Han [6] (Figure 5).

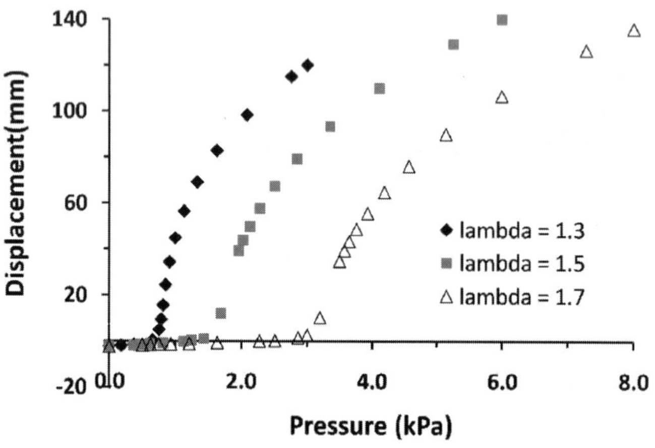

Figure 4. Deflection of an artery plotted against the lumen pressure pre- and post-buckling at three different axial stretch ratios (1.3, 1.5, and 1.7). The deflection starts at the critical pressure and continues to increase nonlinearly post-buckling as the pressure increases. The deflection starts at a higher critical pressure value at a higher axial stretch ratio.

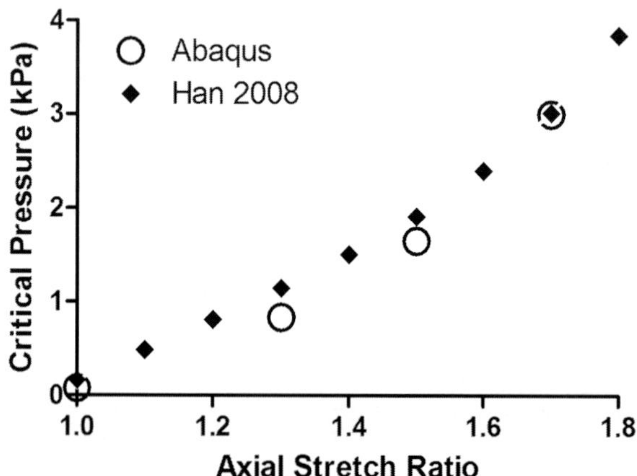

Figure 5. Comparison of critical pressure obtained using FEA and theoretical buckling equations given by Han [6] and plotted as functions of the axial stretch ratio.

3.2. *The Effect of Oval Cross-Section*

Arteries with oval cross sections buckled in the direction of the minor diameter at the basal mode due to a smaller sectional modulus, EI, in this direction (Figure 6). The higher order buckling modes occurred in both directions of the major and minor diameters. Interestingly, when oval arteries buckled in the minor axis direction, the high stress zones were along the neutral axis of the bending where tips of the oval cross section were before loading. This is due to the fact that a high stress zone was generated there when the oval arteries were inflated under lumen pressure. Compared to circular arteries, the oval arteries had higher buckling pressures. Furthermore, the severer the oval shape (the smaller the minor to major diameter ratio), the higher the critical pressure (Figure 7).

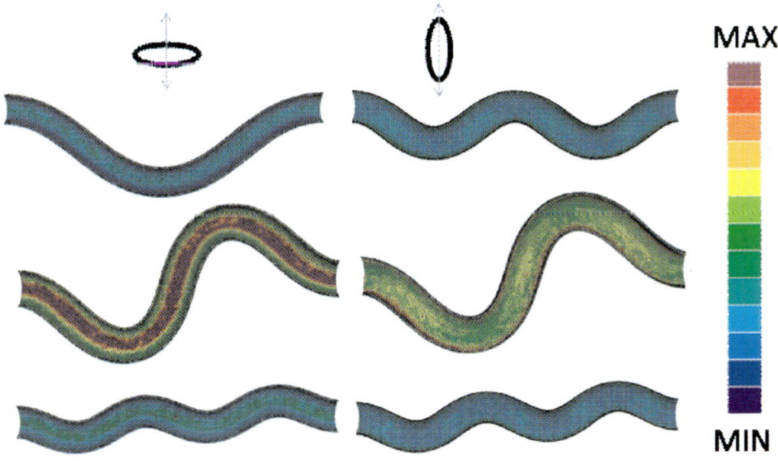

Figure 6. Comparison of buckling modes and stress maps of arteries with oval and circular cross sections (R_2/R_1=1/5). Left panels: Buckle modes (n = 1, 2, 4) in the minor axis direction. Right panels: Buckle modes (n = 3, 2, 4) in the major axis direction. The arrows illustrate the direction of the buckling.

                    A. Northcutt et al.

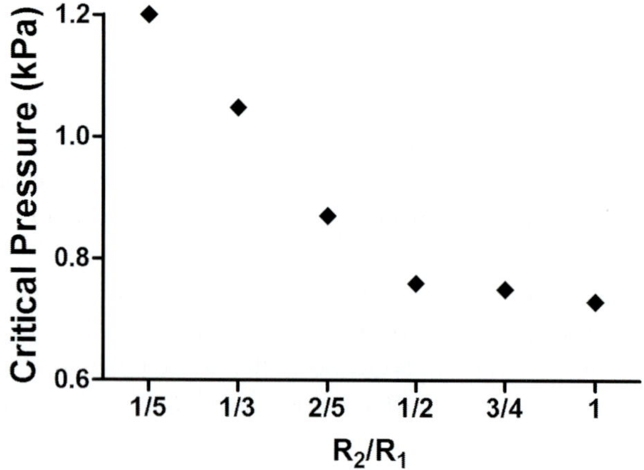

Figure 7. Critical pressure plotted as a function of ovalness level (minor to major axis ratio R_2/R_1).

3.3. *The Effect of Vessel Tapering*

Tapering reduced the critical pressure of the artery as compared to the cylindrical artery of a same diameter of the proximal end. Further, tapering made the buckling mode shape deviate from the sinusoidal shapes and the peak deflection point shifted towards the distal end (Figure 8). In addition, kinking occurred in some of the higher buckling modes.

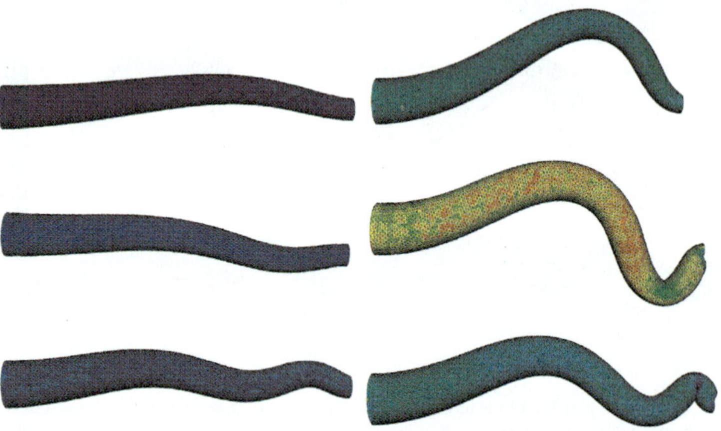

Figure 8. Comparison of buckling modes and stress maps (color coded) of tapered arteries. Top to bottom panels are mode shapes n=1, 2, 3 respectively. The left panels show the initial buckled shape at the critical pressure and the right panels show the post-buckling mode shapes at pressures beyond the critical pressures.

4. Discussion

We studied the buckling of arteries with oval cross sections and tapering using finite element analysis. Our results confirmed that an increased axial stretch ratio significantly increased the critical pressure and vice versa. An oval cross section increased the critical pressure and oval arteries tend to buckle in the short axis direction. In contrast, tapering reduced the critical buckling pressure and shifted the peak deflection point towards the distal end.

An interesting finding is that the ovalness of the cross section actually increases the critical pressure compared to the circular ones and thus improves the stability of the artery. It may seem counter-intuitive but can be explained based on our previous work. Even with an initial oval shape, an artery's cross sectional shape under lumen pressure is nearly as circular as an initially circular vessel. However, the difference is that a "prestress" is developed under the pressure, and as we showed previously, that "prestress" increases the critical pressure [6]. Thus, our results indicate that the slightly oval-shaped cross section is beneficial to the mechanical stability of the arteries.

4.1. *Clinical Relevance*

Buckled arteries demonstrate a tortuous path that greatly resembles the tortuosity of arteries commonly observed in angiography. Arterial tortuosity is seen on 10 to 40% of the population [24-26] and leads to symptoms such as transient ischemic attacks, cerebrovascular symptoms, and stroke [27-30]. The similarity between artery tortuosity *in vivo* and the buckling mode shapes, especially the higher order ones, suggests that artery mechanical buckling could be a possible mechanism for some of the tortuosity we see *in vivo*.

There is a significant taper in the human aorta, from thoracic to abdominal (iliac bifurcation), and tortuosity occurs more often in the abdominal aorta than the thoracic [2]. Our results suggest that the tapering effect on buckling modes might be a possible mechanism for the localized tortuosity in the abdominal region.

4.2. *Limitations*

Though arterial walls are anisotropic with a layered structure, we used a nonlinear isotropic and homogenous model in the analysis. Arterial walls are often considered transversely isotropic by many investigators with the circumferential and longitudinal stiffness close to each other [31]. While errors exist, the isotropic material model provides reasonable results as previously shown [22]. We have previously showed that cylindrical vessels buckle under internal pressure using both linear elastic and nonlinear anisotropic Fung-type material models [5-7]. The current study used a nonlinear isotropic Mooney-Rivlin material model and the results are similar when comparing the critical pressure to axial stretch ratio relationships [6] (see Figure 5). Furthermore, the post-buckling deflection given in Figure 4 are similar to our previous experimental observation measurement of the post-buckling deflection curves of porcine carotid arteries [32]. These agreements indicated that the isotropic approximation used here is reasonable.

On the other hand, tissue tethering and the dynamic effect from the pulsatile flow were neglected in the current study and will be address in our future studies.

4.3. *Significance*

Our results demonstrated the buckling modes of arteries under internal pressure. It also demonstrated the effect of oval cross section and tapering on the buckling behavior of arteries. These models better simulate real arteries than the cylindrical models. These results not only enrich our understanding of the mechanical behavior of blood vessels but also provide a new approach to study vessel tortuosity using computational simulations.

Acknowledgments

This work was supported by a CAREER award (#0644646) from the National Science Foundation. It was also partially supported by a research grant (#0602834) from the National Science Foundation, a seed grant from UTSA, and an MBRS-RISE fellowship (GM60655) from NIH. We thank Dr. Neil Dong for his technical help in ABAQUS computation. We thank the Computational Biology Initiatives at UTSA for the use of the computational facilities and for their technical support.

References

1. H. C. Han and Y. C. Fung, *J Biomech* **28**, 637-41 (1995).
2. W. W. Nichols and M. F. O'Rourke, *McDonald's Blood Flow in Arteries: Theoretical, Experimental, and Clinical Principles. 4th Edition. Chapter 16.* London, Arnold Publisher.(1998).
3. H. C. Han, L. Zhao, M. Huang, L. S. Hou, Y. T. Huang, and Z. B. Kuang, *J Biomech Eng* **120**, 211-6 (1998).

Author (Han) with doctoral advisor Dr. Fung (1990)

4. Z. S. Jackson, D. Dajnowiec, A. I. Gotlieb, and B. L. Langille, *Arterioscler Thromb Vasc Biol* **25**, 957-62 (2005).
5. H. C. Han, *J Biomech* **40**, 3672-8 (2007).
6. H. C. Han, *J Biomech* **41**, 2708-13 (2008).
7. H. C. Han, *Molecular & Cell Biomech.* **6**, 93-100 (2009).
8. Y. C. Fung and S. Q. Liu, *Circ Res* **65**, 1340-9 (1989).
9. H. C. Han and Y. C. Fung, *J Biomech Eng* **113**, 446-51 (1991).
10. D. Fleischmann, T. J. Hastie, F. C. Dannegger, D. S. Paik, M. Tillich, C. K. Zarins, and G. D. Rubin, *J Vasc Surg* **33**, 97-105 (2001).
11. N. F. MacLean and M. R. Roach, *Heart Vessels* **13**, 95-101 (1998).
12. A. R. Zeina, U. Rosenschein, and E. Barmeir, *Coron Artery Dis* **18**, 477-82 (2007).
13. Y. C. Fung, *Biomechanics: Circulation, Chapter 4*. New York, Springer. (1997).
14. L. H. Back, E. Y. Kwack, and D. W. Crawford, *J Biomech Eng* **110**, 310-9 (1988).
15. R. K. Banerjee, L. H. Back, and Y. I. Cho, *Biorheology* **32**, 655-84 (1995).
16. U. Krueger, A. Huhle, K. Krys, and H. Scholz, *Artif Organs* **28**, 623-8 (2004).
17. L. J. Myers and W. L. Capper, *Med Eng Phys* **26**, 147-55 (2004).

18. L. H. Timmins, C. A. Meyer, M. R. Moreno, and J. E. Moore, Jr., *Ann Biomed Eng* **36**, 2042-50 (2008).
19. P. M. Sundell and M. R. Roach, *Atherosclerosis* **139**, 123-9 (1998).
20. J. Sugawara, K. Hayashi, T. Yokoi, and H. Tanaka, *JACC Cardiovasc Imaging* **1**, 739-48 (2008).
21. A. Wolak, H. Gransar, L. E. Thomson, J. D. Friedman, R. Hachamovitch, A. Gutstein, L. J. Shaw, D. Polk, N. D. Wong, R. Saouaf, S. W. Hayes, A. Rozanski, P. J. Slomka, G. Germano, and D. S. Berman, *JACC Cardiovasc Imaging* **1**, 200-9 (2008).
22. M. L. Raghavan, S. Trivedi, A. Nagaraj, D. D. McPherson, and K. B. Chandran, *Ann Biomed Eng* **32**, 257-63 (2004).
23. ABAQUS, *Abaqus Analysis Users Manual.*
24. J. Pfeiffer and G. J. Ridder, *Laryngoscope* **118**, 1931-6 (2008).
25. M. Aleksic, G. Schutz, S. Gerth, and J. Mulch, *J Cardiovasc Surg (Torino)* **45**, 43-8 (2004).
26. F. Paulsen, B. Tillmann, C. Christofides, W. Richter, and J. Koebke, *J Anat* **197 Pt 3**, 373-81 (2000).
27. T. J. Leipzig and G. J. Dohrmann, *Surg Neurol* **25**, 478-86 (1986).
28. K. Okami, J. Onuki, K. Ishida, T. Kido, and M. Takahashi, *Auris Nasus Larynx* **28**, 373-6 (2001).
29. G. Illuminati, J. B. Ricco, F. G. Calio, A. D'Urso, G. Ceccanei, and F. Vietri, *Surgery* **143**, 134-9 (2008).
30. E. S. Zegers, B. T. J. Meursing, E. B. Zegers, and A. J. M. O. Ophuis, *Netherlands Heart Journal* **15**, 191-195 (2007).
31. Y. C. Fung, *Biomechanics: Mechanical Properties of Living Tissues.* New York, Springer Verlag. (1993).
32. C. A. Fierro, K. Johnson, and H. C. Han, ASME Summer Bioengineering Conference, Keystone, Colorado, (2007).

Chapter 7

ROLE OF STRUCTURAL AND SIGNALING MOLECULES IN CARDIAC MECHANOTRANSDUCTION

ANNA M. RASKIN

Department of Bioengineering, University of California at San Diego
La Jolla, CA 92093-0412, U.S.A.

ANDREW D. MCCULLOCH

Department of Bioengineering, University of California at San Diego
La Jolla, CA 92093-0412, U.S.A.

JEFFREY H. OMENS

Department of Bioengineering, University of California at San Diego
La Jolla, CA 92093-0412, U.S.A.

The link between mechanical stimuli and alterations in cellular gene/protein expression is defined by the processes of mechanotransduction. Several cell types, which include osteoblasts, fibroblasts, smooth, cardiac, and skeletal muscle cells, belong to the "mechanocyte" family. These cells are able to detect mechanical signals and transform them into a biological response. Mechanical loads applied to cardiac cells can promote cellular growth (hypertrophy) or degeneration (atrophy). The mechanisms by which cardiac myocytes sense mechanical signals are still being uncovered – structures such as the cell membrane and the internal cytoskeleton and sarcomeres may be involved with stress sensing and mechanotransduction. Recent work has implicated the sarcomeric elastic protein titin and several binding partners in stress-sensing pathways. Defects in mechanotransductive processes may be an independent risk factor underlying cardiovascular maladies including cardiomyopathy and heart failure.

1. Introduction

The cardiac cells within the myocardium are constantly sensing their external environment and altering cellular function based on external cues. Physiologic growth through adulthood is highly dependent on myocyte hypertrophy, and many of the normal growth and remodeling functions are again activated in cardiac diseases in which external loading conditions are abnormal. How myocytes sense external (and possibly internal) stresses is unknown, although much evidence points to involvement of the intercellular cytoskeleton as a key player in stress sensing and mechanotransduction. When these stress-sensing pathways become altered, downregulated cellular responses (such as

65

hypertrophy) can lead to contractile dysfunction at both the cellular and tissue levels, eventually leading to heart failure.

2. Cardiac Hypertrophy

Cardiac hypertrophy is defined as an increase in heart muscle mass due to an increase in the size of constituent myocytes. It is a physiological adaptation of the heart in response to an increase in stress [1]. It is also a major independent risk factor of death, underlying various cardiovascular maladies including heart failure, dilated cardiomyopathy, ischemic heart disease, and sudden death [2, 3]. Myocardial hypertrophy is characterized by an increase in cardiomyocyte size, increased protein synthesis, and re-activation of genes for contractile proteins and natriuretic peptides (atrial natriuretic peptide (ANP) and B-type natriuretic peptide (BNP)) [2, 4]. Currently, stress-induced cardiac hypertrophy is thought to possibly involve growth and neurohumoral factors (endothelin-1, angiotensin-II, insulin-like growth factor, vascular endothelial growth factor), apoptosis, ischemia, and biochemical signaling events induced by mechanical stimuli. Mechanical load is thought to be a primary stimulus for the development of hypertrophy [4]. The focus of this chapter will be to evaluate recent findings on cardiac myocyte mechano-biochemical transduction processes that lead to cellular hypertrophy.

3. Myocardial Stress and Strain Regulate Cardiac Muscle Growth

The most common cause of cardiac hypertrophy in the human population is an increase in blood pressure, which occurs in patients with hypertension or stenosis of the aortic valve [5]. Although a variety of stimuli play a role in the development of hypertrophy in these patients, the increase in myocardial systolic stress, which is commonly associated with an increase in myocardial diastolic stress, is believed to be a major cause for thickening of the ventricular wall. In vivo animal studies further indicate that chronically overloading the left ventricle (LV) by coarctation of the ascending aorta results in an increase in LV mass (hypertrophy), while chronically unloading right ventricular (RV) papillary muscles by transecting the cordea tendinae results in a decrease in tissue mass (atrophy) [6, 7]. In combination these observations suggest that mechanical load plays a critical role in the regulation of myocardial mass and the development of hypertrophy.

Consistent with the role mechanical load plays in regulating cardiac hypertrophy in vivo, models employing stretch of cells and tissues have been used to demonstrate that mechanical load also modulates the development of hypertrophy in vitro. Many labs have utilized a cell culture model and have demonstrated in neonatal ventricular myocytes that a static stretch of 10 to 20% can induce the in vivo hypertrophic phenotype, characterized by an increase in protein synthesis, induction of immediate early genes, and re-expression of genes for contractile proteins and natriuretic peptides [8-10]. Consistent with these findings, the primary response to an increase in tissue stress and/or strain (relative to an unloaded state) of excised right ventricular papillary muscles is a rapid increase in protein synthesis and alterations in gene expression [11]. In chronically cultured trabeculae muscles the application of high mechanical loads results in changes in protein expression patterns and upregulation of myofilament proteins, events which are correlated with an increase in myocyte and tissue size [12]. Together these data suggest that an increase in external loading alone is capable of initiating hypertrophic biochemical events, as well as hypertrophy at the cellular and tissue level.

4. Mechanotransduction

Cardiomyocytes are dynamic systems that respond to mechanical load with mechanotransduction, where physical forces acting on transmembrane and/or intracellular mechanosensors are converted into biochemical information that regulates gene expression, protein synthesis, and eventual development of hypertrophy [4, 13]. In most cases the transduction of physical forces into biochemical signaling is mediated through changes in cell and/or protein conformation [11, 14]. Posttranslational phosphorylation of proteins alters protein structure thereby modulating enzymatic activity and protein-protein interactions involved in biochemical signaling. Identical to this process, force-induced effects on conformation of proteins or cell structures may represent a general mechanism of mechanotransduction [14, 15].

4.1. *Role of the nucleus in mechanotransduction*

The nucleus is the largest and stiffest organelle within the cell, contains the genome and is the site of transcriptional regulation. Intra-cellular and extra-cellular forces alter nuclear shape and structure. For these reasons, the nucleus has been implicated in the processes of mechanotransduction.

The main mechanism by which the nucleus is believed to act as a mechanosensor is mediated by chromatin. Chromatin is the major component of the nuclear interior and functions to package DNA into the small volume of the nucleus. Forces applied onto the nucleus induce changes in nuclear shape, which are associated with changes in chromatin shape and organization. These changes in chromatin are believed to affect transcription and regulate gene expression [16].

4.2. *Role of z-disk and sarcomeric proteins in mechanotransduction*

Structural proteins at various levels within the myocyte have been linked with development of cardiomyopathy, including mutations in genes encoding for actin (ACTC) [17], titin (TTN) [18], myosin binding protein (MYBPC3) [19], myosin heavy chain (MYH7, MYH6) [20], troponin I (TNNI3) [21], troponin T (TNNT2) [22], and desmin [23]. It is likely that defects in these and other structural proteins will alter stress sensing and mechanotransduction processes. The cytoskeleton is an interconnected structure of proteins that contributes to the stabilization of cell shape and cell structure [24]. Because of its interconnected structure, the cytoskeleton is an obvious candidate for mechanotransduction [25, 26]. A mechanical signal applied to the cell membrane can transverse the cytoskeletal network of proteins by inducing molecular deformations and changes in protein conformation, and can reach the nucleus relatively undiminished [27]. The cytoskeleton of cardiac myocytes can be divided into three parts: the force generating sarcomeric cytoskeleton composed mainly of actin and myosin; the intrasarcomeric cytoskeleton containing titin and alpha-actinin; and the extrasarcomeric cytoskeleton which includes desmin, microtubules, and intermediate filaments [28]. Deficiencies in proteins belonging to any one of these cytoskeletal parts, can lead to significant changes in cytoskeletal architecture. For example, a deletion of MLP or CapZ from the intrasarcomeric cytoskeleton leads to defects in z-disc alignment, causes changes in z-disc width and/or length, and contributes to the disorganization of thin filaments and the contractile apparatus which may affect contractile function of the heart [29, 30].

Desmin, an extrasarcomeric protein, functions as a lateral linker of adjacent z-discs and connects myofibrils with the sarcolemma. Removal of desmin results in lateral misalignment of myofibrils as well as decreased attachment of myofibrils to the sarcolemma [30]. Effective diastolic and systolic force transmission from the inside of myocytes (for example) to z-discs, the sarcolemma, the basal lamina, and to the extracellular matrix is dependent on the structure of the cytoskeleton. The cytoskeleton may also play a crucial role in load sensing and cellular signaling by acting as a scaffold for signaling cascades, providing sites where multiple signaling molecules can localize and attach [31]. Hence, a change in cytoskeletal structure may not only affect force transmission and the material properties of the myocardium, but it may also affect the ability of various signaling molecules to translocate and bind to locations within the cytoskeleton where they can mediate protein-protein interactions involved in particular signaling cascades that may influence gene expression and the development of hypertrophy.

4.2.1. *Titin*

Titin is the most abundant protein of the intrasarcomeric cytoskeleton [30] and has multiple functions that contribute to myocardial mechanics and signaling. Titin spans half the sarcomere from z-line to m-line. When sarcomeres are stretched, the extension of titin's I-band gives rise to diastolic force [32]. Titin's extensible region is composed of proximal and distal immunoglobulin (Ig) like domains, the PEVK segment, and the N2B region. When stretched, these different titin regions, having distinct stiffness values, extend sequentially where the Ig segments extend first, followed by the extension of PEVK and N2B segments. When sarcomeres are shortened below slack length, for example, during systolic contraction, the compression of titin's I-band gives rise to the restoring force that pushes z-discs toward their slack positions. Titin is not the only contributor to myocardial diastolic tension. Collagen, which comprises most of the extracellular matrix of cardiac muscle, has a significant contribution to the diastolic tension of myocardial tissue at long sarcomere lengths [33].

Titin also functions as a scaffold for signaling molecules, which traffic between the cytoskeleton and the nucleus. Titin is bound to the z-disc by Tcap, which mechanically connects titin to signaling molecules: PKC, MLP, and calcineurin [34]. At the I-band, titin interacts by way of FHL2 with hypertrophic signaling molecule ERK2 [35]. Titin's I-band region is also associated with FHL1, which we have shown to bind to signaling molecules: Raf, MEK, and ERK . At the C-terminal end of titin, which is mechanically bound to the M-line, titin contains a catalytic serine-threonine kinase domain, which indirectly interacts with MURF2, a signaling molecule involved in myogenesis that can translocate between the cytosol and nucleus under atrophic conditions [36]. Changes in the deformation of specific titin regions, which are accompanied by conformational changes in titin, may occur when titin-associated structural proteins such as MLP or FHL1 are eliminated from the cytoarchitecture and may affect titin function as a contributor to myocardial mechanics and/or mechanosensing. It may also affect force transmission in the fiber direction of cardiac tissue to "mechano-sensors" embedded within the cytoskeleton, leading to alterations in the biochemical response of the tissue to an increase in mechanical load.

In excised, isometrically contracting, papillary muscles diastolic stress and/or strain has been proposed as the primary mechanical stimulus for myocardial remodeling and load-induced hypertrophy [37, 38]. This evidence supports the claim that titin may play a major role in mechanotransduction, since within the physiological sarcomere length range (1.85-2.4μm) titin is the primary contributor to the diastolic tension of cardiac muscle [33].

4.2.2. *Muscle LIM protein*

Muscle LIM protein, MLP, is a member of a unique subclass of LIM proteins that serve several functions in the cytosol as well as in the nucleus [39]. Cytoplasmic MLP is localized at the z-disc and functions as a mechanical stabilizer. At the z-disc MLP binds to Tcap, alpha-actinin, and N-Rap (another LIM domain protein that binds to actin, talin, and vinculin) [34]. In the nucleus MLP may have signaling roles where it has been shown to interact with transcriptional regulators of myogenesis: MyoD and Murf4. MLP knockout mice develop dilated cardiomyopathy with progressive heart failure [29]. The primary affects of MLP deficiency in mice did not alter the development of systolic stresses in the heart [40], but did contribute to disorganized and widened z-disc structure, diastolic myocardial dysfunction, where the diastolic stiffness of heart tissue was affected, as well as a dysfunction in the development of cardiac hypertrophy, possibly due to impairment of mechanotransduction [13].

4.2.3. *Four and a half LIM domain protein*

FHL1 is a member of the four and a half LIM domain protein subfamily [41]. The four identified members of this protein family include: FHL1, FHL2, FHL3, and FHL4. FHL2 and FHL3 are highly expressed in striated muscle [42]. FHL1 is expressed in most tissues. FHL1 expression is significantly upregulated in hearts of mice treated with hypertrophic agonist and in hearts of human patients exhibiting dilated and hypertrophic cardiomyopathy [43, 44]. Our recent studies have shown that FHL1 is localized to titin's N2B region; the direct interaction of FHL1 with titin was supported with co-purification of FHL1 with titin N2B unique sequence. The interaction of FHL1 with signaling molecules, Raf1, MEK1/2, and ERK2, was also determined. These studies further revealed that at baseline (i.e., normal loading conditions), FHL1-deficient cardiac muscle was not phenotypically different than controls. However, with pressure overload FHL1-deficient cardiac muscle had a blunted response to hypertrophic growth, coupled with reduced ANP gene expression, lower ERK2 phosphorylation levels, and preserved cardiac function. Furthermore, when compared to wild-type controls, FHL1-deficient right ventricular papillary muscles exhibit greater myocardial compliance and a reduced hypertrophic response, characterized by lower ANP gene expression levels, to a chronic increase in mechanical load. Our studies revealed that FHL1 plays an essential role in the development of mechanical stress-induced myocardial hypertrophy. Furthermore, these studies have suggested that FHL1 may play some role in mechanotransduction by linking the passive stretch domain of titin to the ERK pathway to modulate cardiac hypertrophy.

In contrast to MLP knockout mice, impairment of mechanotransduction in FHL1-deficient mice did not result in the development of heart failure. This may be a result of differences in biochemical signaling between FHL1 and MLP knockout mice. The Gq-mediated ERK kinase pathway, which may be blunted in FHL1 knockout mice, is among the best established pathways for the development of pathological hypertrophy [45, 46]. Unlike MLP knockout mice, z-disc structure was not affected with a deletion in FHL1. Disruption of z-disc structure is commonly associated with the development of heart failure [34]. The z-disc functions as a junctional complex where various cytoskeletal structures are concentrated. Consequently, the z-disc may have stress levels that are considerably larger than other locations within the cell and may serve as one of the "main sensors" for a multitude of load induced biochemical responses including physiological ones. Comparing differences in the mechanisms of mechanotransduction between the FHL1 transgenic mouse model and the MLP transgenic mouse model would be interesting, but was out of the scope of this study.

4.2.4. *Troponin C*

Troponin C (TnC) is a very highly conserved protein [47] that couples the concentration of Ca^{2+} in the cytoplasm with the generation of force. Upon association of TnC with calcium, it effects a conformational change in the troponin-tropomysin complex that results in opening of actin-myosin binding sites. Mutations in TnC have been implicated in a hereditable form of dilated cardiomyopathy [48].

Experiments in skinned cardiac muscle preparations have shown that rapid length changes result in dissociation of Ca^{2+} from the myofilaments [49]. Ca^{2+} transients measured in intact preparations have also been shown to change when the muscle is allowed to shorten [50]. While some have interpreted these results as evidence for intrinsic force sensing by TnC [51], soluble S1 myosin fragments have been demonstrated to increase affinity of TnC for Ca^{2+} in the absence of force [52-54]. These most recent results suggest that stretch affects Ca^{2+} binding to TnC indirectly by increasing myosin crossbridge detachment and growth.

4.3. *Role of sarcolemmal proteins in mechanotransduction*

The sarcolemma is the cell membrane of myocytes that is composed of two amphipathic phospholipid monolayers. This bilayer contains peripheral proteins attached to the inner surface of the sarcolemma and transmembrane proteins spanning through the thickness of the sarcolemma. Important sarcolemmal proteins that have been implicated in mechanotransduction are G-protein coupled receptors, ion channels, and effector enzymes [55].

The sarcolemma is a likely transmitter of external or internal loads to transmembrane channels or membrane bound sensors. Mechanical stresses may cause deformation of the sarcolemma, which can induce conformational changes in membrane-bound enzymes such as phospholipase C [9, 56], mechanically activated ion channels such as stretch activated channels (SAC), which allow Na^+, K^+, and Ca^{2+} to enter the cell [57, 58], and the Na^+/H^+ exchanger, which functions to regulate intracellular pH and the concentration of Na^+ [59, 60].

4.3.1. *Integrins*

The most popular in vitro experimental model utilized to study mechanotransduction is one where cells adhered to elastic substrates are subject either to a single round of increased strain (static stretch) or repeated cycles of increased strain (cyclical stretch) [8, 9, 61, 62]. Externally applied mechanical loads to these types of in vitro systems are transmitted to intracellular structures via cell contacts, thereby implicating integrins in mechanotransduction. Integrins are transmembrane receptors composed of α and β subunit heterodimers, which bind to the ECM and to cytoskeletal proteins such as α-actinin, talin, tensin, and focal adhesion kinase (FAK) [14, 63]. In myocytes the $\beta1$ integrin isoform ($\alpha1\beta1$, $\alpha3\beta1$, $\alpha5\beta1$, $\alpha6\beta1$, $\alpha7\beta1$, $\alpha10\beta1$, and $\alpha11\beta1$) is predominantly expressed in the postnatal heart [64]. Its role in the development of myocardial hypertrophy has been demonstrated. An increase in $\beta1$ integrin expression resulted in increased ANP levels and protein expression in the mouse heart. The inhibition of $\beta1$ function and signaling resulted in reduced adrenergically mediated hypertrophy.

In vitro experiments utilizing the isolated neonatal stretch system have shown that an increase in strain enhances the localization of FAK, paxillin, cytoskeletal proteins, and signaling molecules (Grb2, Sos, Ras, Raf, PLC,

ERKs, and SAPKs) to focal adhesions forming focal adhesion complexes [95, 101]. It has also been shown that the application of stresses to focal adhesions induce tyrosine phosphorylation of FAK [65]. FAK possesses several tyrosine phosphorylation sites, one of which (Tyr-397) is a binding site to a variety of proteins localized to the focal adhesion, and which has been shown to play a role in activating the extracellular signal regulated kinase ½ (ERK1/2) hypertrophic pathway [66]. Taken together, these results suggest two likely mechanisms by which integrins may play a role in mechanotransduction. First, the formation of focal adhesion complexes may not only contribute to strengthening cell adhesion to the ECM, it may also organize proteins in such a way that signals may be transmitted more efficiently. Second, force vectors acting on integrins in the direction of the focal adhesion protein complex, may alter the conformation of force-sensitive components aligned with the force vector, thereby activating new binding interactions between proteins or activating enzymatic sites. In support of this theory, recent publications have demonstrated a strain vector dependence on FAK phosphorylation [82]. Furthermore, computer simulations have predicted that force-induced conformational changes in specific regions of FAK influence its binding affinity to paxillin [14].

4.3.2. *Phospholipase C*

Phospholipases are enzymes that catalyze the breakdown of phospholipids into fatty acids and other lipophilic substances, which may act as secondary messenger molecules. There are three members in the phospholipase C family: PLCγ, PLCδ, and PLCβ, all of which are membrane coupled and dependent upon calcium for optimal activity [67]. Mechanical stress has been shown to stimulate phospholipase C activity and increase intracellular calcium levels, which result in increased protein kinase C activity (PKC) [9, 56]. Research has suggested that PKCs have played a biochemical role in mechanical stress-induced hypertrophy. PKC translocation from the cytoplasm to cardiac particulate fraction has been shown in vitro and in vivo with hypertrophic agonist treatment [68, 69]. Furthermore, it has been confirmed that the suppression of PKC activity is enough to blunt the development of stretch-induced hypertrophy in cultured neonatal cardiac myocytes [70].

4.3.3. *G-protein coupled receptors*

G-protein coupled receptors (GPCR) are transmembrane-spanning receptors that are coupled to G proteins (G_s, G_i, G_q, or $G_{11/12}$), which function as signal transducers. In the heart, many studies have implicated the GPCR mediated G_q

pathway in the development of pathological cardiac hypertrophy [71]. Studies done on transgenic mice have shown that over expression of the G_q protein in the heart was associated with the development of hypertrophy and heart failure [1], while the inhibition of G_q signaling resulted in a blunted, but beneficial response to hypertrophy following transverse aortic constriction [46, 72]. Endothelin 1 and angiotensin II are two agonists of the G_q GPCR pathway that have been implicated in the development of mechanical load-induced hypertrophy. Endothelin 1 is induced by stretch in neonatal cardiac myocytes and has been shown to stimulate the activation of two kinases implicated in the development of hypertrophy: Raf1 and MAPK [73]. The secretion of angiotensin II from cultured cardiac myocytes was also induced by mechanical stretch [56]. Treatment of neonatal cardiac myocytes with angiotensin II resulted in the activation of Raf1 and MEK (two hypertrophic signaling molecules), which have been shown to activate ERKs in vitro [74]. Furthermore, angiotensin II has been shown to induce cardiomyocyte hypertrophy directly by increasing protein synthesis [75]. Consequently, angiotensin II and endothelin 1 have been suggested to be involved in autocrine/paracrine mechanisms that mediate stretch-induced hypertrophy. In addition to these findings, the angiotensin II type 1 receptor (AT1) has been shown to be activated by mechanical stretch alone [76]. In vitro and in vivo experiments have shown that the hypertrophic response to an increase in mechanical load is not blunted in myocytes of angiotensinogen-deficient mice. In support of these findings, mechanical stretch did not activate ERKs in HEK293 cells and Cos7 cells that did not express the AT1 receptor. But when the AT1 receptor was over expressed in these cells, stretch-activated ERK phosphorylation was detected. This response was inhibited with AT1 receptor blockers.

4.3.4. *Stretch activated ion channels*

Stretch-activated channels (SACs) are ion channels which open their "pore" in response to mechanical deformation, thereby allowing the passage of ions such as Na^+, K^+, and Ca^{2+}. Although the involvement of SACs in mechanotransduction is a controversial topic [61], some evidence exists in the literature supporting the role of SACs in mechanical load-induced hypertrophy. In a study published by Cooper's group, stretch of contracting papillary muscles increased the rate of protein synthesis which was associated with an increase Na^+ uptake most likely mediated by SACs [77]. In confirmation of this statement, the load-induced effect on protein synthesis was inhibited with streptomyocin treatment, a SAC blocker, with no effect on the systolic tension developed by these specimens. An increase in Ca^{2+} influx and intracellular Ca^{2+} concentration through SACs has also been reported in cultured cardiac myocytes [4]. There are several mechanisms by which an increase in intracellular calcium concentration may contribute to the development of hypertrophy. First, PLC, endothelin-1, and calcineurin are three signaling molecules implicated in the development of hypertrophy, which are dependent upon calcium for optimal activity [67, 78]. Second, an increase in intracellular calcium will most likely have a positive effect on the systolic tension developed by cardiac muscle, resulting in greater mechanical loading of the tissue.

4.3.5. *Na^+/H^+ exchanger*

The Na^+/H^+ exchanger is a transmembrane glycoprotein that functions to regulate intracellular pH and the concentration of Na^+ by electroneutrally exchanging intracellular H^+ for extracellular Na^+ [79]. Although the mechanisms by which the Na^+/H^+ exchanger contributes to the development of hypertrophy are still poorly understood, its role in the development of hypertrophy has been demonstrated in literature. In vitro studies have revealed that mechanical loading of cardiomyocytes has resulted in the enhanced activity of the Na^+/H^+ exchanger which is associated with the induction of hypertrophy at the molecular level [79]. Furthermore, treatment of cultured cardiomyocytes with HOE 694, a specific inhibitor of the Na^+/H^+ exchanger, noticeably attenuated stretch-induced activation of the ERK pathway and stimulation of protein synthesis. In vivo, the enhanced activity of the Na^+/H^+ exchanger has been reported in the hypertrophied myocardium of spontaneously hypertensive rats [80]. Additionally, cardiac hypertrophy in rats induced by the continuous treatment with isoproterenol was prevented by the inhibition of the Na^+/H^+ exchanger [81]. Taken together these results suggest a vital role for the Na^+/H^+ exchanger in the development of mechanical stress-induced hypertrophy.

5. Summary and Conclusions

Evidence at the cellular, tissue, and organ levels points to mechanical stress and/or strain as likely stimuli for the development of myocardial hypertrophy. Although many load inducible hypertrophic signaling pathways have been identified in literature, (the three main pathways being: MAPK, JAK/STAT, and the calcineurin dependent pathway), only a limited number of studies have investigated the mechanisms by which physical forces are translated into the biochemical events which lead to the development of myocardial hypertrophy. Most likely a deformation signal, which is accompanied by conformational changes at the molecular level, initiates protein synthesis and myocyte growth. At the tissue level, mechanotransduction signaling may involve the ECM, the sarcolemma, the cytoskeleton, and/or the nucleus. A variety of "mechanosensors" have been proposed, and stress sensing in myocytes may involve multiple molecules including integrins, PLCs, SACs, the Na^+/H^+ exchanger, GPCRs, titin, LIM domain cytoskeletal proteins such as MLP and FHL1, and chromatin. Although many studies have implicated cytoskeletal proteins in the pathogenesis of cardiac disease in humans and mice, only a few have examined the role of cytoskeletal proteins in mechanotransduction. Specifically, the LIM domain MLP protein, having signaling and structural roles, has been speculated as playing a major role in mechanotransduction processes. FHL1 is a newly identified LIM domain cytoskeletal protein which is a binding partner of several downstream hypertrophic signaling molecules, as well as titin. Recent results reveal that FHL1 is critical for the transition to pathological hypertrophy, and functions to regulate diastolic mechanics and mechanical-load induced hypertrophic signaling in the heart.

Acknowledgments

The authors would like to acknowledge the technical assistance of Irina Ellrott, as well as valuable collaborations with Drs. Farah Sheikh and Ju Chen. This work was supported by NIH Grants HL32583 and HL46345.

References

1. J. Heineke and J. D. Molkentin, *Nat. Rev. Mol. Cell. Biol.* **7**, 589 (2006).
2. N. Frey and E. N. Olson, *Ann. Rev. Physiol.* **65**, (2003).
3. M. J. Koren, R. B. Devereux, P. N. Casale, D. D. Savage and J. H. Laragh, *Ann. Intern. Med.* **114**, 345 (1991).
4. C. Ruwhof and A. van der Laarse, *Cardiovasc. Res.* **47**, 23 (2000).
5. C. R. Asher and A. L. Klein, *Cardiol. Rev.* **10**, 218 (2002).
6. R. J. Tomanek and G. Cooper, *Anat. Rec.* **200**, 271 (1981).
7. M. E. Barbosa, N. Alenina and M. Bader, *Methods Mol. Med.* **112**, 339 (2005).
8. S. M. Gopalan, C. Flaim, S. Bhatia, M. Hoshijima, R. Knöll, K. R. Chien, J. H. Omens and A. D. McCulloch, *Biotechnol. Bioeng.* **81**, 578 (2003).
9. I. Komuro, Y. Katoh, T. Kaida, Y. Shibazaki, M. Kurabayashi, E. Hoh, F. Takaku and Y. Yazaki, *J. Biol. Chem.* **266**, 1265 (1991).
10. J. Sadoshima, L. Jahn, T. Takahashi, T. J. Kulik and S. Izumo, *J. Biol. Chem.* **267**, 10551 (1992).
11. G. Cooper, R. L. Kent and D. L. Mann, *J. Mol. Cell. Cardiol.* **21**, 11 (1989).

12. T. Bupha-Intr, J. W. Holmes and P. M. Janssen, *Am. J. Physiol. Heart Circ. Physiol.* **293**, H3759 (2007).

13. R. Knöll, M. Hoshijima, H. M. Hoffman, V. Person, I. Lorenzen-Schmidt, M. L. Bang, T. Hayashi, N. Shiga, H. Yasukawa, W. Schaper, W. McKenna, M. Yokoyama, N. J. Schork, J. H. Omens, A. D. McCulloch, A. Kimura, C. C. Gregorio, W. Poller, J. Schaper, H. P. Schultheiss and K. R. Chien, *Cell* **111**, 943 (2002).

14. R. D. Kamm and M. R. Kaazempur-Mofrad, *Mech. Chem. Biosyst.* **1**, 201 (2004).

15. E.S. Groban, A. Narayanan and M.P. Jacobson, *PLoS. Comput. Biol.* **2**, e32 (2006).

16. K.N. Dahl, A. J. Ribeiro and J. Lammerding, *Circ. Res.* **102**, 1307 (2008).

17. T. M. Olson, V. V. Michels, S. N. Thibodeau, Y.S. Tai and M.T. Keating, *Science* **280**, 750 (1998).

18. M. Satoh, M. Takahashi, T. Sakamoto, M. Hiroe, F. Marumo and A. Kimura, *Biochem. Biophys. Res. Commun.* **262**, 411 (1999).

19. H, Niimura, L. L. Bachinski, S. Sangwatanaroj, H. Watkins, A. E. Chudley, W. McKenna, A. Kristinsson, R. Roberts, M. Sole, B.J. Maron, J. G. Seidman and C. E. Seidman, *N. Engl. J. Med.* **338**, 1248 (1998).

20. W. McKenna, *Cardiologia* **38**, 277 (1993).

21. A. Kimura, H. Harada, J.E. Park, H. Nishi, M. Satoh, M. Takahashi, S. Hiroi, T. Sasaoka, N. Ohbuchi, T. Nakamura, T. Koyanagi, T. H. Hwang, J. A. Choo, K. S. Chung, A. Hasegawa , R. Nagai, O. Okazaki, H. Nakamura, M. Matsuzaki, T. Sakamoto, H. Toshima, Y. Koga, T. Imaizumi and T. Sasazuki, *Nat. Genet.* **19**, 379 (1997).

22. L. Thierfelder, H. Watkins, C. MacRae, R. Lamas, W. McKenna, H. P. Vosberg, J. G. Seidman and C. E. Seidman, *Cell* **77**, 701 (1994).

23. L.G. Goldfarb, *Nat. Genet.* **19**, 402 (1994).

24. S. Hein, S. Kostin, A. Heling, Y. Maeno and J. Schaper, *Cardiovasc. Res.* **45**, 273 (2000).

25. G. Forgacs, *J. Cell. Sci.* **108**, 2131 (1995).

26. A. J. Maniotis, C. S. Chen and D. E. Ingber, *Proc. Natl. Acad. Sci.* **94**, 849 (1997).

27. Y. Shafrir and G. Forgacs, *Am. J. Physiol. Cell. Physiol.* **282**, C479 (2002).

28. J. Chen and K. R. Chien, *J. Clin. Invest.* **103**, 1483 (1999).

29. S. Arber, J. J. Hunter, J. Ross, M. Hongo, G. Sansig, J. Borg, J. C. Perriard, K. R. Chien and P. Caroni, *Cell* **88**, 393 (1997).

30. K. A. Clark, A. S. McElhinny, M. C. Beckerle and C. C. Gregorio, *Annu. Rev. Cell. Dev. Biol.* **18**, 637 (2002).

31. W. G. Pyle, M. C. Hart, J. A. Cooper, M. P. Sumandea, P. P. de Tombe and R. J. Solaro, *Circ. Res.* **90**, 1299 (2002).

32. H. L. Granzier and S. Labeit. *Circ. Res.* **94**, 284 (2004).

33. H. L. Granzier and T. C. Irving, *Biophys. J.* **68**, *1027* (1995).

34. M. Hoshijima, *Am. J. Physiol. Heart Circ. Physiol.* **290**, H1313 (2006).

35. N. H. Purcell, D. Darwis, O. F. Bueno, J. M. Müller, R. Schüle and J. D. Molkentin, *Mol. Cell Biol.* **24**, 1081 (2004).
36. S. Lange, F. Xiang, A. Yakovenko, A. Vihola, P. Hackman, E. Rostkova, J. Kristensen, B. Brandmeier, G. Franzen, B. Hedberg, L. G. Gunnarsson, S. M. Hughes, S. Marchand, T. Sejersen, I. Richard, L. Edström, E. Ehler, B. Udd and M. Gautel, *Science* **308**, 1599 (2005).
37. K. A. Guterl, C .R. Haggart, P. M. Janssen and J. W. Holmes, *Am. J. Physiol. Heart Circ. Physiol.* **293**, H3707 (2007).
38. A. M. Raskin, M. Hoshijima, E. Swanson, A. D. McCulloch and J. H. Omens, **(In Press)** *Mol. Cell. Biomech.* (2009).
39. S. Arber and P. Caroni, *Genes Dev.* **10**, 289 (1996).
40. I. Lorenzen-Schmidt, B. D. Stuyvers, H. E. D. J. ter Keurs, M. Date, M. Hoshijima, K. Chien, A. D. McCulloch and J. H. Omens, *J. Mol. Cell. Cardiol.* **39**, 241 (2005).
41. S. M. Lee, S. K. Tsui, K. K. Chan, M. Garcia-Barcelo, M. M. Waye, K. P. Fung, C. C. Liew and C. Y. Lee, *Gene* **216**, 163 (1998).
42. M. Johannessen, S. Moller, T. Hansen, U. Moens and M. Van Ghelue, *Cell Mol. Life Sci.* **63**, 268 (2006).
43. V. Gaussin, J. E. Tomlinson, C. Depre, S. Engelhardt, C. L. Antos, G. Takagi, L. Hein, J. N. Topper, S. B. Liggett, E. N. Olson, M. J. Lohse, S. F Vatner and D. E. Vatner, *Circulation* **108**, 2926 (2003).
44. D. M. Hwang, A. A. Dempsey, C. Y. Lee and C. C. Liew, *Genomics* **66**, 1 (2000).
45. F. Sheikh, A. Raskin, P. H. Chu, S. Lange, A. A. Domenighetti, M. Zheng, X. Liang, T. Zhang, T. Yajima, Y. Gu, N. D. Dalton, S. K. Mahata, G. Dorn, 2nd, J. Heller-Brown, K. L. Peterson, J. H. Omens, A. D. McCulloch, and J. Chen, *J. Clin. Invest.* **118**, 3870 (2008).
46. G. Esposito, A. Rapacciuolo, S. V. Naga Prasad, H. Takaoka, S. A. Thomas, W. J. Koch and H. A. Rockman, *Circulation* **105**, 85 (2002).
47. A. Roher, N. Lieska and W. Spitz, *Muscle Nerve* **9**, 73 (1986).
48. J. Mogensen, R. T. Murphy, T. Shaw, A. Bahl, C. Redwood, H. Watkins, M. Burke, P. M. Elliott and W. J. McKenna, *J. Am. Coll. Cardiol.* **44**, 2033 (2004).
49. D. G. Allen and J. C. Kentish, *J Physiol* **407**, 489 (1988).
50. P. M. Janssen and P. P. de Tombe PP, *Am. J. Physiol.* **272**, H1892 (1997).
51. P. J. Hunter, A. D. McCulloch and H. E. ter Keurs, *Prog. Biophys. Mol. Biol.* **69**, 289 (1998).
52. J. M. Robinson, W. J. Dong, J. Xing and H. C. Cheung, *J. Mol. Biol.* **340**, 295 (2004).
53. J. P. Davis, C. Norman, T. Kobayashi, R. J. Solaro, D. R. Swartz and S. B. Tikunova, *Biophys. J.* **92**, 3195 (2007).
54. W. J. Dong, J. J. Jayasundar, J. An, J. Xing and H. C. Cheung, *Biochemistry* **46**, 9752 (2007).
55. J. H. Omens. *Prog. Biophys. Mol. Biol.* **69**, 559 (1998).

56. J. Sadoshima, Y. Xu, H. S. Slayter and S. Izumo, *Cell* **75**, 977 (1993).
57. H. Hu and F. Sachs, *J. Mol. Cell. Cardiol.* **29**, 1511 (1997).
58. W. Sigurdson, A. Ruknudin and F. Sachs, *Am. J. Physiol.* **262**, H1110 (1992).
59. N. Matsuda, N. Hagiwara, M. Shoda, H. Kasanuki and S. Hosoda, S, *Circ. Res.* **78**, 650 (1996).
60. D. M. Bers, *Kluwer Academic Pub*, (2001).
61. J. Sadoshima, T. Takahashi, L. Jahn and S. Izumo, *Proc. Natl. Acad. Sci. USA* **89**, 9905 (1992).
62. H. W. de Jonge, D. H. Dekkers, A. B. Houtsmuller, H. S. Sharma and J. M. Lamers, *Cell. Biochem. Biophys.* **47**, 21 (2007).
63. R. L. Juliano and S. Haskill, *J. Cell. Biol.* **120**, 577 (1993).
64. R. S. Ross, *J. Card. Fail.* **8**, S326 (2002).
65. Y. Yano, J. Geibel and B. E. Sumpio, *Am. J. Physiol.* **271**, C635 (1996).
66. R. W. Katz, S. Y. Teng, S. Thomas and R. Landesberg, *Bone* **31**, 288 (2002).
67. S. G. Rhee and K. D. Choi, *J. Biol. Chem.* **267**, 12393 (1992).
68. A. Clerk and P H. Sugden, *Circ. Res.* **89**, 847 (2001).
69. K. Paul, N. A. Ball , G. W. Dorn, 2nd and R. A. Walsh, *Circ. Res.* **81**, 643 (1997).
70. T. Yamazaki, I. Komuro, S. Kudoh, Y. Zou, I. Shiojima, T. Mizuno, H. Takano, Y. Hiroi, K. Ueki, K. Tobe *J. Clin. Invest.* **96**, 438 (1995).
71. N. C. Salazar, J. Chen and H. A. Rockman, *Biochim Biophys. Acta.* **1768**, 1006 (2007).
72. M. Suzuki, K. M. Carlson, D. A. Marchuk and H. A. Rockman, *Circulation* **105**, 1824 (2002).
73. T. Yamazaki, T. Yamazaki, I. Komuro, S. Kudoh, Y. Zou, I. Shiojima, Y. Hiroi, T. Mizuno, K. Maemura, H. Kurihara, R. Aikawa, H. Takanoand Y. Yazaki Y, *J. Biol. Chem.* **271**, 3221 (1996).
74. T. Yamazaki, I. Komuro and Y. Yazaki, *Cell Signal.* **10**, 693 (1998).
75. K. M. Baker and G. W. Booz, *Ann. Rev. Physiol.* **54**, 227 (1992).
76. Y. Zou, H. Akazawa, Y. Qin, M. Sano, H. Takano, T. Minamino, N. Makita, K. Iwanaga, W. Zhu, S. Kudoh, H. Toko, K. Tamura, M. Kihara, T. Nagai, A. Fukamizu, S. Umemura, T. Iiri, T. Fujita and I. Komuro, *Nat. Cell. Biol.* **6**, 499 (2004).
77. R. L Kent, J. K. Hoober and G. Cooper 4th, *Circ Res* **64**, 74 (1989).
78. W. Zhu, Y. Zou, I. Shiojima, S. Kudoh, R. Aikawa, D. Hayashi, M. Mizukami, H. Toko, F. Shibasaki, Y. Yazaki, R. Nagai and I. Komuro, *J. Biol. Chem.* **275**, 15239 (2000).
79. H. E. Cingolani and I. L. Ennis, *Circulation* **115**, 1090 (2007).
80. K. Kusumoto, J. V. Haist and M. Karmazyn, *Am. J. Physiol. Heart Circ. Physiol.* **280**, H738 (2001).

81. I. L. Ennis, E. M. Escudero, G. M. Console, G. Camihort, C. G. Dumm, R. W. Seidler, M C. Camilión de Hurtado and H. E. Cingolani, *Hypertension* **41**, 1324 (2003).
82. S. E. Senyo, Y. E. Koshman and B. Russell, *FEBS Lett.* **581**, 4241 (2007).

Chapter 8

A NOVEL HEMODYNAMIC ANALYSIS OF ECHOCARDIOGRAM

TIN-KAN HUNG

Department of Bioengineering, University of Pittsburgh
Pittsburgh, Pennsylvania, 15668, U.S.A.

This study is to extract hemodynamic information from echocardiogram to complement the left ventricle ejection fraction (*EF*) for cardiac evaluation and diagnosis. The data of velocity vector imaging stored in echocardiogram can be used in a novel fluid mechanics analysis for quantification of left ventricular (*LV*) function. The study is focused on assessing the *LV* contraction, and is patient specific. Cardiac patients with the same *EF* could have different *LV* contraction, which can be identified by the kinetic energy delivered from the wall to blood flow. Both the normal and abnormal wall contraction can be also evaluated by the work done by pressure and shear stresses during systole. The abnormal ventricular contractility can be further assessed by an index for normal velocity effect and by another index for dys-synchrony.

Introduction

Noninvasive evaluation of left ventricular (*LV*) function is vital to an effective diagnosis of heart failure. Velocity vector imaging is a new technology utilizing gray-scale 2D echocardiographic images that are then analyzed by software for frame-by-frame movement of the natural acoustic markers, or speckles, present in ultrasound tissue images over the cardiac cycle. The frame-by-frame location shift of these acoustic markers represents tissue movement or deformation, and provides the spatial and temporal data to calculate velocity and regional strain vectors. This technique extends the ability of echocardiography to yield data on ventricular mechanics noninvasively. Recently, Cannesson et al. [1] reported on the utility of velocity vector imaging (*VVI*) to quantify *LV* dys-synchrony to predict long-term response to cardiac resynchronization therapy. The kinetic energy delivered from the left ventricle to blood flow was recently demonstrated as a useful methodology for clinical application [6]. Other hemodynamic characteristics of *LV* contraction can be obtained by solving the Navier-Stokes equations for pressures and velocities, and studying the work done by pressure and viscous stresses.

                                    T.-K. Hung

Energy Transfer of Cardiac Pumping

The dynamic performance of a left ventricle can be effectively assessed from the work energy equation for blood flowing from the contracting chamber to the aortic root. For a two-dimensional echocardiogram, the equation is

$$\int \frac{\rho V^2}{2}\left(u\frac{\partial x}{\partial n}+v\frac{\partial y}{\partial n}\right)d\ell + \iint\left(\frac{\rho}{2}\frac{\partial V^2}{\partial t}\right)dxdy + \int p\left(u\frac{\partial x}{\partial n}+v\frac{\partial y}{\partial n}\right)d\ell =$$

$$\mu\int\left(\frac{\partial u^2}{\partial x}\frac{\partial x}{\partial n}+\frac{\partial v^2}{\partial y}\frac{\partial y}{\partial n}\right)d\ell + \mu\int\left(v\left(\frac{\partial v}{\partial x}+\frac{\partial u}{\partial y}\right)\frac{\partial x}{\partial n}+u\left(\frac{\partial v}{\partial x}+\frac{\partial u}{\partial y}\right)\frac{\partial y}{\partial n}\right)d\ell$$

$$-\mu\iint\left(2\left(\frac{\partial u}{\partial x}\right)^2+2\left(\frac{\partial v}{\partial y}\right)^2+\left(\frac{\partial u}{\partial y}+\frac{\partial v}{\partial x}\right)^2\right)dxdy \tag{1}$$

in which ρ is the blood density, V the velocity on the wall (with components u and v in the $x-$ and $y-$directions), p the pressure, μ the dynamic viscosity of blood, and n the outer normal direction on the wall. The first integral in Eq. 1 represents the instantaneous kinetic energy influx along the ventricle wall and the out flux at the aortic root. The second integral is the time rate of change of kinetic energy in the ventricle, and the third integral the rate of work done by pressure. The integrals on the right hand side are, respectively, the rate of work done by normal viscous stresses, by shear stress, and the rate of energy dissipation of the flow. The kinetic energy transfer to the blood flow from wall contraction can be calculated directly from the *VVI* data recorded in echocardiogram. The rest of integrals cannot be evaluated until the velocity and pressure fields of the ventricular pumping are obtained from the Navier-Stokes equations. The instantaneous rate of kinetic energy transferred from the *LV* wall to blood flow can be calculated from the line integral of kinetic energy along the inner wall of the ventricle:

$$KEW = \int\limits_{WALL} \frac{\rho V^2}{2}(V_N)d\ell$$

where V_N is normal velocity on the wall. The tangential velocity does not contribute to the energy delivery; it produces shear stresses to the blood flow.

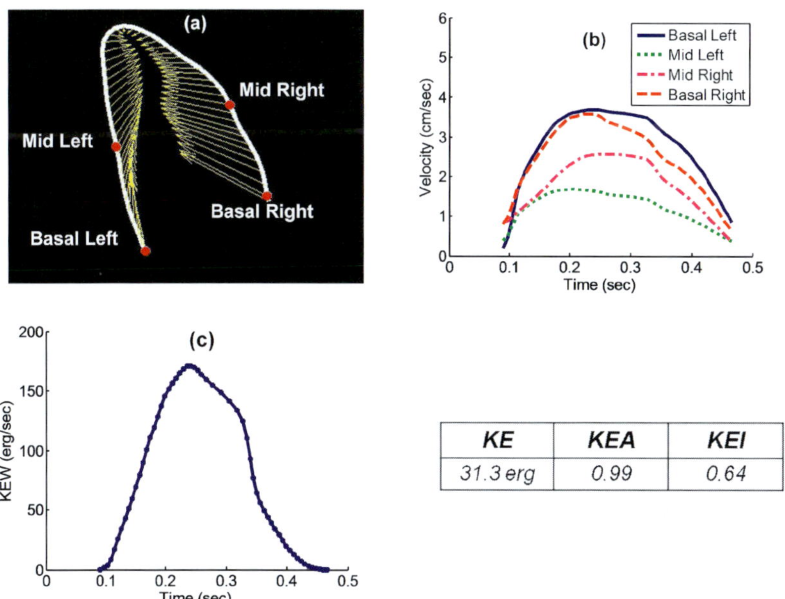

KE	KEA	KEI
31.3 erg	0.99	0.64

Figure 1. (a) Endocardial wall velocity vector at T=0.2 second of 3-chamber echocardiogram. (b) Variation of wall velocity with time at four points in (a). (c) Time variation of kinetic energy wall flux (**KEW**) during systole.

The quantity KEW is the kinetic energy wall flux calculated from an instantaneous velocity distribution (see Fig. 1a) of a 3-chamber echocardiogram. Figure 1b shows the velocity magnitudes at four points of the wall motion. The time variation of KEW during systole is demonstrated in Fig. 1c. In view of optimal function in nature, the tangential velocity is expected to be small for normal cardiac contraction. Integration of KEW with time yields the total kinetic energy (**KE**) delivered to the flow on this section:

$$KE = \int_0^s \int_{WALL} \frac{\rho V^2}{2}(V_N)\,d\ell\,dt$$

in which the superscript "s" represents the duration of systole. The dynamic performance of the *LV* contraction can be further assessed by two indices **KEI** and **KEA**:

$$KEI = \frac{\int_0^S \int_{WALL} \frac{\rho V^2}{2} V_N \, dl \, dt}{\int_0^S \int_{WALL} \frac{\rho V^2}{2} |V| \, dl \, dt} \qquad KEA = \frac{\int_0^S \int_{WALL} \frac{\rho V^2}{2} V_N \, dl \, dt}{\int_0^S \int_{WALL} \frac{\rho V^2}{2} |V_N| \, dl \, dt}$$

The denominator in **KEI** is based on an absolute velocity for a hypothetic maximum **KE**. If V_N in the numerator is equal to $|V|$, **KEI** will be equal to unity. In reality, the value of **KEI** is always less than 1. The index **KEA** is a ratio of kinetic energy (**KE**) to that when the normal velocity in the denominator is based on its absolute value $|V_N|$. In the absence of dys-synchrony, the index **KEA** is equal to unity. Its decrease from unity can be used to identify the seriousness of dys-synchrony.

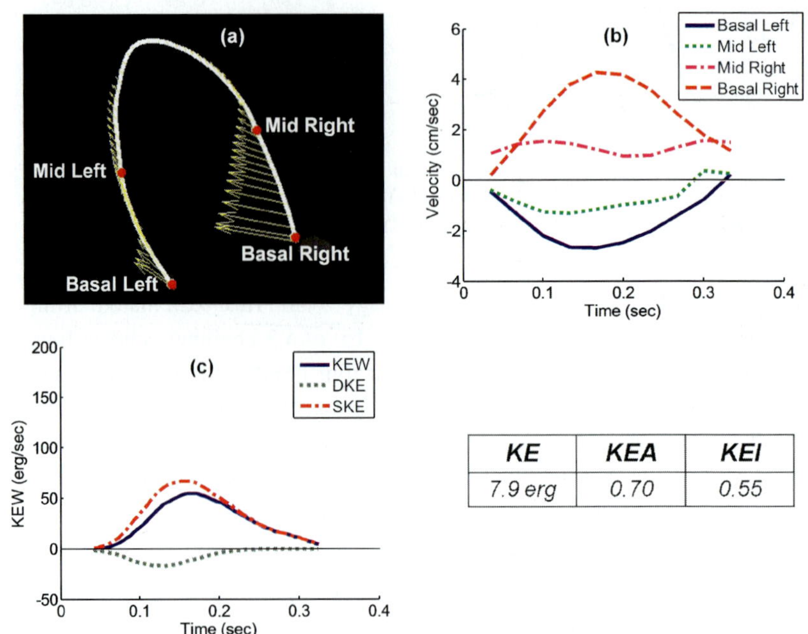

KE	KEA	KEI
7.9 erg	0.70	0.55

Figure 2. (a) Endocardial wall velocity vector at T=0.2 second of 3-chamber echocardiogram. (b) Variation of wall velocity with time at four points in (a). (c) Kinetic energy wall flux (**KEW**) during systole.

The VVI data of echocardiogram shown in Fig. 2a and 2b demonstrate abnormal outward velocities in the mid and basal left. The instantaneous kinetic energy wall flux is the sum of synchronized kinetic energy influx (SKE) and dys-synchronized kinetic energy out-flux (DKE, plotted as negative): KEW=SKE+DKE. The pumping effectiveness is seriously reduced as indicated by KE = 7.9 erg. This value is much lower than 31.3 erg for a normal wall motion shown in Figure 1. The dys-synchrony can be quantified by the dyssynchronized index KEA. As shown in the table in Figure 2, **KEA** is reduced to 0.70 and **KEI** to 0.56.

Computational Approach

The computational flow analysis with moving boundaries is a challenging task. The time-dependent domain of systolic blood flow can be mapped onto a fixed circle (Fig. 3) by transforming the polar coordinates (r,θ,t) to (R,θ,t):

$$R = \frac{r}{a(\theta,t)}, \quad \theta = \theta \quad \text{and} \quad t = t \tag{2}$$

in which $a(\theta,t)$ is the instantaneous radius and is function of angle θ and time t. As shown in Fig. 3, this transformation maps the time dependent domain (left side in Fig. 3) to a fixed circle.

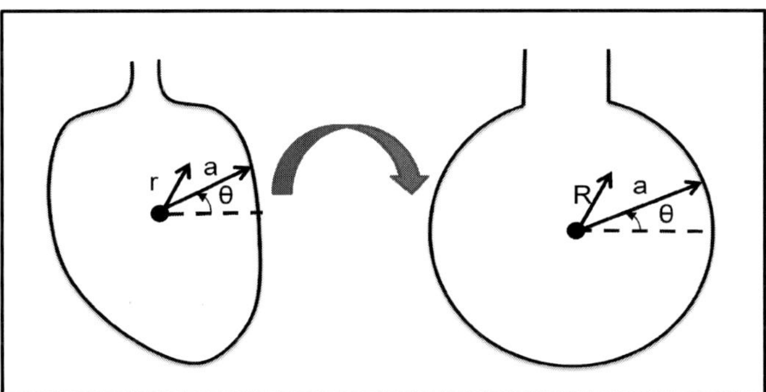

Figure 3. Definition sketch for transformation.

The transformation is similar to that developed in our previous studies [4,5], the continuity equation can be expressed as

$$\frac{u}{Ra} + \frac{1}{a}\frac{\partial u}{\partial R} + \frac{1}{Ra}\left(\frac{\partial v}{\partial \theta} - \frac{R}{a}\frac{\partial a}{\partial \theta}\frac{\partial v}{\partial R}\right) = \frac{1}{\rho c^2}\left(\frac{R}{a}\frac{\partial a}{\partial t}\frac{\partial p}{\partial R} - \frac{\partial p}{\partial t}\right) \tag{3}$$

in which u *is* the radial velocity component and v the velocity component in the θ-direction. The Navier-Stokes equations become

$$\rho\left[\frac{\partial u}{\partial t} - \frac{R}{a}\frac{\partial a}{\partial t}\frac{\partial u}{\partial R} + \frac{1}{a}\frac{\partial u^2}{\partial R} - \frac{1}{a^2}\frac{\partial a}{\partial \theta}\frac{\partial uv}{\partial R} + \frac{1}{Ra}\frac{\partial uv}{\partial \theta} + \frac{u^2-v^2}{Ra}\right] = -\frac{1}{a}\frac{\partial p}{\partial R}$$

$$+\frac{\mu}{R^2a^2}\left[\frac{R^2}{a^2}\left(\frac{\partial a}{\partial \theta}\right)^2\frac{\partial^2 u}{\partial R^2} + \frac{\partial^2 u}{\partial R\partial \theta} + \frac{\partial^2 u}{\partial \theta^2} + \left(\frac{2R}{a^2}\left(\frac{\partial a}{\partial \theta}\right)^2 - \frac{R}{a}\frac{\partial^2 a}{\partial \theta^2}\right)\frac{\partial u}{\partial R}\right]$$

$$+\mu\left[\frac{1}{a^2}\frac{\partial^2 u}{\partial R^2} - \frac{u}{R^2a^2} + \frac{1}{Ra^2}\frac{\partial u}{\partial R}\right] - \frac{2\mu}{R^2a^2}\left[\frac{\partial v}{\partial \theta} - \frac{R}{a}\frac{\partial a}{\partial \theta}\frac{\partial v}{\partial R}\right] \tag{4}$$

$$\rho\left[\frac{\partial v}{\partial t} - \frac{R}{a}\frac{\partial a}{\partial t}\frac{\partial v}{\partial R} + \frac{1}{a}\frac{\partial uv}{\partial R} - \frac{1}{a^2}\frac{\partial a}{\partial \theta}\frac{\partial v^2}{\partial R} + \frac{1}{Ra}\frac{\partial v^2}{\partial \theta} + \frac{2uv}{Ra}\right] = \frac{1}{a^2}\frac{\partial a}{\partial \theta}\frac{\partial p}{\partial R}$$

$$-\frac{1}{Ra}\frac{\partial p}{\partial \theta} + \mu\left[\frac{1}{Ra^2}\frac{\partial v}{\partial R} - \frac{v}{R^2a^2} + \frac{1}{a^2}\frac{\partial^2 v}{\partial R^2}\right] + \frac{2\mu}{R^2a^2}\left[\frac{\partial u}{\partial \theta} - \frac{R}{a}\frac{\partial a}{\partial \theta}\frac{\partial u}{\partial R}\right]$$

$$+\frac{\mu}{R^2a^2}\left[\left(\frac{2R}{a^2}\left(\frac{\partial a}{\partial \theta}\right)^2 - \frac{R}{a}\frac{\partial^2 a}{\partial \theta^2}\right)\frac{\partial v}{\partial R} + \frac{R^2}{a^2}\left(\frac{\partial a}{\partial \theta}\right)^2\frac{\partial^2 v}{\partial R^2} + \frac{\partial^2 v}{\partial R\partial \theta} + \frac{\partial^2 v}{\partial \theta^2}\right] \tag{5}$$

The aortic root is considered as the outlet for the computational flow analysis. As the velocity $V_o(t)$ at the aortic root is uniform (for accelerated converging flow) and time dependent, the continuity equation for the 3-chamber (long axis) plane requires

$$V_o(t) = \frac{1}{D}\int_{Wall} \vec{V}\bullet\vec{n}\,d\ell \tag{6}$$

where D is the diameter at the aortic root. Iso-volumetric contraction will be simulated using a prescribed sudden acceleration on the wall while the pressure at the aortic root is initially based on the end diastolic pressure. The Neumann boundary condition will be used for solving the pressure wave equation. In this formulation [2, 3], not only the normal pressure gradient on the wall will be applied, but also viscous stresses gradients and the convective acceleration on the wall will be satisfied in the determination of the pressure field. The results will be presented elsewhere.

Conclusions

The study is aimed at extracting biomechanical information from *2D* echo-cardiogram for clinical evaluation of abnormal ventricular contraction. The magnitudes of these energy parameters for normal cardiac function will be used as a reference for evaluating cardiac dysfunction of patients with low ejection fraction, dys-synchrony, and cardiac re-synchrony therapy. The ability of analyzing and quantifying ventricular pumping effectiveness during systole could be a powerful tool to advance our knowledge and methodology of this highly prevalent disease.

Tributes

The challenges in bioengineering research and education were seeded with the vision, creativity, dedication and leadership of Professor Y. C. Fung along with his peers in the 1960s. I had an opportunity to join Professor George Bugliarello at Carnegie Mellon University in 1967 for biofluid mechanics and Bioengineering education. The mile stone paper of Dr. Fung and Professor Chia-Shung Yih on nonlinear peristaltic flows led me to study the peristaltic pumping (with T. D. Brown), using the Navier-Stokes equations in orthogonal curvilinear coordinates and the low Reynolds number towing tank of Professor Bugliarello. The wisdom, aspiration and warmth of these giants in engineering science and education remain inspirational to me and in my interaction with my family, friends and students. To your health and happiness, Professor and Mrs. Fung, we salute you.

                                    T.-K. Hung

Dr. Fung visited President Hsia of the National Cheng Kung University in 1987. From left: You-Li Chou, Yuan-Cheng Fung, Han-Min Hsia, Tin-Kan Hung, Ming-Te Tseng and Ting-Cheng Hung.

Dr. Fung received an honorary doctorial degree from President Kao of the National Cheng Kung University in 2004. From left: Fong-Chin Su, Jhing-Fa Wang , Chiang Kao, Shu Chien, Yuan-Cheng Fung, Jin Wu, Tin-Kan Hung, Shan-Hwei Ou and You-Li Chou.

Shunan Cho Hung, Luna and Yuan-Cheng Fung, and Tin-Kan Hung at Professor Yuan Cheng Fung's 90th Birthday Symposium.

References

1. M. Cannesson, M. Tanabe, M.S. Suffoletto, D. Schwartzman and J. A. Gorcsan, *Am J Cardiology*, **98**, 949 (2006).
2. T.K. Hung, *Journal of Engineering Mechanics Division*, ASCE, EM1,13 (1973).
3. T.K. Hung and T.D. Brown, *Journal of Computational Physics*, **23**, 343 (1977).
4. T.K. Hung, *Journal of Engineering mechanics Division*, **107**, 643-648 (1981).
5. T.K. Hung and T.M.C. Tsai, *Journal of Engineering Mechanics*; **123(2)**, 247 (1997).
6. T.K. Hung, S. Balasubramanian, M.A. Simon, M. Suffoletto, H.S. Borovetz and J.A. Gorcsan, *Journal of Mechanics in Medicine and Biology* **8**, 87 (2008).

Chapter 9

IN VITRO BIOMECHANICAL STUDIES IN AGING HUMAN LUNGS

SHERVIN MAJD

Biomedical Engineering Department, The University of Memphis,
330 Engineering Technology bldg., Memphis, TN, 38152, U.S.A.

MICHAEL YEN

Biomedical Engineering Department, The University of Memphis,
330 Engineering Technology bldg., Memphis, TN, 38152, U.S.A.

Effects of aging on mechanical properties of human lung are studied. Finite element modeling of human lung tissue under both equi-biaxial and three dimensional (uniform) loading is performed. Experimental results are utilized in building both two and three dimensional models in ABAQUS 6.5 for three age groups: 20, 40 and 60 years old. A computed tomography image of a human lung is simplified to build a three dimensional model. Material properties are defined using linear incremental elasticity. Lagrangian stress-strain relationship is estimated for under various loading conditions in all three age groups. Lagrangian Von-Mises stress trends and values analyzed in the equi-biaxial model appear to be in agreement with biaxial experiments performed by Fung et al. and by Gao et al. This validates that parenchyma can be considered as the main source of lung mechanical properties. Effect of aging on lung stress-strain relationship was similar in both two and three dimensional models. For similar stretch ratios, older age samples are experiencing higher stress values as compared to the younger age samples. This study indicates that finite element simulation utilizing experimental results may be a viable approach to build a virtual testing platform in forecasting the change of lung elasticity with aging at different loading conditions and stretch ratios.

1. Introduction

Pressure-volume curves are often used as a major source of information for lung mechanical behavior, specifically to study uniform expansion of lung [1-4]. More recently, investigators focused on obtaining the stress-strain relationship of lung parenchyma [1, 3, 5]. However, characterizing the stress-strain relationship of lung parenchyma is complicated. It involves non-linear stress-strain relationship, blood circulation and gravitational effect, surfactant's role on loading and deformation, and effects of neighboring organs such as diaphragm and sternum [1-2, 4]. In addition, human data is difficult to obtain in-vivo, especially with short post-mortem time. These difficulties make simplification crucial to quantify lung mechanical properties. Triaxial testing was used by Hoppin et al. for measuring the lung mechanical properties [6]. However, edge effects appear to significantly affect the characterizations. To address this issue, biaxial testing was recommended by Fung et al. to characterize the lung parenchyma [3]. This technique was used widely in animal studies. However, there are just only a few experimental data available for human lungs. Fung et al. and Gao et al. performed biaxial tests on human lung tissue samples of two different age groups of 26.20±8.20 (years old, mean ± sd) and 68.30±14.70 (years old, mean ± sd.), respectively [1,3]. Comparing the results of these two studies shows the effects of aging on stress-strain relationship of parenchyma in biaxial testing.

On the other hand, Lai-Fook investigated three dimensional mechanical properties of human excised lung under uniform expansion for three different age groups 20, 40 and 60 years old lungs. He measured the bulk modulus of elasticity and Poisson's ratio by applying local loops of Pressure-Volume curves and small indentations, respectively, from the state of equilibrium at different transpulmonary pressures of 4, 8, 12 and 16 cm H_2O [2]. His experiments represent the aging effects on human excised lung elasticity.

While experimental results on human lung stress-strain relationship are limited, mathematical modeling methods were shown to be very useful in characterizing the stress-strain relationship of lung tissues [7-9].

The aim of this study is to investigate the age dependant stress-strain relationship of human lung under two and three dimensional uniform loading. A numerical approach is exemplified using commercial finite element software, ABAQUS 6.5. First, aging effects on equi-biaxial (2D) mechanical properties of human lung parenchyma is explored. An equi-biaxial model was adapted from Lai-Fook's 3D experiment on excised lungs. Stress-strain relationship of human lung parenchyma is evaluated for three different age groups of 20, 40 and 60 years old models and compared to biaxial testing results.

In the second phase of this study, aging effects on 3D uniform expansion of excised lungs are studied. A 3D model of lung is built in ABAQUS 6.5 by simplifying a computed tomography image of a human lung. Mechanical properties are defined from Lai-Fook's 3D experiments using linear incremental elasticity. Uniform expansion of human excised lung is simulated for three age groups of 20, 40 and 60 years old.

Results of this study suggest that finite element simulation using linear incremental elasticity is a viable method to simulate the aging effect on stress-strain relationship of human lungs.

2. Methods

2.1. *Equi-biaxial modeling*

A 5x5x0.4 (cm) solid biaxial model is built in ABAQUS assuming equi-biaxial loading of sample tissue to simulate equi-biaxial expansion of parenchyma. Lung is considered isotropic [10]. Material properties were defined from the 3D experimental study by Lai-Fook et al. They obtained the bulk modulus and Poisson ratio at specific lung transpulmonary pressures of 4, 8, 12 and 16 cm H_2O for three different age groups (20, 40 and 60 year old samples). Linear incremental elasticity was used to estimate the lung nonlinear behavior [2,10]. Bulk modulus was converted to Young's modulus using Equation-1,

$$E = 3K(1-2\nu)$$

(1)

where K is bulk modulus and ν is Poisson's ratio. E represents the Young's modulus. Stretch ratios of equi-biaxial model were estimated as cubic root of volume ratios measured at specific transpulmonary pressures of 4,8,12 and 16 cm H_2O by Lai-Fook et al. as seen in Equation-2,

$$\lambda = \frac{r_2}{r_1} \propto \sqrt[3]{(V_2/V_1)}$$

(2)

where λ, r_1 and r_2 represent stretch ratio, initial sample length and sample length after expansion, respectively. V_1 and V_2 represent initial lung volume and lung volume after expansion. Lai-Fook et al. showed at very low transpulmonary pressure P_L ($P_L < 1$), lung seems to enlarge rather than stretch from residual volume. In addition, estimation of lung elasticity from collapsed state (very low P_L) would be flawed since extra mechanical forces are needed to reopen the alveoli and airways. Therefore, Lai-Fook's first estimate of the elasticity of the lung was performed at P_L of 4 cm H_2O to ensure all alveoli and airways are open [2]. In this study, P_L of 4 cm H_2O was considered as a reference state for all models in our simulations from which stretching starts.

Knowing the stretch ratios (from Equation-2) and their corresponding bulk modulus and Poisson ratio, ν, incremental linear elasticity of lung sample can be formulated for equi-biaxial modeling of all age groups. As seen in Figure 1. and Figure 2., change of elastic modulus and Poisson ratio in versus stretch ratios of 20 years old sample is calculated. Linear interpolation (arrows) is also performed to further divide the stretch increments into smaller steps and

                         S. Majd & M. Yen

increase the accuracy of the simulation. In addition, large deformation effects within each step are accounted for by dividing that step into infinitesimal steps.

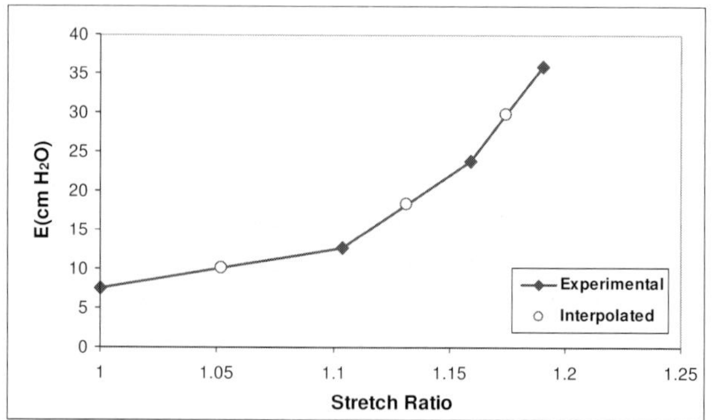

Figure 1. Elastic Modulus versus stretch ratio (20 years old).

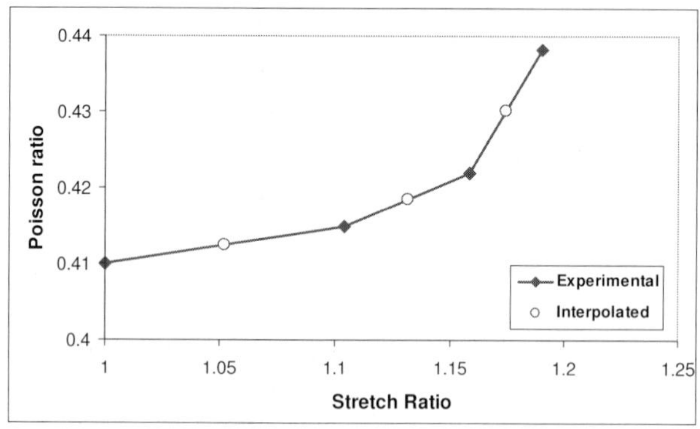

Figure 2. Poisson ratio versus stretch ratio (20 years old).

Element mesh is generated using 10,000 linear hexahedral elements (C3D8R) and 13005 nodes. The estimated displacements (from Equation-1) are applied incrementally to the finite element model and stress values are computed based on static analysis in ABAQUS 6.5 using displacement-control static analysis [1,3]. Results represent the Lagrangian stress values versus nominal strains as defined by Equation-3,

$$\varepsilon n = \lambda - 1 \tag{3}$$

where ε_n is the nominal strain and λ is the stretch ratio. Similar approach is implemented for 40 and 60 year old models to determine the aging effects on equi-biaxial stress-strain distribution of human lung parenchyma. Finally, the simulation result of 20 and 60 year olds models are compared with the biaxial testing characterizations of parenchyma by Fung and by Gao using regression analysis.

2.2. *3D simulation (uniform expansion)*

Lai-Fook showed that 20 year old excised lung volume is 55% of total lung capacity (TLC), when P_L equals 4 cm H_2O. Thereby, to build a 3D model of 20 year old lung, a computational tomography image of lung at end of normal tidal volume (53% TLC) is used to estimate the 3D excised lung geometry. This model is considered as our reference geometry for 20 year old lung model at P_L of 4 cm H_2O. 60,877 tetrahedral elements and 11,799 nodes are used to mesh the model as seen in Figure 3.

Lung is assumed isotropic and incremental elasticity is considered (Figure 1 and Figure 2) [10]. Initial lung volumes are modified considering the enlargement of lung initial volume with aging. Therefore, initial volume in 40 and 60 years old models are estimated to be 65% and 75% TLC, respectively. Force-control static analysis of lung expansion is performed in increments of 4 cm of H_2O. Thereby, the model is loaded from 4 to 25 cm H_2O (100% TLC) to simulate Lai-Fook's experiment in several steps. Similar approach is applied to 40 and 60 years old models. Results are compared to see the effect of aging on excised lung's mechanical behavior.

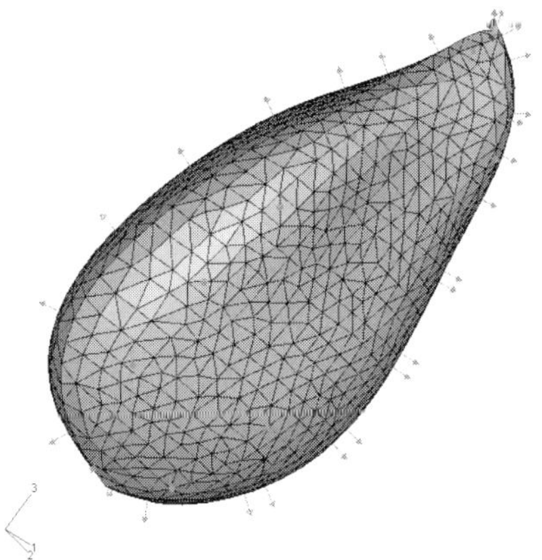

Figure 3. 3D model of meshed lung under pressure of 4 cm H_2O.

3. Results

3.1. *Equi-biaxial simulation*

Computed Von-Mises Lagrangian stress values for 20, 40 and 60 years old samples using simulated biaxial model are obtained and presented in Figure 4. Nominal strains (calculated from Equation-3) are selected instead of strain values, which are suitable for comparison with experimental results. Results suggest that for similar change in stress values, stretch ratio decreases with aging (Figure 3.). Non-linear mechanical behavior can be observed for all age groups. Aging increases lung stiffness under equi-biaxial loading condition.

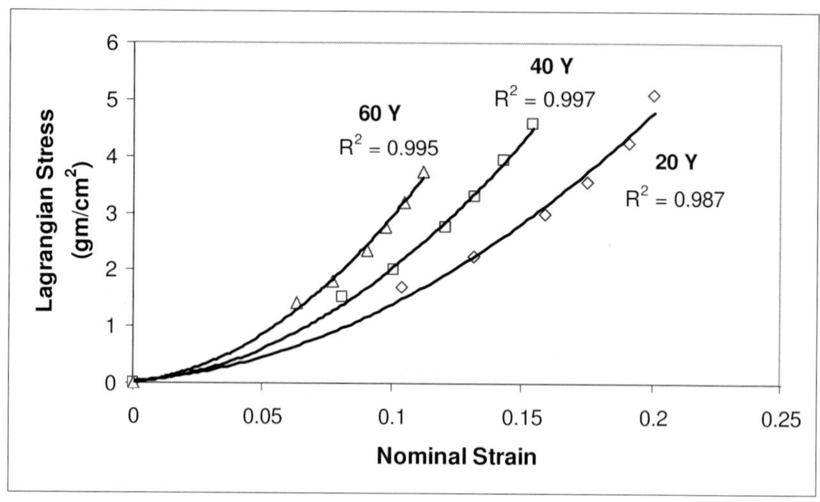

Figure 4. Aging effects on lung stress-strain relationship (Equi-biaxial loading).

For validation, the simulation results of 20 years old model are statistically compared with Fung's biaxial testing results (Mean= 26 years old) as seen in Figure 5.

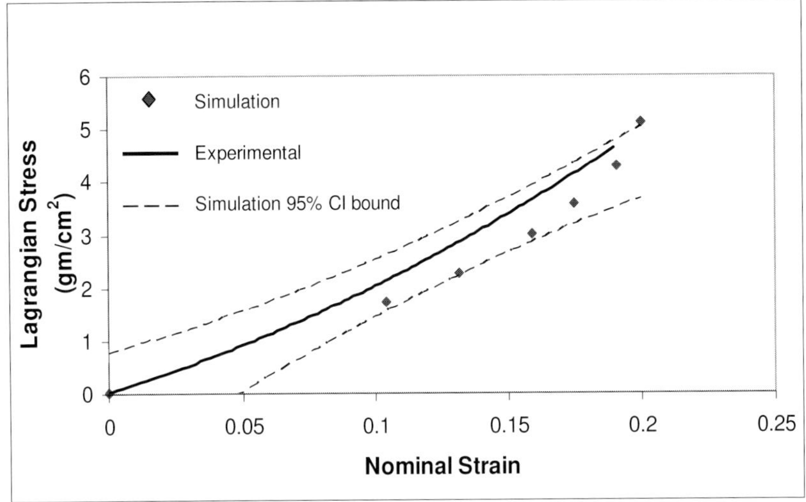

Figure 5. 20 year olds sample biaxial properties with 95% CI bounds.

Square points show the simulated results, whereas dashed lines represent 95% confidence interval (CI) higher and lower bounds (dashed lines). Experimental results (solid line) are compared to simulation data (solid points). Statistical regression analysis shows that equi-biaxial experimental results are located within 95% CI bounds of simulated results for 20 years old model.

3.2. *Uniform expansion (3D)*

Effect of aging on stress-strain relationship of excised lung during uniform expansion from 4 to 25 cm H_2O is estimated. Computed maximum principal Lagrangian stress versus nominal strain values of excised lungs during uniform expansion from 4 to 25 cm H_2O are presented in Figure 6.

Since initial lung volume is estimated when PL was 4 cm H2O, initial Lagrangian stresses (prestress values) are similar for all age groups and equals to 4 cm H_2O. Nominal strain values assumed to be zero at resting state for all age groups (when PL equals 4 cm H_2O). Exponential regression lines estimate the stress-strain relationship of 3D lung for each age group. Aging increases the stiffness of lungs. For similar stress values, younger lungs stretch more than older lungs.

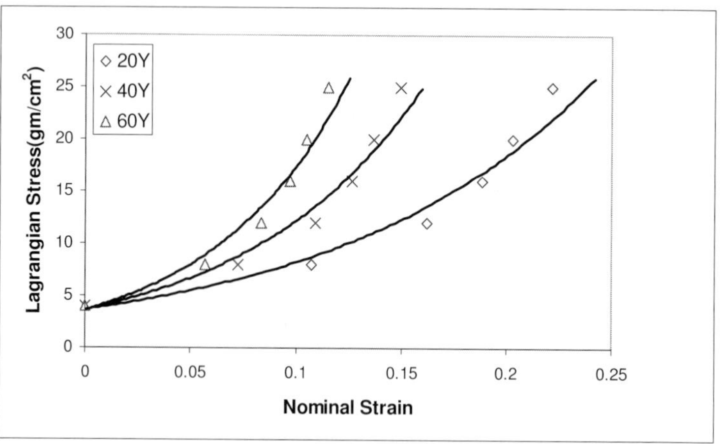

Figure 6. Stress-strain relationship of human excised lung during uniform expansion for three age groups: 20, 40 and 60 years old. Regression lines predict the stress-strain relationship of lung during uniform expansion.

4. Discussion

4.1. *Equi-biaxial modeling*

Simulation showed that aging increases the parenchyma stiffness under equi-biaxial loading. Simulation results show that 20 year olds lung experiences 80% and 30% more stretch values as compared to 60 and 40 years old lungs, respectively. This trend is in agreement with experimental works comparing the results of biaxial testing performed by Fung and by Gao on young and older human parenchyma, respectively [1-3].

One limitation in this study is that material property was derived from average experimental values. Thus, it does not account for sample variations. In addition, lung was considered isotropic at macroscopic level. Tai et al. showed that this assumption is valid for small distortions [10]. Dividing each analysis step into several linear steps ensures that this assumption is effective for our simulation.

Lai-Fook et al. characterized the mechanical properties of human excised lung which includes larger airways, smaller airways and parenchyma. Therefore,

2D adaptation of their results represents the equi-biaxial loading of lung and not only the parenchyma. However, previous studies showed that parenchyma is the main source of lung mechanical support at macroscopic level [1, 4, 7-8]. In addition, Stamenovic showed that bulk modulus and Poisson ratio are mainly determined by tension in alveolar walls [11]. Regression analysis validated this hypothesis and showed that 2D adaptation of Lai-Fook's study can replicate the equi-biaxial testing of parenchyma.

Linear incremental elasticity was used to define material properties as an accepted approach in previous researches [2, 10]. Using this method, stress values are underestimated as compared to nonlinear analysis. Further divisions of analysis step into smaller ranges are expected to increase the accuracy of computation [10]. That is why; interpolation of mechanical properties is performed.

Regression analysis showed that biaxial experimental results for 20 years old sample lung tissues are located within 95% CI bounds of simulated results. This validates that equi-biaxial mechanical behavior of parenchyma can be characterized from 3D mechanical properties of lung uniform expansion using linear incremental elasticity.

Similar approach can be used to compare the equi-biaxial testing results of Gao et al. and simulated 60 years old model. However, standard deviation in the experiments is high mainly as a result of testing various cadaver groups with different post mortem periods. In addition, simulation of 60 years old model was performed based on 3D experimental results by Lai-Fook et al., which was derived from limited sample numbers with high standard deviation. Therefore, the validation was not performed.

4.2. *3D simulation results*

3D uniform expansion of lung is simulated using an estimated geometry of human lung obtained from a computed tomography image at 53% TLC. Lung neighboring organs and chest wall are excluded to represent the excised lung geometry. Stress-strain relationship of lung is obtained as lung was loaded from negative pressure of 4 to 25 cm H_2O (load control static analysis). This analysis is performed for three different age groups of 20, 40 and 60 years old to assess the effect of aging on 3D uniform expansion of excised lungs. Results show that excised lung becomes stiffer with aging. Maximum nominal strain of 20 years old lungs (when P_L increases from 4 to 25 cm H_2O) was found to be 91% and 47% higher as compared to 60 and 40 years old lungs, respectively. Previous studies showed that change of lung mechanical properties with aging can be attributed to elastin and collagen fibers in lung structure [12-13]. Crapo found that elastic fiber network is rearranged mainly after 50 years of age [13]. This change of fiber network leads to enlargement of alveolar space and thereby, reduction of alveolar surface from 75 cm^2 at age of 30 to 60 cm^2 at age of 75 [13]. This phenomenon justifies that lung volume at FRC is increased with aging as reported in previous studies [2, 14]. That is probably why it is repeatedly mentioned in the literature that lung is more flaccid in aged groups [13]. On the other hand, Turner et al. showed that aging reduces the static recoil pressure of human intact lung [14]. This is mainly caused by enlarged resting volume in the aged lung, while the slope of P-V curve, which represents the lung compliance reduces with aging (for $P_L> 4$ cm H_2O) as shown by Lai-Fook et al. [2]. Therefore, lung stretches less in the aged lung under similar applied pressure values (at $P_L>4$ cm H_2O). This is in agreement with our simulation results in which aged lungs experience higher values of stress under similar stretch ratios.

As for limitations, zero strain reference was defined as P_L equals to 4 cm H_2O for all three age groups for uniform expansion simulations. This assumption is essential since mechanical testing of lung is not valid without preconditioning [1,3]. It assures that characterization of stress-strain relationship of lung expansion does not include the required forces to reopen the airways and alveoli from a collapsed state.

Acknowledgement

The authors would like to thank Dr. Benjamin M. W. Tsui and Dr. William P. Segars and the division of medical imaging physics at Johns Hopkins for providing the geometrical model of human lung and Dr. J. Gao, for providing biaxial testing results and helpful discussions.

This chapter is a tribute to Dr. Y. C. Fung's 90th-year-old Birthday. I wish to congratulate him and acknowledge his contributions in the health related sciences, especially in the field of Biomechanics. He is regarded as the "father" of modern Biomechanics. Happy Birthday and thanks a million for the inspiration and guidance.

References

1. J. Gao, W. Huang and RT. Yen, *Biomed Sci Instrum.* **42**, 172 (2006).
2. S.J. Lai-Fook and RE. Hyatt, *J Appl Physiol.* **89**, 163 (2000).
3. Y.J. Zeng, D. Yager and YC. Fung, *J Biomech Eng.* **109**,169 (1987).
4. Y.C. Fung, *Circ Res.* **37**, 481 (1975).
5. Y.C. Fung, P. Tong and P. Patitucci, *ASCE Journal of Engineering Mechanics* **104**, 201 (1978).
6. F.G. Hoppin, G.C. Lee and S.V. Dawson, *J Appl Physiol.* **39**, 742 (1975).
7. J.B. West and F.L. Matthews, *J Appl Physiol.* **32**, 332 (1972).
8. D.L. Vawter, Y.C. Fung and J.B. West, *J Appl Physiol.* **45**, 261 (1978).
9. A. Al-Mayah, J. Moseley and KK. Brock, *Phys Med Biol.* **53**, 305 (2008).
10. R.C. Tai and G.C. Lee, *J. Biomech.* **14**, 243 (1981).
11. D. Stamenovic, *Physiol Rev.* **70**, 1117 (1990).
12. J.A. Pierce, J.B. Hocott, *J Clin Invest.* **39**, 8 (1960).
13. R.O. Crapo, in *Pulmonary Disease in the Elderly Patient.* Ed. D. Mahler (Marcel Dekker, New York, 1993).
14. J.M. Turner, J. Mead, M.E. Wohl, *J Appl Physiol.* **25**, 664 (1968).

Chapter 10

MODELING THE OXYGEN UPTAKE IN PULMONARY ALVEOLAR CAPILLARIES

CHENG-JEN CHUONG

Bioengineering Department, University of Texas at Arlington
Arlington, TX 76019, U.S.A.

After a review on the modeling of oxygen uptake in pulmonary alveolar capillaries, a summary of our modeling studies developed to represent the sheet flow characteristics of pulmonary capillaries is presented. With the model, we determined the overall pulmonary diffusing capacity D_L, and the diffusing capacities D_M and D_E of the alveolar membrane and the red blood cell (RBC) segments of the diffusional pathway for O_2. Results showed the membrane segment contributing the major resistance, with RBC segment resistance increases as oxygen saturation rises during the RBCs transit: RBC contributed 7% of the total resistance at the capillary inlet ($S_{O2} = 75\%$) and 30% towards the capillary end ($S_{O2} = 95\%$). Both D_M and D_L increased as the hematocrit increased but began approaching a plateau near a Hct of 35%, due to competition between RBCs for oxygen influx. Both D_M and D_L were found to be relatively insensitive (2~4%) to changes in plasma protein concentration (28~45%). The model correlated reasonably with experimental data and can better represent the oxygen uptake of the pulmonary capillary bed.

1. Introduction

The oxygen uptake within the lungs, from the alveolar air space to a heme binding site on a hemoglobin molecule inside a red blood cell (RBC), is dictated by the diffusion characteristics of this pathway and the chemical reactions within the RBC. The total resistance for this pathway ($1/D_L$) can be written as the sum of the resistance due to a membrane segment ($1/D_M$) and a RBC segment ($1/(\theta \cdot V_c)$). It was originally defined by Roughton and Forster [1] as:

$$\frac{1}{D_L} = \frac{1}{D_M} + \frac{1}{\theta \cdot V_c} \tag{1}$$

where D_L denotes the apparent pulmonary diffusing capacity, D_M the membrane diffusing capacity, both in the unit of ml/(min·mmHg), θ the specific rate of gas uptake by RBCs in blood with a normal hematocrit in ml/(min·mmHg·[ml of blood]), and V_c is the pulmonary capillary blood volume in ml. The product of θ and V_c is also referred to as the RBC diffusing capacity D_E. The diffusive transport across the membrane segment, which accounts for both the blood-gas

                                 C.-J. Chuong

tissue barrier and the intracapillary plasma, can be described as a simple passive diffusion process. However, oxygen transport within the RBC segment involves passive diffusion of oxygen, as well as binding of oxygen to hemoglobin, and diffusion of oxy-hemoglobin, i.e. facilitated diffusion of oxygen.

The lumped parameter representation of Eq. 1 was a conceptual milestone, allowing the membrane diffusing capacity D_M and the pulmonary capillary blood volume V_c to be quantified from experimental measurement of D_L for CO at different oxygen tensions [1]. Known as the RF method, these data could be translated into O_2 diffusing capacities. Equation 1, however, does not address the spatiotemporal distributed nature of the oxygen transport and the role of blood plasma in its membrane resistance. Based on models that study oxygen release in systemic capillaries, Hellums [2] first pointed out that blood plasma could contribute to as much as 50% in the total resistance to oxygen delivery to tissue bed. Federspiel and Popel [3] modeled the two-phase nature of the blood on oxygen release in systemic microcirculation. They considered RBCs as evenly-spaced discrete hemoglobin containing spheres in a capillary tunnel and took the Hb-O_2 reaction kinetics inside the RBC into consideration. They calculated mass transfer coefficient, a measure of O_2 flux out of capillary segment divided by the driving force for O_2 release, to show it depends on particle spacing and clearance; but not on the O_2 tension at the capillary wall. Federspiel [4] later applied this model to study oxygen transport in pulmonary capillaries. He considered spherical cells in cylindrical capillary surrounded by a uniform annulus of alveolar tissues. This geometry may not accurately reflect the blood-gas tissue barrier and cell geometry across which gases transport; it could lead to an overestimation in diffusing capacities. Wang and Popel [5] considered the effects of RBC shapes in oxygen release in systemic capillaries. They showed the RBC shapes could affect capillary O_2 flux and hemoglobin saturation.

A different approach for diffusing capacity estimate was proposed by Weibel *et al.* using morphometric data [6,7]. The method assumes molecular diffusion paths are linear and that all available alveolar and capillary membrane surfaces participate in gas exchange. It was however known to yield conflicting results from other methods [8,9,10]. To explore possible sources in discrepancies, Hsia, Chuong and Johnson [11] used a simple geometric model of a pulmonary capillary segment containing a variable number of RBCs to calculate segmental D_L and D_M to CO as a function of hematocrit (Hct). By comparing D_M predictions derived from modeling that follows the paths of molecular diffusive transport with those from morphometric approach, they concluded that morphometric method seriously overestimated D_M because the tissue-plasma resistance to diffusion is underestimated and the effective membrane utilized for diffusion is overestimated, particularly in the low Hct range. They further showed the use of RF method could modestly overestimates D_M and underestimates V_c at high Hct; because the conductance of tissue-plasma membrane for CO varies with alveolar O_2 tension. In another study, Frank, Chuong and Johnson [12] incorporated the geometric features of pulmonary alveolar sheet flow [13,14,15] and the Hb-O_2 chemical reactions within RBCs in their model studies. They showed the spatiotemporal distributed nature of both the membrane and RBC segmental resistances in alveolar oxygen transport. It suggested the membrane segment contributes the major resistance to O_2 flux and that the RBC segmental resistance increases as O_2 saturation rises through the transit. Reasonable agreement was found when they compared extrapolated model segmental membrane diffusing capacities with whole body membrane diffusing capacity measurements. Later, Hsia *et al.* [16,17,18] examined the effect of RBC distribution uniformity in gas diffusive transport and concluded that uniform distribution of RBC in a given capillary enhances the amount of gas uptake. Using isolated rabbit lungs perfused with washed and re-suspended human RBCs, Betticher *et al.* [19] demonstrated RBC deformability and subsequent deformed shapes enhances O_2 diffusing capacity and reduced pulmonary arterial pressure drop. Based on histological sections from rapidly frozen rat lungs, Nabors *et al.* [20] showed, at high alveolar volume or reduced micro-vascular pressure, both the RBCs and capillary cross-section deform which could lead to increased diffusing capacity because of shorten diffusive paths and enhanced RBC surface area exposure that facilitate oxygen diffusion transport from alveolar air.

With these backgrounds, I would like to present a summary of our modeling study [12] for the transient of oxygen transport in pulmonary alveolar capillaries, from the alveolar air space to the RBCs. With this model, the spatiotemporal distributed nature of both the membrane and RBC segmental resistance in oxygen transport are examined. Effects of varying plasma protein concentrations on the overall oxygen transport were also examined.

2. Formulation

2.1. *Geometric Model of an Alveolar Capillary Segment*

To incorporate the geometric features of the pulmonary alveolar sheet flow [13,14,15] we represented the alveolar capillary as a modified 2D model with parallel-sided channel containing parachute-shaped RBCs. The model geometry consists of a cross-section through the longitudinal axis of a typical pulmonary capillary segment (100 μm length) containing a variable number of equally spaced RBCs depending on the segmental Hct (Fig. 1A). The parachute shape of the RBC was digitized from a photograph of Skalak and Branemark [21]. Human RBCs are known [22,23,24] to have a mean volume of ~97 μm^3 and a mean surface area of ~137 μm^2. We thus modified the 2D planar model to have an effective depth of 4.92 μm, resulting in an RBC volume of 103 μm^3 and an effective surface area of 125 μm^2. The effective surface area considers that used for gas transport at the RBC perimeter along the peripheral surface. The model thus consists of three different regions (Fig 1B): 1) the blood-gas tissue barrier, 2) the plasma fluid within the capillary, and 3) the RBC, each with respective gas diffusive properties. The effect of varying Hct was evaluated by adjusting the "unit length" of the capillary segment, which effectively changes the volume of plasma (Fig 1B).

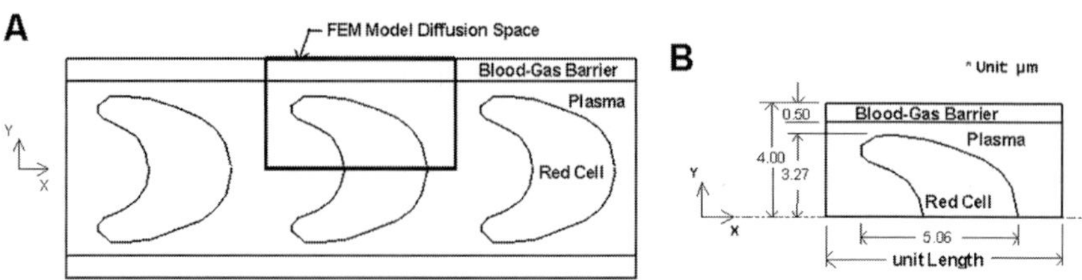

Figure 1. The model consists of parachute-shaped RBCs in a parallel-sided channel (100 μm) was used to represent the sheet flow characteristics in pulmonary alveolar septum. *A:* Geometric model of a pulmonary capillary segment with 3 "unit length" segments containing 3 RBCs. *B:* Respective geometric dimensions where "unit length" is adjusted to account for varying Hct values. From Frank, Chuong, Johnson [12], reproduced by permission.

2.2. *Governing Equations*

Due to the low Peclet number in pulmonary capillary blood flow, we neglected the convective transport [4,25,26]. Neglecting the differences in the plasma fluid and RBC traveling speed, the passive diffusive transport in the membrane segment (blood-gas barrier and plasma) is described by

$$\alpha \frac{\partial P_{O2}}{\partial t} = \alpha \, d_{O2} \, \nabla^2 P_{O2} \tag{2}$$

where P_{O2} denotes the oxygen partial pressure, t the time, α the solubility coefficient for oxygen, and d_{O2} the diffusion coefficient for oxygen. The diffusive transport within the RBC is described by a more complicated facilitated diffusion due to the chemical binding of oxygen to hemoglobin. Under typical conditions in the microcirculation, the facilitated diffusion can increase the rate of oxygen transport by almost two fold [26]. Passive diffusion of oxygen, oxy-hemoglobin, and the reaction between oxygen and hemoglobin must be considered. With the assumption of instantaneous equilibrium for the oxy-hemoglobin reaction, they can be written as [12]:

$$\left([Hb] \frac{d\,S_{O2}}{d\,P_{O2}} + \alpha \right) \frac{\partial\,P_{O2}}{\partial\,t} = \left([Hb]\,d_{Hb} \frac{d\,S_{O2}}{d\,P_{O2}} + \alpha\,d_{O2} \right) \nabla^2\,P_{O2} \qquad (3)$$

where $\dfrac{d\,S_{O2}}{d\,P_{O2}}$ is the differential form of the Hill equation written as:

$$\frac{dS_{O2}}{dP_{O2}} = \frac{n\,\left(P_{50}\right)^{n}\,\left(P_{O2}\right)^{n-1}}{\left(P_{50}^{\ n} + P_{O2}^{\ n}\right)^2} \qquad (4)$$

with P_{50} denoting the oxygen partial pressure at 50% oxy-hemoglobinsaturation. Considering normal human blood at 37°C and PH = 7.4, the empirically derived constants of P_{50} (= 26 mmHg) and n (= 2.7) were used. Equations 2, 3 and 4 are solved for P_{O2} distribution through the transit. All physical properties used are summarized in Table 1.

Table 1. Physical properties used in the model.

	PARA-METERS	VALUES	UNITS	SOURCES (Ref)
Blood-gas barrier	α_{O2}	1.4	nano-moles/(cm³-mmHg)	3
	d_{O2}	2.4 x 10⁻⁵	cm²/sec	3
Plasma	α_{O2}	1.4	nano-moles/(cm³-mmHg)	3
Protein conc. level				
Low (5 gm/100 ml)	d_{O2}	2.49 x 10⁻⁵	cm²/sec	28
Average (6.9 gm/100 ml)	d_{O2}	2.40 x 10⁻⁵	cm²/sec	28
High (10 gm/100 ml)	d_{O2}	2.25 x 10⁻⁵	cm²/sec	28
Red blood cell	α_{O2}	1.40	nano-moles/(cm³-mmHg)	3
	d_{O2}	2.4 x 10⁻⁵	cm²/sec	3
	$[Hb]$	2.0 x 10⁴	nano-moles/cm³	3
	d_{Hb}	1.4 x 10⁻⁷	cm²/sec	3

2.3. *Calculations of D_L, D_M, and D_E*

Through the transient the O_2 flux distribution across RBC membrane can be calculated from

$$O_2 \text{ flux} = \alpha\, d_{O2}\, \frac{\partial P_{O2}}{\partial n} \tag{5}$$

where $\dfrac{\partial P_{O2}}{\partial n}$ denotes P_{O2} gradients at a local surface normal. Once the total oxygen flow, calculated as

$$\text{Flow} = \int \alpha\, d_{O2}\, \frac{\partial P_{O2}}{\partial n}\, d(Area) \tag{6}$$

is determined, D_L, the diffusing capacity for the capillary segment, can be calculated from

$$D_L = \frac{\text{Flow}}{P_{alv} - \overline{P_{rbc}}} \tag{7}$$

where P_{alv} is the P_{O2} at the air-tissue barrier and $\overline{P_{rbc,V}}$ is the volume weighted mean of P_{O2} in the RBC when it is in equilibrium with S_{O2}. D_M, the diffusing capacity for the membrane segment, can be calculated as:

$$D_M = \frac{\text{Flow}}{P_{alv} - \overline{P_{rbc,S}}} \tag{8}$$

where $\overline{P_{rbc,S}}$ is the surface weighted mean of P_{O2} at the RBC wall membrane. D_E, the diffusing capacity for the RBC segment, is calculated as:

$$D_E = \frac{\text{Flow}}{\overline{P_{rbc,S}} - \overline{P_{rbc,V}}} \tag{9}$$

Values for $\overline{P_{rbc,V}}$ and $\overline{P_{rbc,S}}$ are determined at each time step through the RBC transit by considering the average of P_{O2} weighted according to volume or surface. Mean values in diffusing capacity $\overline{D_L}$ and $\overline{D_M}$ were calculated for the range of hemoglobin saturation (S_{O2}) from 74% to 96%, approximately corresponding to the normal physiological values [27].

3. Results

3.1. *Transient Values in D_L, D_M, and D_E*

Values for D_E decreased during the RBCs transit along the capillary because of the progressive fall in the number of hemoglobin binding sites available for oxygen binding. Nevertheless, D_E remains relatively large throughout most of the capillary transit compared to D_M. The transient variations for all three diffusing capacity parameters are plotted for increasing hemoglobin saturation S_{O2}, for the case of average plasma protein concentration at an effective Hct of

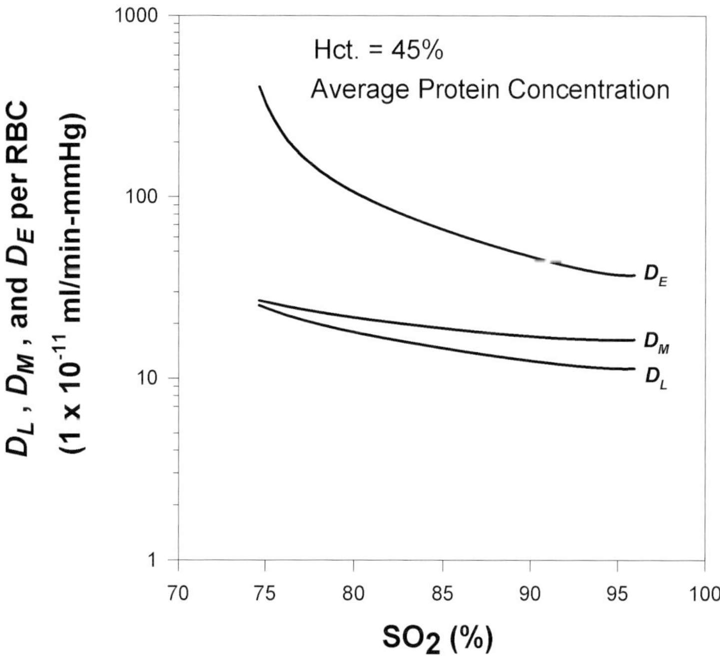

Figure 2. Transient changes in diffusing capacity parameters D_E, D_M, and D_L per RBC at varying S_{O2} (Hct = 45%; plasma protein concentration level = average). From Frank, Chuong, Johnson [12], reproduced by permission.

45% (Fig. 2). The value of D_L per RBC diffusion space was seen to be 24.3 x 10^{-11} ml/(min·mmHg) at the beginning of the transient ($S_{O2} = 75\%$); which became 11.3 x 10^{-11} ml/(min·mmHg) near the end of the transient ($S_{O2} = 95\%$). The % contribution in the total resistance ($1/D_L$) from the RBC segment ($1/D_E$) was found to be only 7% at the capillary inlet ($S_{O2} = 75\%$); which rose to 30% towards the end of the capillary transit ($S_{O2} = 95\%$). Thus there is a progressive increase in the RBCs resistance to oxygen transport through the transit.

3.2. *Changing O_2 Flux Across RBC Membrane*

The changing O_2 fluxes distributions across RBC membrane through the transient are given in Fig. 3 with a reference plot for Hill's equation (HbO_2 vs. P_{O2}). In early stage ($S_{O2} = 75\%$), with the low oxygen saturation in the RBC, O_2 diffuses across the RBC membrane with little resistance. Oxygen enters the RBC most rapidly at its wall membrane closest to the capillary surface, since it is the path offering the least resistance. Much less flux is seen at the concave part of the RBC, which is relatively hidden from the alveolar surface, suggesting that the blood plasma in this vicinity does not contribute to D_M and the utilization of the plasma fluid in membrane segment is not uniform. At 85% S_{O2}, due to the gradual increase in the relative resistance of the RBC segment with respect to the membrane segment, the distribution of O_2 flux over different regions of the RBC wall membrane becomes less non-uniform. At the same time, there is an increased level of O_2 entering the RBC through intracapillary plasma, indicating better utility of the plasma fluid in oxygen transport. Additionally, the spatial distribution of O_2 flux within the plasma is seen to become more uniform. Finally at 95% S_{O2}, total flux into the RBC becomes so low that there is little regional preference in O_2 flux through the RBC wall membrane or plasma fluid. Thus, there is a gradual progression during the RBCs transit towards a more uniform distribution in the O_2 flux across the RBC wall membrane and in the utilization of the plasma fluid with respect to O_2 transport. This results in the gradual decrease in D_M (Fig 2), since effectively the diffusion of O_2 from the air-tissue barrier into the RBC takes a longer path due to the increased regional resistance of the RBC. Therefore, the regional changes in D_E occurring throughout the transient cause the utilization of the RBC wall membrane and intracapillary plasma to be altered, which affects D_M.

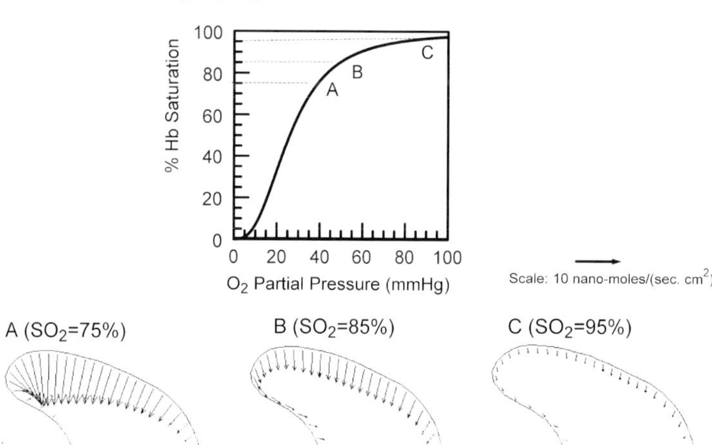

Figure 3. Changing O$_2$ flux distribution across RBC membrane at different stages of the transient: S_{O2} = 75, 85, and 95% and a reference plot for Hill's equation (HbO_2 vs. P_{O2}).

3.3. *Effects of HCt Changes on D_L, D_M*

Results of mean diffusing capacities during the RBCs transit, i.e. $\overline{D_L}$ and $\overline{D_M}$ per RBC diffusion space and per 100 μm of capillary, are given in Fig 4A-D. For the average plasma protein concentration, $\overline{D_L}$ per RBC diffusion space was 17.5 x 10^{-11} ml/min·mmHg at a Hct of 10% but it decreased to 12.2 x 10^{-11} ml/min·mmHg at a Hct of 50% (Fig 4A). The decrease at higher Hct is due to the competition among cells for the O$_2$ influx. Total $\overline{D_L}$ for the entire 100 μm capillary blood volume was seen to increase from 58.5 x 10^{-11} to 204 x 10^{-11} ml/min·mmHg as the Hct increases from 10% to 50%. A progressively decreasing slope is seen after 35% Hct, indicating a gradual approach towards a plateau at higher Hcts (Fig 4B). Similar trends were seen for $\overline{D_M}$ per RBC, it decreases from 25.8 x 10^{-11} to 16 x 10^{-11} ml/min·mmHg as the Hct varies from 10% to 50% because of the competition among cells (Fig 4C). Total $\overline{D_M}$ for the entire 100 μm capillary blood volume was seen to increase from 86.2 x 10^{-11} to 268 x 10^{-11} ml/min·mmHg as the Hct increases from 10% to 50%. A progressively decreasing slope is seen beyond 35% Hct, indicating a gradual approach towards a plateau (Fig 4D).

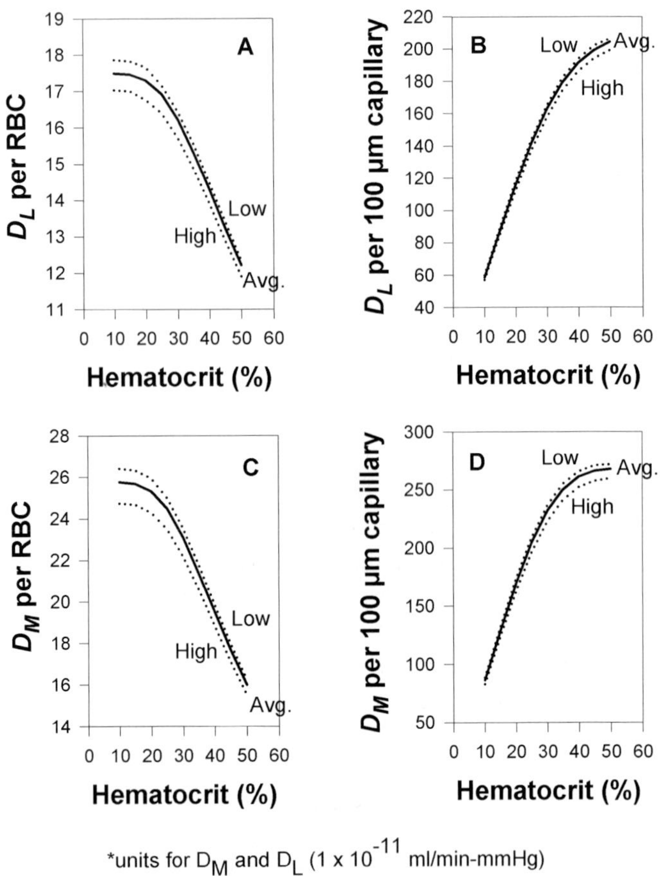

*units for D_M and D_L (1×10^{-11} ml/min-mmHg)

Figure 4. Mean D_L ($\overline{D_L}$) and D_M ($\overline{D_M}$) at varying Hct values for three different levels of plasma protein concentration. **A:** D_L per RBC. **B:** Total D_L for 100 µm capillary segment. **C:** D_M per RBC. **D:** Total D_M for 100 µm capillary segment. Low, average and high refer to the level of plasma protein concentration change, see values in Table 1. From Frank, Chuong, Johnson [12], reproduced by permission.

3.4. *Effect of Plasma Protein Concentration Changes*

The model was applied to examine the effect of plasma protein concentration change on D_L, D_M, and D_E. Three levels of plasma protein concentration (5, 6.9, and 10 gm/100 ml), all within physiological range, were considered. Their respective gas diffusion coefficients [28] are included in Table 1. Results showed an increase in protein concentration causes a fairly uniform reduction in both D_M and D_L; whereas a decrease in protein concentration leads to a uniform elevation. The % change in D_M and D_L due to protein concentration changes however are small (2~4%) relative the actual % changes in protein concentration (28-45%). Similar results apply to all Hct values (Fig 4).

3.5. *Comparing Model Predictions with Total Lung D_M Measurements*

To compare with available experimental measurements, we calculated the mean D_M at the total lung volume using the following extrapolation:

$$\overline{D_{M,\,Total\ lung}} = Hct \cdot \frac{\overline{D_{M,\,rbc}}}{V_{rbc}} \cdot V_C \tag{10}$$

where $\overline{D_{M,\,rbc}}$ is the mean D_M per RBC calculated by the model throughout the transient, V_{rbc} is the model volume of a RBC (103 μm^3), and Vc is the total pulmonary capillary blood volume. Note that $\overline{D_{M,\,rbc}}$ is an implicit function of *Hct*. Using experimentally determined data for V_C of 100.3 ml at a resting state (cardiac index ~ 7 l/min) and 162.8 ml at an exercise state (cardiac index ~ 15.5 l/min) [29], we approximated $\overline{D_{M,\,Total\ lung}}$ at resting and exercise states according to Eq. 10, see Fig 5. These extrapolated D_M values for the total lung showed a linear increase up to a Hct of ~ 35% before gradually approaching a plateau. Measurements of $D_{M,\,CO}$ based on the rebreathing technique[29] were multiplied by 1.24 to yield estimated values for $D_{M,\,O2}$, shown as constant values in Fig 5. Shaded regions represent ranges from a mean resting state of 54.1 ml/min·mmHg, to a moderate exercise state of 69.9 ml/min·mmHg, and to a peak exercise state of 115.7 ml/min·mmHg [29]. A visual comparison indicates that the current model predicted total $D_{M,\,O2}$ within a reasonable range, more so for the resting state.

The model has slightly overpredicted $D_{M,\,O2}$ because it assumed an ideal condition of uniform RBC distribution and spacing within the capillaries, which offers the optimal use of the membrane segment, i.e. highest values in D_M [16,18]. It is known that there is spatial and temporal fluctuation in both RBC distribution and spacing in the pulmonary capillary bed under physiological conditions. For the resting state, the corresponding Hct was in the range of 18-30% (Fig 5), somewhat lower than the physiological values of 28~37% [30,31] because of the over-prediction in $D_{M,\,O2}$. For the exercise state, this source of error is further exaggerated and the deviation in the estimated Hct is larger.

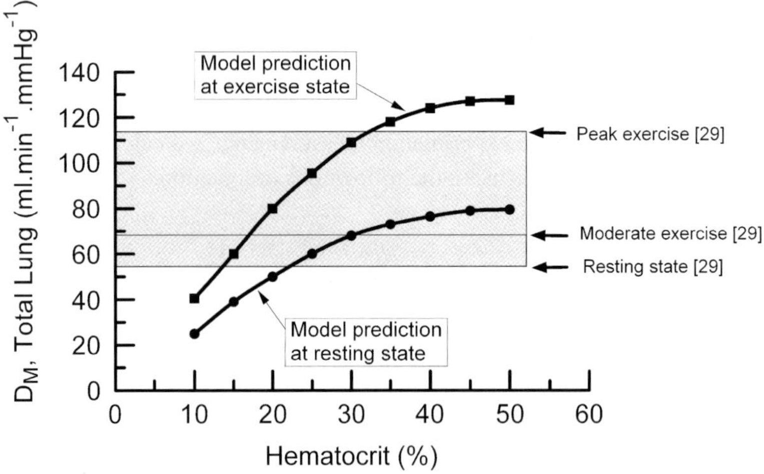

Figure 5. A comparison of model predictions vs. measurements: Extrapolated $D_{M,\,Total\,lung}$ are plotted as a function of Hct for resting and exercise states. D_M derived from experimentally determined D_M are shown for a range from mean resting state, moderate exercise, to peak exercise states. Adapted from ref. [12].

3.6. *Comparing with RBC Conductance θ*

Using the mean $\overline{D_E}$ per RBC throughout the entire transient, we calculated an equivalent θ value of 5.9 ml/min·mmHg per ml of whole blood for a Hct of 45%. This was obtained on the assumption of infinite reaction velocity between oxygen and hemoglobin. To calculate it, we have considered a transient analysis from 0 to 80% S_{O2} saturation to simulate the experimental conditions of rapid-reaction techniques [28,32], which indicated nearly constant θ values for this S_{O2} saturation range. Above 80% saturation, the RBC conductance θ progressively falls as the saturation rises due to the progressive decrease in the number of binding sites provided by the last heme on the hemoglobin molecule [32]. Measured values of θ have been reported to be 2.8 and 3.8 ml/min·mmHg per ml of whole blood [28,32]. However, the rapid-reaction technique may never completely eliminate the unstirred plasma layer around the RBC and hence tends to underestimate the θ value; whereas the present work tends to overestimate the θ value due to the assumed infinite reaction velocity.

4. Summary

A review on the modeling of the oxygen uptake in pulmonary alveolar capillaries is made. A model that incorporates the geometric features of alveolar sheet flow [13,14,15] is presented. Results showed the spatiotemporal nature of both RBC and membrane segmental resistance in oxygen transport through the transit. Membrane segment contributed the major resistance; the RBC segmental resistance increases as the oxygen saturation rises through the RBC transit. Both D_M and D_L increased as the Hct was increased but gradually approached a plateau as the Hct exceeded 35%. Both D_M and D_L were found to be relatively insensitive to changes in plasma protein concentration. The model correlated reasonably with experimental data and can better represent the oxygen uptake at the pulmonary capillary bed.

References

1. F. J. W. Roughton and R. E. Forster, *J Appl Physiol.* **11**: 290 (1957).
2. J. D. Hellums, *Microvasc. Res.* **13**, 131 (1977).
3. W. J. Federspiel and A. S. Popel, *Microvasc. Res.* **32**, 164 (1986).
4. W. J. Federspiel, *Respir. Physiol.* **77**: 119 (1989).
5. C. H. Wang and A. S. Popel, *Math Biosciences.* **116**: 89 (1993).
6. E. R. Weibel, *Respir. Physiol.* **11**: 54 (1970).
7. E. R. Weibel, W. J. Federspiel, F. Fryder-Doffey, C. C. W. Hsia, M. Konig, V.Staider-Navarro, and R. Vock, *Respir. Physiol.* **93**: 125 (1993).
8. J. D. Crapo and R. O. Crapo, *Respir. Physiol.* **51**: 183 (1983).
9. J. D. Crapo, R. O. Crapo, R. L. Jensen, R. R. Mercer, and E. R. Weibel, *J. Appl. Physiol.* **64**: 2083 (1988).
10. E.R. Weibel, C.R. Taylor, J. J. O'Neal, D.E. Leith, P. Gehr, H. Hoppeler, V. Langman, and R. V. Baudinette, Respir. Physiol. **54**: 173 (1983).
11. C. C. W. Hsia, C. J. C. Chuong, and R. L. Johnson, Jr., *J. Appl. Physiol.* **79(3)**: 1039 (1995).
12. A. O. Frank, C. J. C. Chuong, and R. L. Johnson, Jr., *J. Appl. Physiol.* **82(6)**: 2036 (1997).
13. Y. C. Fung and S. S. Sobin, *J Appl. Physiol.* **26**, 472 (1969).
14. Y. C. Fung and S. S. Sobin, *Circ. Res.* **240**, 470 (1972).
15. Y. C. Fung, *Biomechanics: Circulation* (2nd edition, Springer-Verlag, Inc. New York 1997).
16. C. J. Chuong and R. L. Johnson, Jr., Proceeding of the 14th Annual Houston Conference on Biomedical Engineering Research, pp. 98 (1996), also presented at the 1995 American Thoracic Society Meeting, Seattle WA (1995).

17. C. C. W. Hsia, C. J. C. Chuong, and R. L. Johnson, Jr., *J. Appl. Physiol.* **83(4)**: 1397 (1997).
18. C. C. W. Hsia, R. L. Johnson, Jr., and D. Shah, *J. Appl. Physiol.* **86(5)**: 1460 (1999).
19. D. C. Betticher, W. H. Reinhart, and J. Geiser, *J. Appl. Physiol.* **78(3)**: 778 (1995).
20. L. K. Nabors, W. A. Baumgartner, Jr., S. J. Janke, J. R. Rose, W. W. Wagner, Jr., and R. L. Capen, *J Appl Physiol* **94**: 1634 (2003).
21. R. Skalak and P. T. Branemark, *Science*, **164**: 717 (1969).
22. E. Evans and Y. C. Fung, *Microvasc. Res.* **4**, 335 (1972).
23. P.C.Y. Chen and Y. C. Fung, *Microvasc. Res.* **6**, 32(1973).
24. Y. C. Fung, *Biomechanics: Mechanical Properties of Living Tissues.* (Springer-Verlag, Inc. New York, 2nd edition, 1993).
25. J. Aroesty and J. F. Gross, *Microvas. Res.* **2**: 247 (1970).
26. K. Groebe and G. Thews, *Adv. Exp. Med. Biol.* **248**: 175 (1989).
27. J. West, *Respiratory Physiology: The Essentials.* (Lippincott Williams & Wilkins; 8th ed. 2008).
28. K. Yamaguchi, P. D. Nguyen, P. Scheid, and j. Piiper, *J. Appl. Physiol.* **58(4)**: 1215 (1985).
29. C. C. W. Hsia, D. G. McBrayer and M. Ramanathan, *Am. J. Respir. Crit. Care Med.* **152**: 658 (1995).
30. L.H. Brudin, S. O. Valind, C. G. Rhodes, D. R. Truton and J. M. B. Hughes, *J Appl. Physiol*, **60(4)**: 1155 (1986).
31. J. Dupuis, C. A. Goresky, J. L. Rouleau, G. G. Bach, A. Simard, and A. J. Schwab, *J. Appl. Physiol.* **80(1)**: 30 (1996).
32. N. C. Staub, J.M. Bishop, and R. E. Forster, *J. Appl. Physiol.* **16**: 511 (1961).

Chapter 11

TWO BIOENGINEERING SOLUTIONS FOR A PULMONARY CIRCULATION

JOHN B. WEST

Department of Medicine, University of California at San Diego
La Jolla, CA 92093-0623, U.S.A.

Dr. Fung has made important contributions to our understanding of the mammalian pulmonary circulation particularly on the sheet-flow concept, distensibility of small pulmonary blood vessels, and fluid dynamics in zone II of the lung. Recently we have become interested in the avian pulmonary circulation where the physiology is very different from the mammalian counterpart, and a comparison of the two is rewarding. In contrast to the situation in mammals, the pulmonary capillaries of birds show extraordinary rigidity in spite of large changes in capillary transmural pressure. The explanation is the support that the blood capillaries receive from the surrounding air capillaries. The epithelial bridges that are attached to the outside of the blood capillaries are able to support a substantial load in spite of being exceptionally thin. The avian pulmonary circulation also shows zone II behavior under some conditions. This is surprising because the sluice-like behavior in zone II is traditionally ascribed to narrowing of the capillary bed, and this does not occur in birds. Possibly the flow characteristics are caused by compression of small veins, and indeed this may be the case in the mammalian lung as well.

1. Introduction

On a personal note, I have probably known Y. C. Fung longer than most of the contributors to this volume. We first met in 1968 when UCSD Medical School was largely a gleam in the eye, and I have been privileged to have him as a colleague ever since. He has made many major contributions to respiratory physiology, and some of the most important have been on the pulmonary circulation. Three of these can be listed as follows:

1. The sheet flow concept of the capillary circulation [1,2].
2. Distensibility of small pulmonary blood vessels [3].
3. Fluid dynamics of zone II, that is the sluice-like behavior that occurs when alveolar pressure exceeds venous pressure [4,5].

All these studies have contributed greatly to our understanding of the mammalian pulmonary circulation

Recently we have been studying the avian pulmonary circulation and to our surprise the physiology is very different from that in mammals. A comparison of

117

the two is especially rewarding and I shall take this opportunity to summarize some of the main findings.

As a background, it was about 300 million years ago that the ancestors of the present reptiles emerged from water and made a commitment to air breathing. However they were poikilothermic exotherms (cold-blooded) and could not sustain high levels of exercise. But from them came the two great evolutionary lines that produced the mammals and birds, both capable of high, sustained levels of oxygen consumption. A fascinating aspect of these two evolutionary paths is that although the physiology of many organ systems have many similarities, the lungs are radically different. Indeed it has been argued that from some points of view, the avian lung is superior to the mammalian lung [6].

Figure 1 is a simplified diagram of the two types of lung. In the familiar mammalian pattern, inspired air is drawn into the alveoli and an almost equal amount is expired. In other words, ventilation is reciprocating. By contrast, the avian lung has bellows-like air sacs that receive the inspired air and then pump this through the gas-exchanging tissue known as the parabronchi. The flow through the parabronchi is unidirectional and continues both during inspiration and expiration. This is a flow-through system of ventilation.

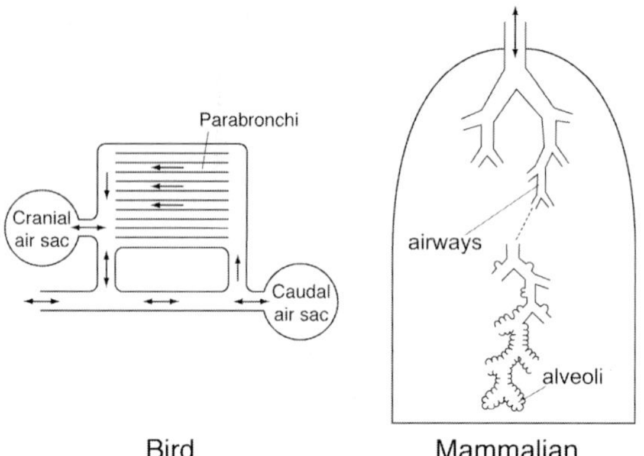

Figure 1. Simplified diagram of the bird and mammalian lungs. In the bird, gas exchange occurs in the parabronchi, and this is a flow-through system. The mammalian lung has reciprocating ventilation. From [6].

There are major differences between the two types of lungs:

1. The bird has separated the ventilatory and gas-exchanging functions of the lung. This makes a lot of sense. The gas-exchanging tissue needs to be very delicate with extremely thin walls so that gas exchange by a passive

diffusion can easily occur. By contrast, ventilation is best done by robust bellows-like structures such as the air sacs. To a bioengineer, combining these two functions in the alveolar tissue of the mammalian lung seems to invite problems. In fact one of the most common, serious diseases of the human lung is emphysema which is characterized by breakdown of alveolar walls.

2. As indicated above, the mammalian lung uses a reciprocating method of ventilation whereas the bird lung has a flow-through system rather like a car radiator. This means that in the avian lung, the terminal air spaces in the gas-exchanging region can be very small. In fact they are "air capillaries" which are only slightly larger than the blood capillaries. By contrast, in the mammalian lung, the terminal air spaces, that is the alveoli, are very much larger. The reason for this is that the small inspired tidal volume of say 0.5 l (in humans) is distributed into a much larger volume of perhaps 3.5 l. In order for the inspired gas to reach the most peripheral alveoli by a combination of convection and diffusion, the terminal air spaces have to be relatively large. Figure 2 shows typical histological appearances of the gas-exchanging tissue in avian and mammalian lungs where the contrast in size between the air capillaries and alveoli is very striking.

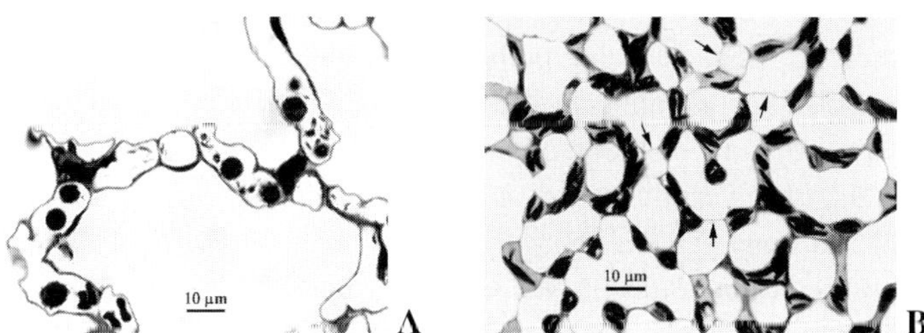

Figure 2. Light micrographs of the gas-exchanging tissue in a typical mammal (A) and bird (B). The terminal air spaces in the mammal, that is the alveoli, are very much larger than that the air capillaries in the bird. From [6].

2. Distensibility of Small Pulmonary Blood Vessels

Sobin *et al.* [3] reported that in the cat, the small pulmonary blood vessels (that is the capillaries or alveolar sheet vessels) showed marked distensibility when their transmural pressure (difference between the pressures inside and outside the capillary) was increased. Their data are replotted in Figure 3. The measurements are similar to those found by Mazzone [7] in dog lung which are also shown in the figure, and confirmed earlier data on distension of pulmonary capillaries by Glazier *et al.* [8].

J. B. West

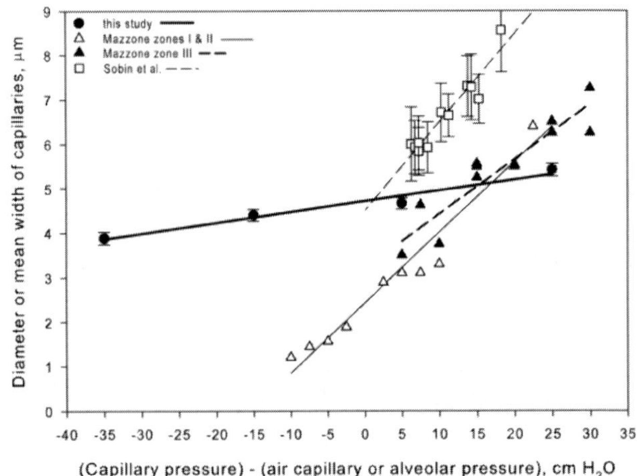

Figure 3. Comparison of bird and mammalian (dog and cat) pulmonary capillary diameter changes under different vascular and alveolar pressures. Note that the bird pulmonary capillaries are almost rigid. Dog data from [7]; cat data from [3]. From [9].

However when we made similar measurements in chicken lung, we were astonished to find that the pulmonary capillaries were almost completely indistensable. For example in the study by Watson *et al.* [9], increasing the transmural pressure of the capillaries from -10 to +25 cm H$_2$O increased the mean width of the capillaries by only 19 percent, By contrast, the increase in the study of Mazzone [7] was about fivefold. The distensibility of the capillaries in the study by Sobin *et al.* [3] was similar to that of Mazzone [7] as can be seen from the figure.

These morphometric data are from chickens are consistent with the pressure-flow relationships obtained when the vascular pressures are increased and pulmonary vascular resistance is measured. For example, in a study on perfused chicken lung *in situ*, large increases in either pulmonary arterial or pulmonary venous pressure resulted in no consistent change in pulmonary vascular resistance [10]. By contrast, measurements in mammalian lungs such as those by Borst *et al.* [11] show striking falls in pulmonary vascular resistance when either pulmonary arterial or pulmonary venous pressure is raised, the other pressure being held constant. Our findings in chicken lung are consistent with the report by Powell *et al.* [12] showing that temporary occlusion of one pulmonary artery in anesthetized ducks caused almost no change in the pulmonary vascular resistance of the unoccluded lung in spite of the fact that the blood flow through it was doubled.

What is the reason for this extraordinary rigidity of avian pulmonary capillaries? The answer is that the capillaries are supported by the honeycomb-like structure of the air capillaries surrounding them as shown in Figure 2. Close examination of the figure shows fine struts or bridges that are attached to the outside of the blood capillaries. These bridges are composed of two sheets of epithelial tissue as shown in Figure 4. They are so thin (as little as little as 0.1 µm in some places) that at first sight it seems unlikely that they could provide the necessary mechanical support to cause such rigidity of the pulmonary capillaries in the face of large changes in capillary transmural pressure. Presumably the explanation is that the loads are shared by the network of bridges rather like the spokes of a bicycle wheel share the load imposed on them. It should be pointed out that while the epithelial bridges look like spokes in Figure 4, they are in fact thin epithelial sheets.

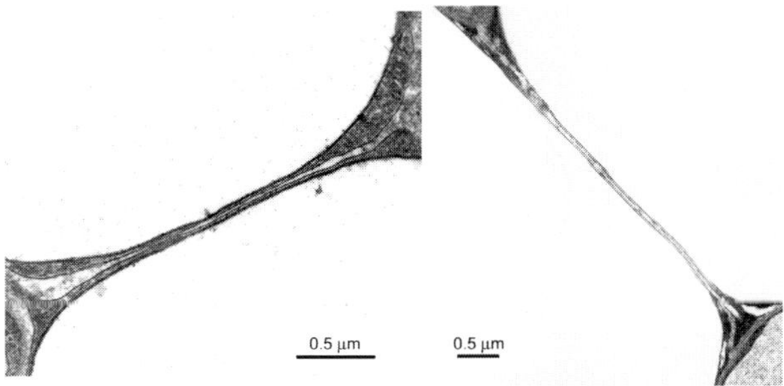

Figure 4. High-powered electron micrographs of two epithelial bridges that surround and support the pulmonary capillaries in chickens. Note the extreme thinness.

3. Pressure-Flow Relations in Zone II

Zone II of the lung is that region where pulmonary artery pressure exceeds alveolar pressure, but alveolar pressure is greater than pulmonary venous pressure [13]. A feature of Zone II is that the blood flow is determined by the difference between arterial and alveolar pressure, not arterial and venous pressure as might be expected. The traditional explanation is that because alveolar pressure exceeds venous pressure, a collapse point occurs at the downstream end of the pulmonary capillaries thus limiting flow. The result is that the alveolar pressure becomes the operative downstream pressure [14].

During our studies of the pressure-flow relationships in the chicken lung in situ, we found a pattern similar to zone II when the pressure in the air capillaries was higher than pulmonary venous pressure. In this preparation, the pressure outside the blood capillaries is the pressure in the air capillaries, so that air capillary pressure is equivalent to alveolar pressure in the mammalian lung. Figure 5 shows typical results. The upper panel shows the pressure-flow relations in the chicken and the lower panel in the dog. In both cases it can be seen that at the highest pulmonary venous pressure, there is no flow. The reason is that venous pressure is then equal to the inflow arterial pressure. Initially, as the venous pressure is reduced (moving from right to left in the figure), flow increases, but after the venous pressure is reduced below the point shown as zero, no further consistent change in flow is seen. In fact in the lower panel, flow tends to fall somewhat and this is seen in a couple of tracings in the upper panel as well. The zero point on the horizontal axis is that where venous pressure is equal to either alveolar pressure (in the lower panel) or air capillary pressure (in the upper panel). In other words these two panels show that when venous pressure is lowered below either air capillary or alveolar pressure, no further change in blood flow occurs.

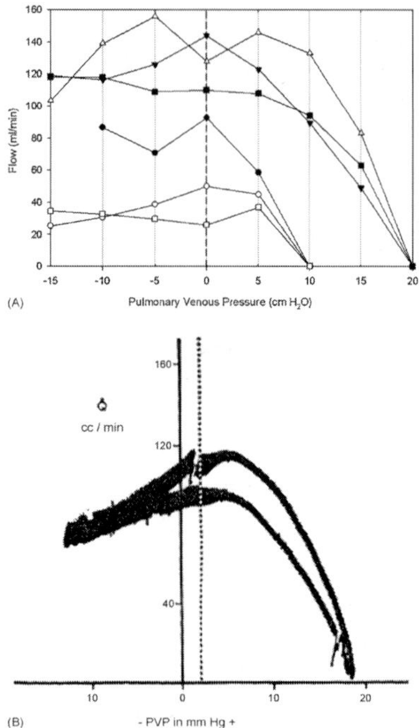

Figure 5. (A) Effects of changing pulmonary venous pressure on flow rate in chickens. Each series of symbols represents an individual chicken. When the venous pressure was reduced below air capillary pressure (indicated by the vertical broken line) there was no consistent change in flow rate under these zone II conditions. From [10]. (B) Similar experiments carried with a dog lung. When the venous pressure was reduced below alveolar pressure (indicated by the vertical broken line), flow no longer increase and in fact tended to decrease. From [14].

The pattern shown in the upper panel of Figure 5 was very surprising to us because, as stated above, the explanation for the zone II phenomenon has always been compression of the capillaries by alveolar pressure. Indeed this is part of the analyses by Fung and Yen [4], and Fung and Zhuang [5]. However as we saw earlier, there is very strong evidence that the blood capillaries in the chicken lung cannot collapse even when the air capillary pressure is considerably higher than venous or even arterial pressure. This is shown in Figure 3 where on the extreme left of the figure, air capillary pressure exceeds arterial pressure by as much as 35 cmH2O. The different behavior of the capillaries in the two types of lung is shown dramatically in Figure 6 where both lungs are in zone I, that is alveolar (or air capillary) pressure is greater than arterial pressure. In the left panel it can be seen that the chicken capillaries are widely patent, but in the right panel from the dog, the capillaries are completely collapsed.

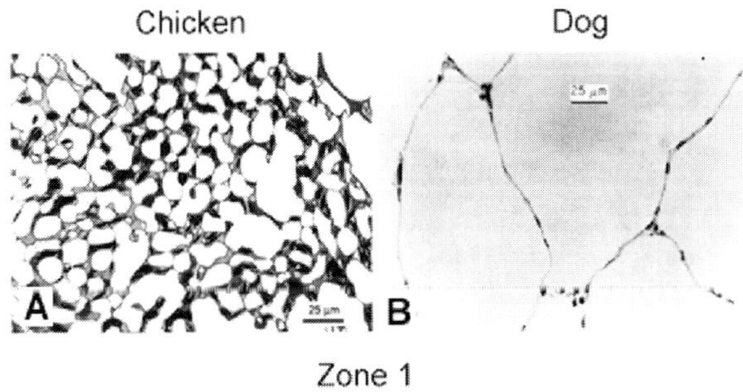

Figure 6. Light micrographs showing the differences in capillary filling in chicken and dog in zone I, that is when alveolar pressure is raised about pulmonary arterial pressure. Note that in the chicken the capillaries remain widely patent, while in the dog they are completely collapsed.

What is the explanation for the zone II behavior in the chicken lung when it is clear that the capillaries cannot collapse? One possibility is that the small pulmonary veins which are not held open by the epithelial bridges are able to collapse when venous pressure falls below the air pressure in the parabronchi. Indeed this may even be the explanation of the zone II phenomenon in mammalian lungs. There is no experimental evidence against it. For example, no one has been able to obtain histological evidence to support the fact that it is the collapse of the pulmonary capillaries in mammalian lungs that is responsible for the zone II pressure-flow relations. The results in the chicken lung certainly suggest that the small veins may be responsible.

                    J. B. West

4. Extremely Thin Blood-Gas Barrier in the Avian Lung

One of the fascinating features of the bird lung is that the blood-gas barrier is often extremely thin. Figure 7 shows the thickness of the barrier in birds, mammals, reptiles, and amphibia based on a large series of measurements tabulated in Maina and West [15]. In fact some birds have thinner blood-gas barriers than any mammals [16]. Furthermore not only is the total thickness of the blood-gas barrier extraordinarily thin in birds but the thickness of the extracellular matrix which is thought to be responsible for the strength of the barrier is also extremely thin. This is a paradox because flying is very energetic and presumably the pulmonary vascular pressures rise substantially. Therefore it is remarkable that the capillary wall is able to withstand the resultant high stresses if the extracellular matrix is so thin [17].

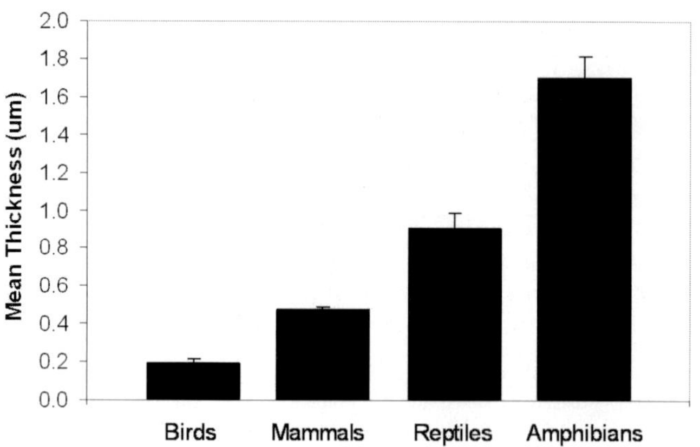

Figure 7. Harmonic mean thicknesses ±SD for the birds, mammals, reptiles, and amphibia included in Table 1 of [15]. This includes 34 species of birds, 37 of mammals, 16 or reptiles, and 10 of amphibia. Note the progressive increase in mean thickness. From Maina and West [15].

How can it be that the blood-gas barrier in the bird can be so thin, and in particular how can the extracellular matrix that is responsible for the strength of the barrier be so thin? Again we believe that the explanation is the environment of the capillaries. As Figure 2A shows, the capillaries in mammalian lung are strung out along the alveolar wall like a string of beads, and they are essentially unsupported at right angles to the wall. By contrast, as Figure 2B shows, the blood capillaries in the bird lung are supported by the epithelial bridges of the surround air capillaries. This explains both the rigidity of the pulmonary capillaries, and also the thinness of their walls.

Acknowledgments

The work was supported by NIH Grant R01 HL60968.

References

1. Y.C. Fung and S.S. Sobin. *J. Appl. Physiol.* **26**, 472 (1969).
2. S.S. Sobin, H.M. Tremer, and Y.C. Fung. *Circ. Res.* **26**, 397 (1970).
3. S.S. Sobin, H.M. Tremer, and Y.C. Fung. *Circ. Res.* **30**, 440 (1972).
4. Y.C. Fung and R.T. Yen. *J. Appl. Physiol.* **60**, 1638 (1986).
5. Y.C. Fung and F.Y. Zhuang. *J. Biomech. Eng.* **108**, 175 (1986).
6. J.B. West, R.R. Watson, and Z. Fu. *Eur. Resp. J.* **29**, 11 (2007).
7. R.W. Mazzone. *Microvasc. Res.* **20**, 295 (1980).
8. J.B. Glazier, J.M.B. Hughes, J.E. Maloney, and J.B. West. *J. Appl. Physiol.* **26**, 65 (1969).
9. R.R. Watson, Z. Fu, and J.B. West. *Resp. Physiol. Neuro.* **160**, 208 (2008).
10. J.B. West, R.R. Watson, and Z. Fu. *Resp. Physiol. Neuro.* **157**, 382 (2007).
11. H.G. Borst, M. McGregor, J.L. Whittenberger, and E. Berglund. *Circ. Res.* **4**, 393 (1956).
12. F.L. Powell, R.H. Hastings, and R.W. Mazzone. *Am. J. Physiol.* **249**, R34 (1985).
13. J.B. West, C.T. Dollery, and A. Naimark. *J. Appl. Physiol.* **19**, 713 (1964).
14. S. Permutt, B. Bromberger-Barnea, and H.N. Bane. *Med. Thorac.* **19**, 239 (1962).
15. J.N. Maina and J.B. West. *Physiol. Rev.* **85**, 811 (2005).
16. P. Gehr, S. Sehovic, P.H. Burri, H. Claasen, and E.R. Weibel. *Respir. Physiol.* **40**, 33 (1980).
17. R.R. Watson, Z. Fu, and J.B. West. *Am. J. Physiol. Lung Cell Mol. Physiol.* **292**, 769 (2007).

Chapter 12

FLUID FLOW INDUCED CALCIUM RESPONSE IN BONE CELL NETWORK[*]

BO HUO[1,2], XIN L. LU[1], X. EDWARD GUO[1]

[1]Bone Bioengineering Laboratory, Department of Biomedical Engineering
Columbia University, New York, NY 10027, U.S.A.

[2]Institute of Mechanics, Chinese Academy of Sciences, Beijing 100080, P. R. China

In vitro bone cell networks with controlled pattern and intercellular gap junctions were successfully established by using microcontact printing and self assembled monolayers technologies. The intracellular calcium responses of bone cell networks exposed to a steady fluid flow and the underlying signaling pathways were further investigated. The cell network samples were separated into eight groups for treatment with specific pharmacological agents that inhibit pathways significant in bone cell calcium signaling. The calcium transients of the network were recorded and quantitatively evaluated with a set of network parameters. The results showed that gap junction blocker, extracellular ATP pathway inhibitor, and thapsigargin (depleting intracellular calcium stores) significantly reduced the occurrence of multiple calcium peaks, which were visually obvious in the untreated group. The number of responsive peaks also decreased slightly yet significantly when either the COX-2/PGE$_2$ or the NOS/nitric oxide pathway was disrupted. In the absence of calcium in the culture medium, the intracellular calcium concentration decreased slowly with fluid flow without any calcium transients observed. These findings have identified important factors in the flow mediated calcium signaling of bone cells within an *in vitro* network.

1. Introduction

Osteocytes are interconnected through numerous intercellular processes, forming extensive cell networks throughout the bone tissue [1-3]. It has been shown that osteocyte density is an important physiological parameter, which decreases with age and microdamage accumulation [4]. However, most previous studies on mechanotransduction in osteocytes and osteoblasts were performed on confluent or sub-confluent uncontrolled monolayers of cells [5, 6]. In order to examine the roles of the cellular connectivity between bone cells in

---

[*] This was published in part in Cellular and Molecular Bioengineering, **1(1)**:58, 2008 and presented at International Symposium on Genomic Biomechanics: Frontier of the 21st Century on the Occasion of the 90th birthday of Professor Yuan-Cheng "Bert" Fung.

127

mechanotransduction, a controlled bone cell network is necessary for quantitative studies. In our previous study [7], a two-dimensional patterned bone cell network was successfully established to mimic the *in vivo* osteocyte network by using microcontact printing and self assembled monolayers (SAM) technologies. Each individual bone cell in the network was connected with four neighboring cells via functional gap junctions. Gap junctions are membrane-spanning channels, where each pair of connexons (*i.e.* hemichannels) forms a cylinder with a pore in the center through which small molecules (<1 kDa) can pass from one cell to another cell [8]. It is widely accepted that this intercellular connection plays a significant role in coordinating bone cell network activities.

Calcium signaling is essential for numerous bone cell functions, such as proliferation and differentiation [9-13]. It has been demonstrated that mechanical stimuli such as fluid flow can induce a robust intracellular calcium ($[Ca^{2+}]_i$) response in bone cells [14, 15]. The fluid flow induced elevation of cytosolic calcium comes mainly from two sources: intracellular stores (*e.g.*, endoplasmic reticulum, ER) and the extracellular environment [16, 17]. $[Ca^{2+}]_i$ signaling is mediated by various molecular pathways, *e.g.*, inositol trisphosphate (IP$_3$), adenosine 5'-triphosphate (ATP), prostaglandin E$_2$ (PGE$_2$), and nitric oxide (NO) (Fig. 1).

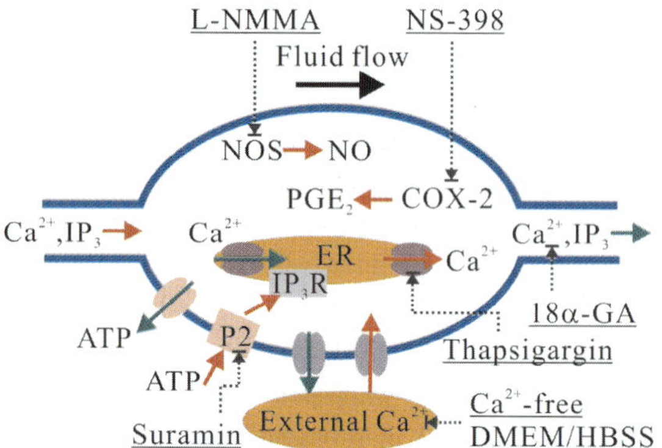

Figure 1. A schematic drawing of calcium signaling pathways in bone cells and corresponding inhibitors employed in the current study. Red arrow: influx or upregulation activity, green arrow: efflux from cytosol, dash arrow: inhibition or blocking with a specific inhibitory intervention noted and underlined.

IP_3 can lead to a rapid release of calcium stored in the ER via binding to the ER membrane receptor [9, 11]. When a bone cell is under fluid flow induced shear stress, the activation of gap junction hemichannels (connexin 43) induces ATP efflux from the cytosol to the pericellular environment [18, 19]. Extracellular ATP can elicit a significant $[Ca^{2+}]_i$ response by binding with the purinergic membrane receptors [20]. Fluid shear stress can also prompt the induction of COX-2 protein and further PGE_2 release in MC3T3-E1 osteoblast-like cells [21]. Nitric oxide modulates $[Ca^{2+}]_i$ signaling via a cyclicguanosine monophosphate (cGMP) dependent pathway [22] or nitrosylation of proteins [23].

In the present study, *in vitro* bone cell networks with controlled pattern and functional intercellular gap junctions were built with MC3T3-E1 osteoblast-like bone cells. The cell networks were divided into 8 groups to test the effects of treatment with a battery of pharmacological agents to interrupt or inhibit $[Ca^{2+}]_i$ signaling. The spatiotemporal characteristics of the $[Ca^{2+}]_i$ responses were quantitatively characterized with a set of network parameters. The inter/intra cellular calcium signaling mechanisms and the roles of different molecular pathways in $[Ca^{2+}]_i$ signaling were further analyzed.

2. Materials and Methods

2.1. *Bone Cell Network*

Microcontact printing and SAM surface chemistry technologies were used to prepare the bone cell networks as described previously [7]. In brief, a chromium mask containing the desired square grid of connected circles was fabricated. The pattern was replicated to a master made of Shipley 1818 positive photoresist (MicroChem Corp, Newton, MA) by exposing it to UV light through the mask. The geometric parameters of the pattern adopted here were optimized for osteoblast-like cells [7]. A polydimethylsiloxane (PDMS, *Sylgard 184,* Dow Corning, Midland, MI) stamp with the desired micro-pattern was then generated by curing PDMS elastomer on the master. On the day before the fluid flow experiment, the stamp was coated with an adhesive SAM (octadecanethiol; Sigma-Aldrich Co., St. Louis, MO), and pressed onto a gold coated glass slide (150 Å Au using an e-beam evaporator, SC2000, SEMICORE Inc., Livermore, CA) for 60 seconds. The stamped glass slide was immediately immersed in a non-adhesive ethylene glycol terminated SAM solution (HS-C11-EG3; Prochimia, Sopot, Poland) for 3 hours to cover areas that were not patterned with the adhesive SAM. To facilitate cell adhesion, the patterned slide was further incubated in 10 µg/mL fibronectin (Sigma-Aldrich Co., St. Louis, MO) in PBS for one hour, which can only be absorbed by the adhesive SAM coated region. Osteoblast-like MC3T3-E1 cells with controlled density (1.0×10^4 cell/cm^2) were seeded on the patterned slide and cultured in α-MEM medium supplemented with 2% charcoal-stripped fetal bovine serum (CS-FBS, Hyclone Laboratories Inc., Logan, UT) for 24 hours before further testing. Figure 2 shows the snapshots of a typical bone cell network in the untreated group with a 70 µm separation distance.

2.2. *Steady Fluid Flow and $[Ca^{2+}]_i$ Response*

To indicate $[Ca^{2+}]_i$, patterned bone cells on the glass slide were loaded with 5 µM Fluo-4 AM (Molecular Probes, Eugene, OR) in α-MEM medium for 1 hour at 37°C. After calcium dye loading, the slide was mounted on a custom-built parallel plate flow chamber, which was further attached to an inverted fluorescence microscope (Olympus IX71, Melville, NY) with a 10X objective [16]. A magnetic gear pump was connected to the chamber to generate a steady fluid flow with a constant shear stress of 40 dyne/cm^2 on the cell surface. The $[Ca^{2+}]_i$ response of the bone cell network under fluid flow stimulation was recorded with a high-speed CCD camera (ORCA-ER-1394, Hamamatsu Photonics K.K., Hamamatsu City, Japan) for a total 10-minute period, one minute for baseline and 9 minutes after the onset of fluid flow. The intensity of $[Ca^{2+}]_i$ for each cell was later normalized by its corresponding baseline.

2.3. *Experimental Groups*

Besides the untreated group (12 slides, 130 cells analyzed), seven specific biochemical blocking agents were employed to identify the sources and the contributions of molecular pathways to the network characteristics of $[Ca^{2+}]_i$ response. *(1) Extracellular calcium depletion:* In this group, calcium-free Dulbecco's Modified Eagle Medium (DMEM, Invitrogen Corporation, Carlsbad, CA) and calcium-free Hank's Balanced Salt Solution (HBSS, Invitrogen Corporation) were introduced to replace the regular medium after Fluo-4 AM loading (8 slides, 99 cells analyzed). *(2) ER calcium store depletion*: After Fluo-4 staining, the cell-seeded slides (6 slides, 90 cells analyzed) were incubated in 1 µM Thapsigargin (TG) (Sigma-Aldrich Co.) medium for 30 min prior to flow [20]. TG depletes calcium from the ER store. *(3) PGE$_2$ blocking*: The patterned cells were pretreated with 10 µM NS-398 (EMD Chemicals Inc., San Diego, CA) for 24 hours before flow study. NS-398 selectively inhibits the COX-2 enzyme activity, which in turn blocks the PGE$_2$ release [21]. 123 cells on 11 slides were analyzed in this group. *(4) NO blocking*: In this group, 100 µM NG-monomethyl-L-arginine (L-NMMA; EMD Chemicals Inc.) was introduced into the cell culture medium one day before seeding the cells on slides and continuously presented in mediums afterwards. L-NMMA inhibits the production of nitric oxide via competitively inhibiting all three isoforms of nitric oxide synthase (NOS) [24]. In this group, 189 cells on 16 slides were analyzed. *(5) Gap junction blocking*: To study the role of intercellular gap junctions on calcium responses of the bone cell network, 75 µM 18α-Glycyrrhetinic acid (18α-GA, Sigma-Aldrich Co.), a reversible gap junction blocker which binds to membrane proteins and causes disassembly of gap junction plaques [25], was supplied in Fluo-4 AM loading medium and fluid flow medium [5]. In this group, 61 cells on 6 slides were studied. *(6) ATP blocking:* Suramin (EMD Chemicals Inc.), a P2 purinergic receptor blocker, was employed to investigate the effect of ATP release on the $[Ca^{2+}]_i$ response. 100 µM suramin was applied to the cells 30 minutes before exposing to shear flow [26]. 50 cells from 3 slides were analyzed. *(7) Vehicle DMSO control:* Among all the previously employed chemicals, NS-398, 18α-GA, and TG were dissolved into DMSO. Therefore vehicle control testing was performed with 0.3% v/v DMSO presented in medium (3 slides, 51 cells analyzed).

2.4. *Data analysis*

To quantitatively analyze the network characteristics of the $[Ca^{2+}]_i$ responses, a set of network parameters were defined from the time-lapse of $[Ca^{2+}]_i$ response (Fig. 2).

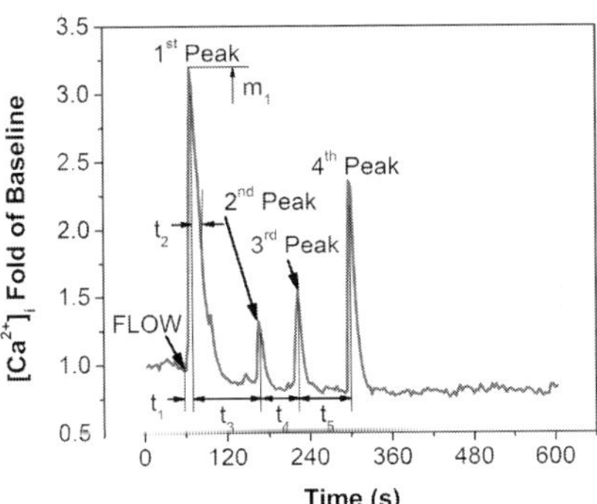

Figure 2. A typical calcium response of bone cells in an untreated network under applied steady fluid flow. The characteristic temporal parameters are also illustrated. t_1 denotes the time from flow onset to the maximum value of the first responsive peak; t_2 is the time from peak value to the 50% relaxation; t_3, t_4, and t_5 are the time intervals between different peaks. m_1 represents the normalized magnitude of the first peak.

For each individual cell, the total number of $[Ca^{2+}]_i$ responsive peaks during the flow stimulation period was counted. t_1 denoted the time between the onset point of fluid flow and the maximum value of the first responsive spike, while t_2 was the time from first peak to its 50% relaxation point. One-way analysis of variance (ANOVA) with Bonferroni's post hoc analysis was performed to determine statistical differences between mean values of different treatments, $p<0.05$. The condition of equal population variances of treatments was tested by Levene's method.

3. Results

A typical set of intracellular calcium images during the testing period are shown in Fig. 3, clearly demonstrating multiple responses of bone cells within the network under fluid flow. A set of typical $[Ca^{2+}]_i$ traces of individual cells from the untreated group and those with different treatments are shown in Fig. 4. In each group, $[Ca^{2+}]_i$ profiles of three cells from a single glass slide were plotted.

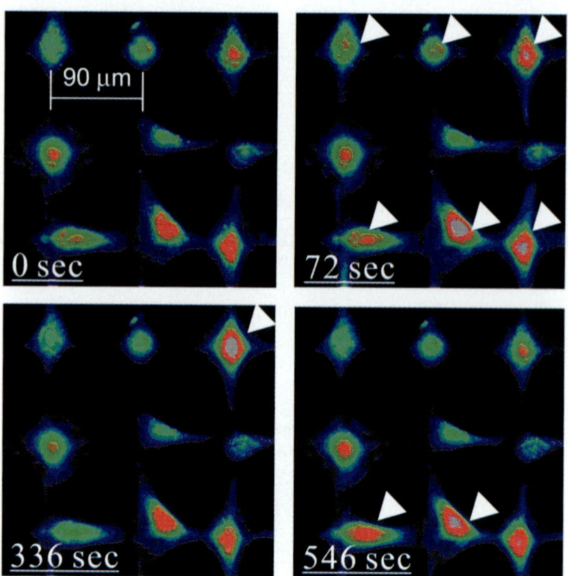

Figure 3. Pseudo-color images of typical intracellular calcium responses of a bone cell network from untreated group. Fluid flow was applied at 60 seconds. Arrowheads highlight those cells responding. Some cells were able to respond multiple times.

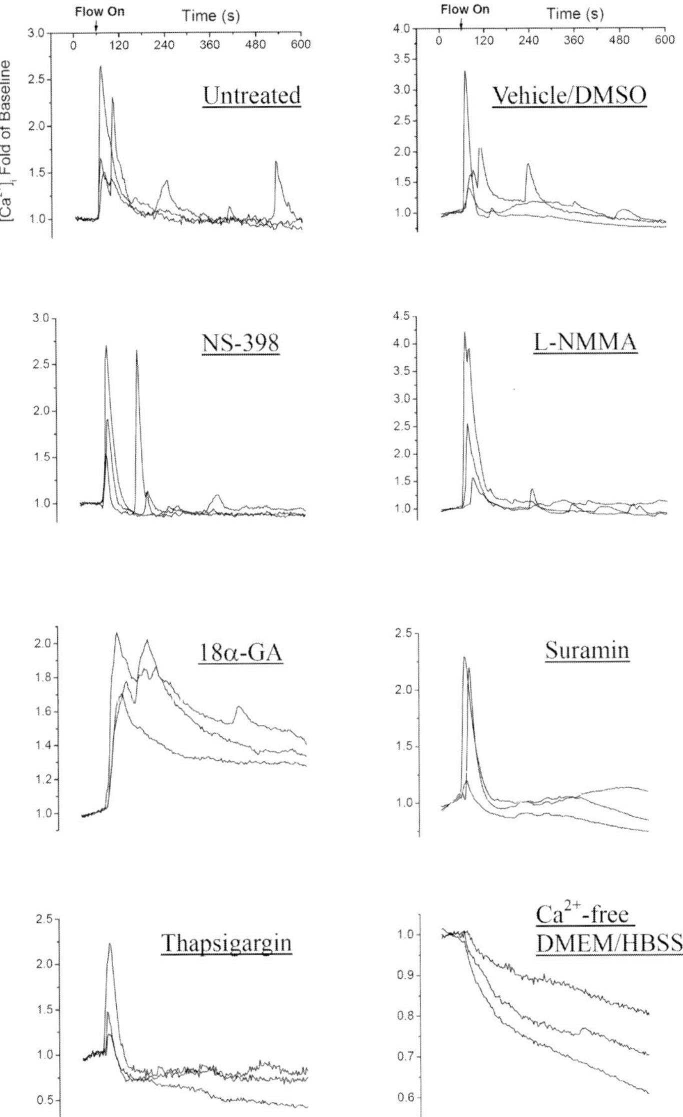

Figure 4. A set of typical intracellular calcium traces of individual cells with or without pharmacological blocker treatment. The calcium images were recorded for 10 minutes (1 minute for baseline and 9 minutes for flow stimulation). The intensity of intracellular calcium concentration shown here was normalized to the baseline.

For untreated, vehicle control, NS-398 (PGE_2 blocking), and L-NMMA (NO blocking) groups, multiple $[Ca^{2+}]_i$ responsive peaks were quite obvious even by visual inspection. However, 18α-GA (gap junction blocking) and TG (ER calcium depletion) groups tended to exhibit only a single $[Ca^{2+}]_i$ peak. In the absence of extracellular calcium, unlike the $[Ca^{2+}]_i$ response in any other group, $[Ca^{2+}]_i$ decreased slowly after the onset of fluid flow without any $[Ca^{2+}]_i$ peak.

To quantitatively and statistically analyze the calcium waves, the frequencies of multiple responding peaks are shown in Fig. 5a, while the mean value of the number of responding peaks are plotted in Fig. 5b.

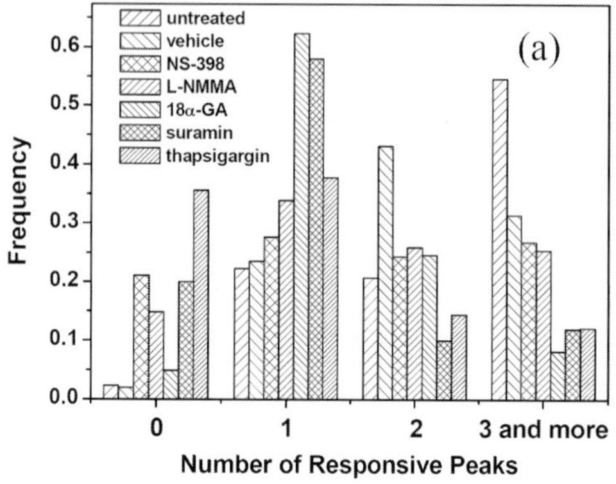

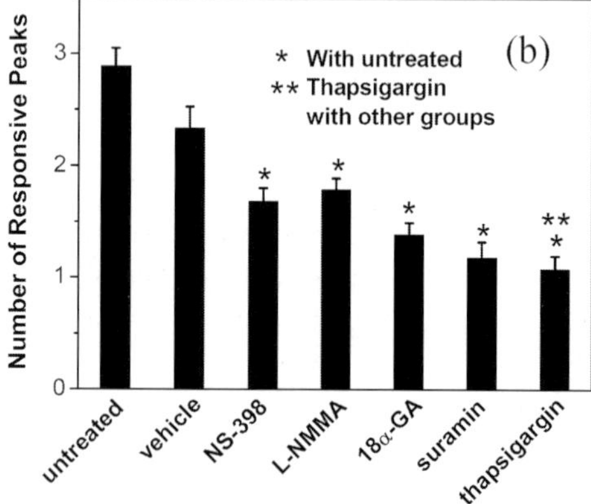

Figure 5. (a) The frequency distribution of the number of responsive peaks of cells in different groups and (b) the average of the number of responsive peaks (the error bars are standard errors of the means (SEM) while * and ** indicate statistical significance of p<0.05). The group of Ca^{2+} free medium had no response to fluid flow and was not included.

In the untreated group, over 55% of cells were able to respond three or more times with an overall mean value as 2.89 ± 0.16 times of baseline. In the vehicle control group, this value was 2.33 ± 0.19 but no significant difference was detected when compared with the untreated group. Both NS-398 (PGE_2 blocking) and L-NMMA (NO blocking) resulted in fewer multiple responses, only 50% of cells exhibited two or more calcium transients. Furthermore, 15% and 22% of cells had no response to fluid flow in these two groups, respectively, which was significantly higher than those of the untreated or vehicle control group. 18α-GA (gap junction blocking) and suramin (ATP blocking) significantly reduced the number of calcium peaks, while cells with a single response comprised approximately 60% of the total population. ANOVA testing revealed no significant difference between the mean values of these two groups, but they were both significantly different from untreated and vehicle control groups. The TG-treated group (ER depleted) generated the least number of responding calcium peaks, which was significantly different from any other group. The removal of extracellular Ca^{2+} completely abolished the $[Ca^{2+}]_i$ response.

As shown in Fig. 6, cells in the gap junction blocked group took the longest time to reach the first $[Ca^{2+}]_i$ peak ($t_1 = 61 \pm 7$ seconds). Moreover, the relaxation time of this group was nearly 5 times that of all other groups ($t_2 = 222 \pm 23$ seconds). The untreated and TG groups shared a similar t_1 value (40 ± 4 *vs.* 34 ± 2 seconds, P>0.05). The vehicle control, NS-398, L-NMMA, and suramin groups all had significantly shorter first peak response times when compared with the unteated group, but there was no significant difference among these four groups. Aside from the 18α GA group, all the other six groups had no significant difference in the 50% relaxation time.

4. Discussion

The network characteristics of $[Ca^{2+}]_i$ response of bone cells in a controlled cell network have been quantified, and several biochemical pathways related to $[Ca^{2+}]_i$ signaling have been examined. Notably, multiple $[Ca^{2+}]_i$ peaks, a signature of network circuits, have been observed in bone cell networks under applied steady fluid flow. Intracellular calcium signaling in bone cells involves both internal and external sources of calcium [9]. The internal calcium stores are

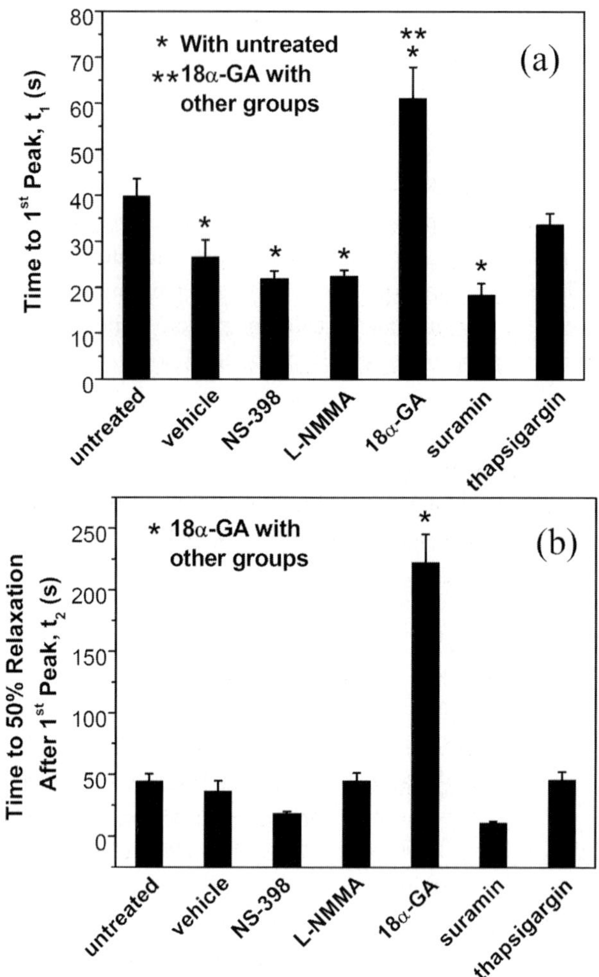

Figure 6. Network characteristic parameters of calcium response in the bone cell network under steady fluid flow from different treatment groups (the error bars are SEM while * and ** indicate statistical significance of $p<0.05$): (a) time to the first peak t_1 and (b) time to 50% relaxation after the first peak t_2. The group of Ca^{2+} free medium had no response to fluid flow and was not included.

mainly held within the membrane systems of the ER [27, 28]. Current experimental results from calcium-free media and ER depletion groups have identified the calcium source underlying the $[Ca^{2+}]_i$ responses in the bone cell networks. The cells in calcium-free media showed absolutely no $[Ca^{2+}]_i$ increase due to fluid flow, indicating the critical role of extracellular calcium. A few researchers have noted that the influx of extracellular calcium is required to evoke the opening of the calcium release channels in the ER [29, 30]. In fact, it has been shown that the gap junction dependent propagation of intercellular calcium signals in rat osteoblast cells requires influx of external calcium through L-type calcium channels [30]. Moreover, the cytosolic calcium concentration was significantly reduced in the absence of extracellular calcium, which was consistent with a previous report [20]. On the other hand when calcium was depleted from ER stores, more than 60% of cells exhibited a calcium peak under the stimulation of fluid flow, similar with those observed in randomly cultured 80-90% confluent MC3T3-E1 cells under an oscillatory fluid flow [31]. The magnitude of the first responsive peak was not significantly altered. This confirmed that extracellular calcium influx was responsible for the initial calcium peak. However, the average number of responsive $[Ca^{2+}]_i$ peaks in the cell network was significantly reduced to barely more than one. Thus ER stores appears to be the major calcium source responsible for the lower-magnitude $[Ca^{2+}]_i$ peaks following the initial calcium transient in the network of bone cells [29, 31].

The $[Ca^{2+}]_i$ responses of the networked bone cells consist of a large initial peak followed subsequently by multiple lower-magnitude peaks, which is similar to those observed in an 80-90% confluent bone cell-culture study [32]. In the untreated group, over 50% of the cells were able to generate three or more calcium peaks within 9 minutes. In the ER calcium depleted group, the ATP activation blocked group, and the gap junction blocked group, the average number of calcium peaks during 9 minutes was significantly reduced, and the number for the first two groups was very close to one (TG and suramin treated, respectively). The importance of ER calcium stores in the multiple responsive $[Ca^{2+}]_i$ peaks [29, 31] has been confirmed in the current networked bone cells. This new bone cell network study suggests the critical importance of the ATP pathway in mediating multiple $[Ca^{2+}]_i$ peaks of bone cells under steady fluid flow. This observation is consistent with literature where calcium wave propagation in bone cells induced by cell poking was influenced by the ATP pathway [30, 33, 34]. We also noted that 18α-GA (gap junction blocking) did not completely extinguish the multiple transients of bone cells, as 32% of cells responded more than once. This implied that a different pathway other than gap junction signal communication also plays a significant role in the multiple calcium peaks in the bone cell network. Indeed, inhibiting COX-2/PGE$_2$ or nitric oxide pathways also slightly reduced the multiple calcium peaks of individual osteoblasts, which was similar to previous findings in other cell types [22, 23, 35, 36]. A previous study by Li et al. has shown that blocking the nitric oxide pathway can significantly reduce the refilling capability of intracellular calcium store [37], thereby decreasing the occurrence of multiple transients.

In summary, a thorough investigation of the spatiotemporal characteristics of $[Ca^{2+}]_i$ signaling of bone cells in a controlled cell network under steady fluid flow was conducted. The effects of several candidate pathways potentially involved in inter/intra cellular calcium signaling were also studied. We confirmed that the major source for the rapid initial calcium transient under mechanical stimulation was attributable to the influx of extracellular calcium. The ER calcium stores and ATP pathways are responsible for the subsequent multiple calcium peaks. The intercellular gap junctions in the cell network demonstrated a significant influence on the $[Ca^{2+}]_i$ response. Bone cells in a gap junction-blocked network exhibited a sustained high calcium concentration after the triggering from fluid flow. It was also shown that the PGE_2 and NO pathways were both involved to a lesser extent in the calcium signaling of the bone cell network.

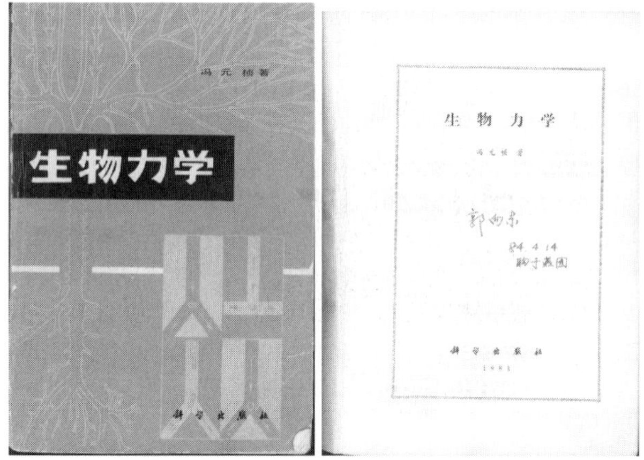

The first Biomechanics book published in China is by Dr. Y.C. Fung and Science Press (China) in 1983, entitled "Biomechanics". Dr. X. Edward Guo's journey in biomechanics research started from reading this book in 1984 when Dr. Guo was a senior in Peking University. His story is not unique and many researchers currently in the field of biomedical engineering who grew up in China share the same experience.

Dr. X. Edward Guo (left) met Dr. Y.C. Fung (center) for the first time at the Symposium of Inauguration of the Department of Biomedical Engineering at Columbia University in 2000. At right is another faculty member in the new Department of Biomedical Engineering at Columbia University, Dr. Clark T. Hung.

Dr. X. Edward Guo with Dr. Y.C. Fung at the Genomics on Biomechanics Symposium in celebration of Professor Y.C. Fung's 90th Birthday at UCSD.

References

1. S. Weinbaum, S.C. Cowin, and Y. Zeng, *J Biomech*, **27**(3), 339 (1994).
2. S.C. Cowin, *Bone*, **22**(5 Suppl), 119S (1998).
3. L.D. You, S. Weinbaum, S.C. Cowin, and M.B. Schaffler, *Anat Rec A Discov Mol Cell Evol Biol*, **278**(2), 505 (2004).
4. D. Vashishth, O. Verborgt, G. Divine, M.B. Schaffler, and D.P. Fyhrie, *Bone*, **26**(4), 375 (2000).
5. C.E. Yellowley, Z. Li, Z. Zhou, C.R. Jacobs, and H.J. Donahue, *J Bone Miner Res*, **15**(2), 209 (2000).

6. G. Zaman, A.A. Pitsillides, S.C. Rawlinson, R.F. Suswillo, J.R. Mosley, M.Z. Cheng, L.A. Platts, M. Hukkanen, J.M. Polak, and L.E. Lanyon, *J Bone Miner Res*, **14**(7), 1123 (1999).

7. X.E. Guo, E. Takai, X. Jiang, Q. Xu, G.M. Whitesides, J.T. Yardley, C.T. Hung, E.M. Chow, T. Hantschel, and K.D. Costa, *Mol Cell Biomech*, **3**(3), 95 (2006).

8. F. Shapiro, *Calcif Tissue Int*, **61**(4), 285 (1997).

9. M.J. Berridge, P. Lipp, and M.D. Bootman, *Nat Rev Mol Cell Biol*, **1**(1), 11 (2000).

10. M. Zayzafoon, K. Fulzele, and J.M. McDonald, *J Biol Chem*, **280**(8), 7049 (2005).

11. J. Iqbal and M. Zaidi, *Biochem Biophys Res Commun*, **328**(3), 751 (2005).

12. K. Sato, A. Suematsu, T. Nakashima, S. Takemoto-Kimura, K. Aoki, Y. Morishita, H. Asahara, K. Ohya, A. Yamaguchi, T. Takai, T. Kodama, T.A. Chatila, H. Bito, and H. Takayanagi, *Nat Med*, **12**(12), 1410 (2006).

13. M. Zayzafoon, *J Cell Biochem*, **97**(1), 56 (2006).

14. A. Liedert, D. Kaspar, R. Blakytny, L. Claes, and A. Ignatius, *Biochem Biophys Res Commun*, **349**(1), 1 (2006).

15. J. Rubin, C. Rubin, and C.R. Jacobs, *Gene*, **367**, 1 (2006).

16. C.T. Hung, F.D. Allen, S.R. Pollack, and C.T. Brighton, *J Biomech*, **29**(11), 1411 (1996).

17. N.X. Chen, K.D. Ryder, F.M. Pavalko, C.H. Turner, D.B. Burr, J. Qiu, and R.L. Duncan, *Am J Physiol Cell Physiol*, **278**(5), C989 (2000).

18. D.C. Genetos, D.J. Geist, D. Liu, H.J. Donahue, and R.L. Duncan, *J Bone Miner Res*, **20**(1), 41 (2005).

19. D.C. Genetos, C.J. Kephart, Y. Zhang, C.E. Yellowley, and H.J. Donahue, *J Cell Physiol*, **212**(1), 207 (2007).

20. C.T. Hung, F.D. Allen, K.D. Mansfield, and I.M. Shapiro, *Am J Physiol*, **272**(5 Pt 1), C1611 (1997).

21. S.M. Norvell, S.M. Ponik, D.K. Bowen, R. Gerard, and F.M. Pavalko, *J Appl Physiol*, **96**(3), 957 (2004).

22. D.K. Looms, K. Tritsaris, B. Nauntofte, and S. Dissing, *Biochem J*, **355** (Pt 1), 87 (2001).

23. T. Volk, K. Mading, M. Hensel, and W.J. Kox, *J Cell Physiol*, **172**(3), 296 (1997).

24. S.H. Ralston, D. Todd, M. Helfrich, N. Benjamin, and P.S. Grabowski, *Endocrinology*, **135**(1), 330 (1994).

25. Y. Guo, C. Martinez-Williams, K.A. Gilbert, and D.E. Rannels, *Am J Physiol*, **276**(6 Pt 1), L1018 (1999).

26. C.E. Yellowley, C.R. Jacobs, and H.J. Donahue, *J Cell Physiol*, **180**(3), 402 (1999).

27. M.J. Berridge, *Nature*, **361**(6410), 315 (1993).

28. D.E. Clapham, *Cell*, **80**(2), 259 (1995).

29. H. Mogami, A.V. Tepikin, and O.H. Petersen, *Embo J*, **17**(2), 435 (1998).

30. N.R. Jorgensen, S.C. Teilmann, Z. Henriksen, R. Civitelli, O.H. Sorensen, and T.H. Steinberg, *J Biol Chem*, **278**(6), 4082 (2003).
31. N.N. Batra, Y.J. Li, C.E. Yellowley, L. You, A.M. Malone, C.H. Kim, and C.R. Jacobs, *J Biomech*, **38**(9), 1909 (2005).
32. S.W. Donahue, H.J. Donahue, and C.R. Jacobs, *J Biomech*, **36**(1), 35 (2003).
33. N.R. Jorgensen, S.T. Geist, R. Civitelli, and T.H. Steinberg, *J Cell Biol*, **139**(2), 497 (1997).
34. N.R. Jorgensen, Z. Henriksen, C. Brot, E.F. Eriksen, O.H. Sorensen, R. Civitelli, and T.H. Steinberg, *J Bone Miner Res*, **15**(6), 1024 (2000).
35. L. Soldati, A. Terranegra, B. Baggio, R. Biasion, T. Arcidiacono, G. Priante, D. Cusi, and G. Vezzoli, *Prostaglandins Leukot Essent Fatty Acids*, **75**(2), 91 (2006).
36. N.S. Krieger, K.K. Frick, K. LaPlante Strutz, A. Michalenka, and D.A. Bushinsky, *J Bone Miner Res*, **22**(6), 907 (2007).
37. N. Li, J.Y. Sul, and P.G. Haydon, *J Neurosci*, **23**(32), 10302 (2003).

Chapter 13

ANALYSIS OF THE MODELS FOR CYTOSKELETAL RHEOLOGY

ROGER D. KAMM

*Departments of Mechanical Engineering and Biological Engineering
Massachusetts Institute of Technology, Cambridge, MA 02139, U.S.A.*

TAEYOON KIM

*Department of Mechanical Engineering, Massachusetts Institute of Technology
Cambride, MA 02139, U.S.A.*

WONMUK HWANG

*Department of Biomedical Engineering, Texas A&M University
College Station, TX 77843, U.S.A.*

Spanning a period of a few decades, there have been several models proposed for the elasticity or viscoelasticity of the cytoskeleton. In addition, there have been numerous experiments using cells and reconstituted actin gels, with or without cross-linkers, designed to characterize the rheology of actin networks under various conditions. Still, debates continue regarding the contributions of various potential mechanisms (filament bending stiffness, thermal fluctuations, filament extensional stiffness, cross-linker stiffness, cross-link binding and unbinding) to cell viscoelasticity, and which model (semiflexible polymer network, tensegrity network, cellular solids) is best able to capture the measured characteristics. Here, we use a computational model of a thermally-active cross-linked network to probe the important factors and discuss these in the context of the different models. We conclude that thermal fluctuations are not important in the cytoskeleton, in contrast to the semiflexible polymer network theory. However, both the cellular solids and tensegrity models fail to capture some of the salient mechanical features. For cross-linked networks under prestress, the tensegrity model appears to be the most consistent with the results of our simulations. However, currently no single model satisfactorily captures viscoelastic behaviors of actin networks over a wide range of conditions. Our simulation scheme provides a basis for delineating multiple mechanisms involved in this complex system.

1. Introduction

In this tribute to YC (Bert) Fung, we wish to provide a contribution that was inspired by his amazing propensity to introduce new ideas that would provoke a stimulating discussion and facilitate further studies. Never has he shied away from controversy, but rather, seems always to thrive on it. Although he never published on cytoskeletal viscoelasticity, he has provided considerable insight into cell mechanics over the years. We consequently chose this topic for our contribution, in the hopes that this, too, might stimulate further discussion, and present some new perspectives on a long-standing problem. Over the past several decades, three major models have emerged, each of which has been applied to the cytoskeleton (Fig. 1).

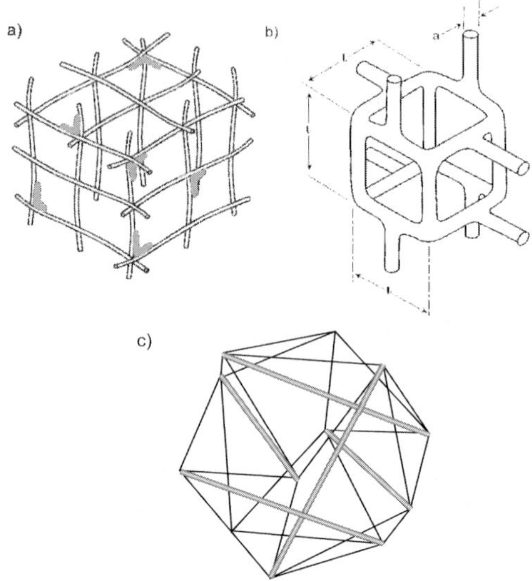

Figure 1. Schematic representations of three models for cellular mechanics. (adapted from [18]) A) *Biopolymer model.* The network is comprised of thermally active filaments with varying degrees of cross-links (light gray). B) *Cellular solids.* The cytoskeleton is assumed to consist of unit cells constructed from elastic beams with rigid connections. C) *Tensegrity.* The network is presumed to be composed of elements that are maintained in a state of tension either by interactions with other compressive members (as shown) or by tethering to an external matrix.

Each differs fundamentally, however, in terms of the primary mechanism(s) that determine the elastic or viscoelastic properties of the network. In the cellular solids model [1,2], the material is depicted as an athermal network of beams with rigid connections, and stiffness is determined primarily by the bending of these individual members. The second model, based on the concepts of tensegrity [3], proposes that the network consists of a balance between elements in tension and other elements in compression. In this model, the properties of the individual elements are relatively unimportant; network stiffness is determined instead by the level of tension in the system. The tensegrity model also neglects thermal effects. Finally, based on the assumption that cross-links or entanglements occur over a length scale comparable to or smaller than the persistence length of thermally active filaments in the network, the cytoskeleton has been modeled as a semiflexible polymer network [4]. The fourth model, based on the concepts of "soft glassy rheology" [5], will not be discussed here, in part because the fundamental physics giving rise to this behavior continues to be a source of some debate and uncertainty. (For comprehensive reviews of these models, see [6] and [7].)

It is likely that each of these models is valid in certain regimes or under specific conditions that occur in biology. Indeed, each has been supported by experimental evidence, either in cells or in reconstituted actin networks. Still, it is surprising that one model has not yet emerged as a clear leader in terms of providing the closest approximation to available experimental data.

One of the complicating factors is that, as with the models, there remains some variability in the experimental data for cytoskeletal networks. While there are numerous publications presenting experimental data, we can cite two classes of measurement that provide a sense of the range of observations. One method used extensively is magnetic twisting cytometry (MTC) [8]. In this case, a magnetic bead is attached to the cell via adhesion receptors (typically one of the integrins), and the beads are oscillated at different frequencies. This type of measurement has led to a weak power law behavior characterized by [6]. Some assumptions are required to be made in order to interpret the observed bead oscillations in terms of a shear modulus (G), and the estimated values for G lie in the range of ~10–300 Pa. In contrast to this, measurements have been made with particles that are internalized by the cell, and the measured thermal motions of the particles were used to determine the frequency-dependent shear moduli (10–10^3 Pa) [9]. Recent data using multiple particle tracking (MPT) methods employing particles introduced into the cell by ballistic methods [10] yielded lower values (< 100 Pa).

2. Methods

In order to probe the mechanisms responsible for the unique viscoelastic behavior of living cells, a computational model was developed and utilized to explore the relative importance of various factors and mechanisms. The model simulates the behavior of a simple system that consists of a single type of microfilament and a single cross-linking agent. While the parameter values were chosen to match as close as possible those of actin and filamin, or actin and α-actinin networks, the results should be viewed more generally as representing any cross-linked network of semi-flexible filaments.

Our approach was to first generate a cross-linked network from a collection of free monomers and cross-linking proteins in solution and then to probe the network for its viscoelastic character. Polymerization and cross-linking were accomplished by allowing each monomer, cross-linker, and partially-formed filament to undergo thermal fluctuations and to bind to each other as governed by close proximity in combination with specified rate constants. Although the results presented here are based on irreversible binding, the model also permits unbinding events according to the Bell equation [11]. Once the network is formed, it is reduced to a coarse-grained form, replacing the initial spherical monomers by cylinders, each representing 10 monomer-segments of the actin filaments [12].

Elastic properties of the network are governed by several critical parameters of its constituents: bending and extensional stiffness of the filaments and cross-linking agents. Of these the characteristics of the cross-linking proteins have recently been characterized by computational studies [13], but reasonable values of these parameters were estimated based on other well-known parameter values.

To test the viscoelasticity of the network formed in this way, two complementary methods were used. In what we term *segment-tracking rheology*, the thermal motion of the individual cylindrical segments was tracked in time to obtain the mean square displacement and analyzed in the manner of [14] to estimate the real (G') and imaginary (G'') parts of the shear modulus, G. This method parallels that employing the thermal motion of microbeads injected into the cell. Since this approach relies on a number of assumptions that may not be valid in the analysis of filament fluctuations, a second method, bulk rheology, was used, more akin to the macroscopic measurements of gel rheology. The entire computational domain is subjected to an oscillatory strain by displacement of the ends of the filaments on one boundary while those on the opposite boundary are fixed. To simulate the movement of the solvent, a linear fluid velocity is imposed between the two surfaces. Shear stress is computed by dividing the forces required to generate the oscillatory strain by the surface area, and G is obtained from the time-varying shear stress and strain.

As a means of identifying the main mechanisms contributing to G' and G'', each of several parameters was varied by the same fractional amount, and the effect of each variance on network modulus was determined.

3. Results

Linear behavior. The frequency dependence of shear moduli calculated either from the response to small-amplitude oscillations of the entire network or from thermal fluctuations of the individual filament segments, matches well the experimental data that were obtained from reconstituted actin networks cross-linked with filamin (Fig. 2) [15], suggesting that, at least for this range of parameters, the thermal fluctuations of networks provide a useful, non-invasive means of determining network viscoelasticity.

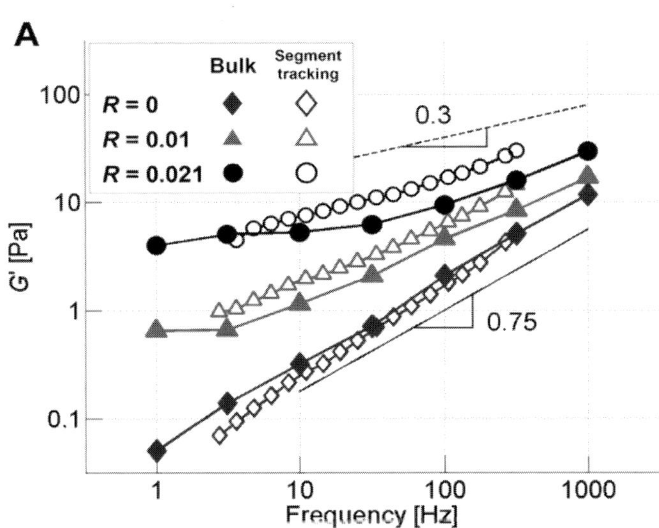

Figure 2. The real (G') and imaginary (G'') parts of the shear modulus as a function of frequency for the two different analyzing methods. Actin concentration (C_A) of the network is 508 µg/ml. Results are shown for three degrees of cross-linking ($R = 0$, 0.01, and 0.021), where R is the ratio of cross-linker concentration to C_A. Note the transition from a shallow slope at low frequency to a slope approaching 0.75 at high frequency, particularly at the high cross link density ($R - 0.021$).

Isolated effects of cross-links, filaments, and thermal effects. By reducing each of the parameters describing filament or cross-linker stiffness by a factor of 25, the relative importance of the various structural components in the network was assessed (Fig. 3A). At low prestrains, bending of the cross-linkers and the filaments are comparably important. As prestrain increases, the influence of filament bending stiffness gradually diminishes, however, while cross-linker bending remains important. At the same time, increasing prestrain leads to important contributions of extensional stiffness, both of the actin itself and the cross-linkers. It should be noted, however, that in these simulations no cross-link unbinding or unfolding were allowed, which would likely become important in reality at high prestrains.

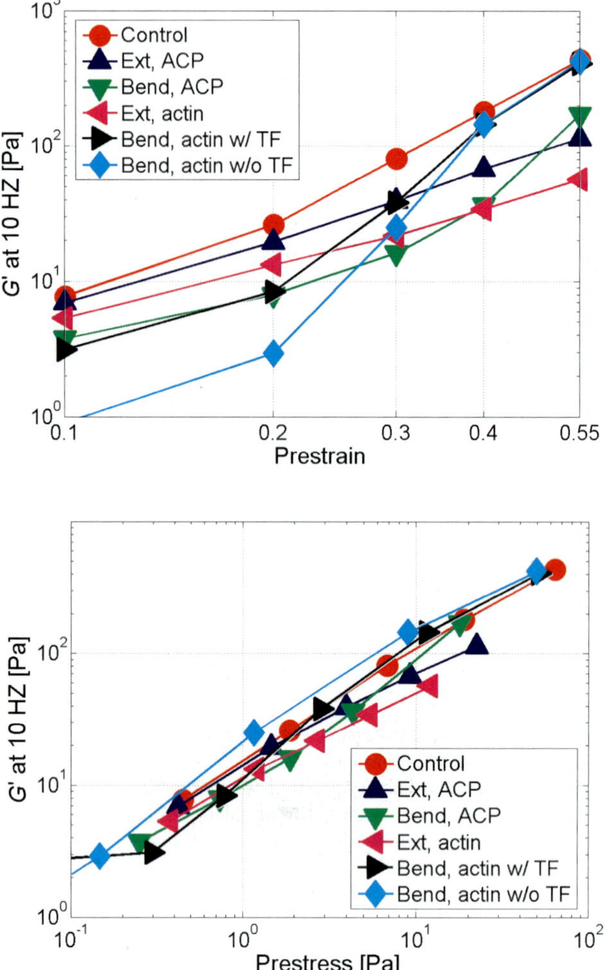

Figure 3. Values of shear storage modulus (G') evaluated at 10 Hz with normal parameter values (control, red)) compared to G' calculated when individual structural parameters are reduced 25-fold: filament bending stiffness, filament extensional siffness, cross-linker bending stiffness, and cross-linker extensional stiffness. In addition, when filament bending stiffness is reduced 25-fold (thereby reducing persistence length), thermal fluctuations are negated (cyan). A) Plotted against prestrain. B) Plotted against prestress (black solid line has a slope of 1).

Strain-stiffening. Focusing on the interplay between thermal fluctuations and prestress, a 3-dimensional plot can be generated that examines the combination of these effects, and leads to an interesting, but intuitively simple idea (Fig. 4). Thermal effects tend to be manifest under the combined effects of low prestrain and short persistence length of the filaments. These conditions are contrary to what has been thought to be the case for the cytoskeleton [16]. While the degree of prestrain is difficult to identify *in vivo*, it is generally believed that cells normally exist in a prestressed state due to actomyosin contractions balanced by forces from the extracellular matrix.

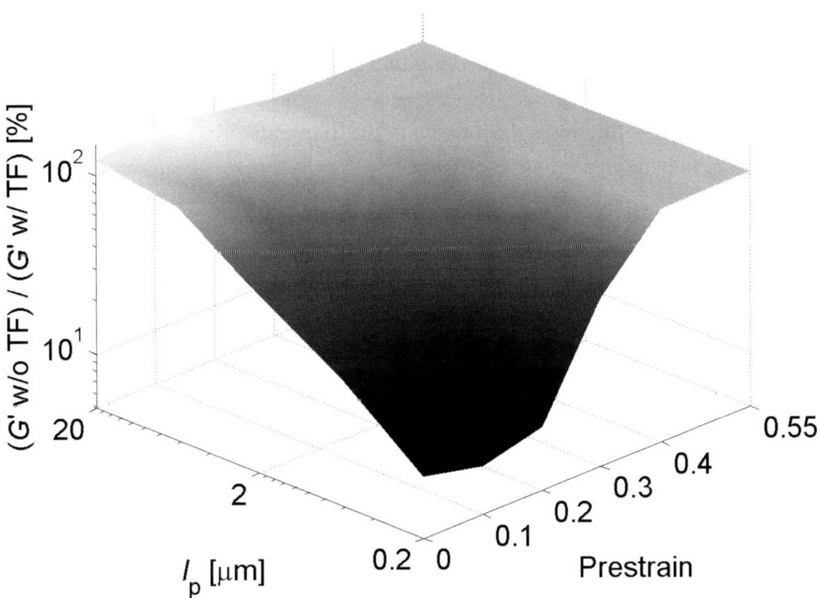

Figure 4. Shear storage modulus (*G'*) computed in the absence of thermal fluctuation (TF) divided by *G'* with TF at different prestrain levels and persistence length. A value of 100% indicates that TF has no influence on *G'*. Note that only at low prestrain and low persistence length do TF exert a substantial effect. In the case of actin filaments, persistence length is approximately 17μm.

4. Discussion

While there is still much to learn, several of the results presented above have the potential to help resolve some of the outstanding controversies regarding cellular rheology.

The importance of thermal effects. One of the most surprising conclusions that can be drawn from these simulations is that under conditions that are selected to best replicate those of cells, thermal effects apparently exert little effect on stiffness, as demonstrated by the observation that the complete absence of thermal motions had a negligible effect on G' (Fig. 4). This appears to be true at zero prestress, whenever the persistence length is greater than the distance between cross-links. Given the persistence length of actin is ~17 μm, this condition is likely to be satisfied in most biological and physiological situations.

If thermal effects can generally be neglected, what then determines the network elasticity? It appears that the answer to this question depends upon the degree of prestrain imposed, and that the situation is somewhat complicated. Also, thermal fluctuations may be important during formation of the network, since residual stress may develop while actin filaments cross-link under thermal fluctuations.

Which model is the most consistent with these results? When these results are interpreted in the context of the different models, each is found to have certain advantages and disadvantages depending on the level of prestrain in the cytoskeletal network. In the case of tensegrity, it is only fair to draw comparisons at elevated prestrains, since without them, and lacking any compressive elements such as the microtubules, tensegrity would predict a zero modulus. To the extent that tensegrity is a useful model, one would expect two salient features [17]: that the modulus would would be a linear function of prestress, and that it would be nearly independent of the elastic properties of the network constituents, in this case, the actin filaments and cross-linking proteins. Indeed, both of these features of the tensegrity structure were exhibited by the model, as can be seen by plotting the same data as in Fig. 3A but in terms of prestress vs. G' on a log-log scale. While there is some dependence on constituent properties when reduced by a factor of 25 with a slight deviation from a linear slope, the degree of consistency in the results is really quite remarkable. However, when tensegrity is used to predict the *magnitude* of the modulus, at least in its most simple form:

$$G' \approx \tfrac{1}{3}\sigma_s$$

(1)

where σ_s is the prestress in the system, the values are significantly lower, by at least an order of magnitude, than those observed in the simulation.

At prestrains close to zero, the cellular solids model would be the most appropriate, given the lack of thermal effects noted above and the inability of the tensegrity model to capture the modulus of a cross-linked actin network at zero strain. Despite these seeming advantages, however, the moduli predicted by cellular solids are somewhat higher than those calculated in the model. The following equation provides estimates of ~316 Pa, about 80-fold higher than that observed in simulation at 1 Hz:

$$G' \approx \tfrac{3}{8} E_f \Phi^2 \tag{2}$$

where E_f is the Young's modulus of the filaments, and Φ is the volume fraction of the filaments. This is likely due to the fact that G' shows a strong dependence on the degree of cross-linking; the cellular solids model, as currently formulated, assumes a maximal degree of cross-linking, and that these cross-links are rigid. Our model shows that both of these effects can cause a significant increase in modulus. Consequently, while cellular solids would appear to be based on a reasonable physical representation of the system, its lack of these two critical features makes it difficult to use for direct quantitative predictions.

Biopolymer models, as stated earlier, have the advantage that they alone account for thermal fluctuations. However, the results presented here suggest that for conditions corresponding to the cytoskeleton, and for many of the studies of cross-linked actin gels, thermal effects seem to be of relatively little importance. Nonetheless, the shear modulus predicted by this theory:

$$G' \approx \frac{\kappa_b l_p}{\xi^2 l_c^3} \tag{3}$$

where κ_b is the filament bending stiffness, l_p the persistence lenth, ξ the mesh length, and l_c the distance between cross-links, yield numerical values somewhat closer to observation, ≈ 180 Pa. Moreover, only the biopolymer models have the capability of predicting the frequency scaling of the moduli, and it is likely that the ¾ power law behavior at high frequency is accurately captured by these models

One important feature observed in cells and not captured by any current model is the effect of unfolding or unbinding event of cross-link under high stress. These presumably account for the softening (fall in modulus) at high strain, and will require a more sophisticated model to accurately capture. The present model has the potential to perform this, and the required extensions are currently being incorporated.

Acknowledgements

Support of the NIH (GM076689) and the Singapore MIT Alliance for Research and Technology is gratefully acknowledged.

References

1. R. L. Satcher, C. F. Dewey, *Biophysical Journal* **71**, 109-118 (1996).
2. L. J. Gibson, M. F. Ashby, Cellular Solids: Structure and Properties, Oxford: Pergamon Press, 357 p, (1988).
3. D. E. Ingber, *Journal of Cell Science* **116**, 1157-1173 (2003).
4. F. C. Mackintosh, J. Kas, P. A. Janmey, *Physical Review Letters* **75**, 4425-4428 (1995).
5. B. Fabry, J. J. Fredberg, *Respiratory Physiology & Neurobiology* **137**, 109-124 (2003).
6. B. D. Hoffman, J. C. Crocker, *Annu Rev Biomed Eng* 11, (in press).
7. P. A. Janmey, C. A. McCulloch, *Annu Rev Biomed Eng* **9**, 1-34 (2007).
8. G. N. Maksym, B. Fabry, J. P. Butler, D. Navajas, D. J. Tschumperlin, J. D. Laporte, J. J. Fredberg, *J Appl Physiol* **89**, 1619-1632 (2000).
9. S. Yamada, D. Wirtz, S. C. Kuo, *Biophys J* **78**, 1736-1747 (2000).
10. J. S. H. Lee, P. Panorchan, C. M. Hale, S. B. Khatau, T. P. Kole, Y. Tseng, D. Wirtz, *Journal of Cell Science* **119**, 1760-1768 (2006).
11. G. I. Bell, *Science* **200**, 618-627 (1978).
12. T. Kim, W. Hwang, R. D. Kamm, *PLoS Comp Biol* (submitted).
13. M. H. Zaman, M. R. Kaazempur-Mofrad, *Mech Chem Biosyst* **1**, 291-302 (2004).
14. T. G. Mason, K. Ganesan, J. H. vanZanten, D. Wirtz, S. C. Kuo, *Physical Review Letters* **79**, 3282-3285 (1997).
15. Y. Tseng, K. M. An, O. Esue, D. Wirtz, *Journal of Biological Chemistry* **279**, 1819-1826 (2004).
16. D. A. Head, A. J. Levine, F. C. MacKintosh, *Physical Review E* **68**, 061907 (2003).
17. D. Stamenovic, D. E. Ingber, *Biomech Model Mechanobiol* **1**, 95-108 (2002).
18. T. Kim, W. Hwang, R. D. Kamm, *PLoS Comp Biol* submitted, (2009).

Chapter 14

Y. C. FUNG AND THE BIOMECHANICS OF HEARING

RONG ZHU GAN[†]

School of Aerospace and Mechanical Engineering and OU Bioengineering Center
University of Oklahoma, Norman, Oklahoma 73019, U.S.A.

Y. C. Fung introduced me to the biomechanics field in 1979. He advised my Ph.D. research in lung biomechanics (pulmonary blood flow) in 1992, and led me to biomechanics of hearing in 1999. During the past 30 years, I developed my research career in biomechanics based on Fung's mentoring. In this paper, I wish to convey my deep appreciation for his guidance and to recognize his influence in my career, which has taken me to my current research in hearing, both theoretical modeling and experimental measurements of sound transmission through the ear.

1. Introduction

1.1. *Meeting Y. C. Fung*

I first met Dr. Y. C. Fung when he led a group of biomedical engineers and research scientists in biomechanics (including my Ph.D. advisor, Michael Yen) from the University of California San Diego to Huazhong University of Science and Technology (HUST) in Wuhan, China in 1978. It was the first time I heard the word "Biomechanics". Soon after his historical visit to HUST, when he introduced Biomechanics to China, I became one of the first three students in China to graduate with Master's degrees in Biomechanics in 1981. Under Dr. Fung's leadership, the first "China-Japan-USA Biomechanics Conference" was hosted by HUST in 1983, and I had the opportunity to present my first scientific paper on rheological properties of pulmonary arteries at the conference. Since then, my professional career has changed from traditional mechanical engineering to more applied biomedical engineering and Dr. Fung has been there to guide me through several important milestones in my life.

The first milestone was in June 1988 when Dr. Fung and I met at a Fluid Mechanics Conference in Ottawa, Canada (see picture below with Dr. and Mrs. Fung behind the Canadian Parliament Building on June 18, 1988). At that time, I was a Ph.D. student in the Department of Mathematics at the University of

† Work supported by grants NIH R01DC006632, NSF/CMS 0510563, and OCAST HR01-45, AR02.2-026.

153

Alberta studying wave propagation in the arterial system, and I wanted to return to biomedical engineering. Dr. Fung was very supportive, and recommended me to Dr. Michael Yen who has just moved to Memphis, Tennessee and started a new Biomedical Engineering Department at Memphis State University (now known as University of Memphis). This change in my Ph.D. direction became the starting point for my future career in biomedical engineering. My Ph.D. research in pulmonary blood flow resulted in three papers published in Journal of Applied Physiology in the early 1990s that were directly supervised by both Dr. Fung and Dr. Yen.

The second milestone occurred between 1998 and 1999 when I was involved in preparing a grant application for the Whitaker Foundation's Special Opportunity Award at the University of Oklahoma. Dr. Fung gave me the most important advice on grant proposal: "be focused on one major theme". I developed the proposal with that in mind, and we won the Whitaker award to establish the bioengineering program at the University of Oklahoma (OU). With Whitaker funding, the OU Bioengineering Center was established in 1999, and OU became the first school in the Midwest to start the Bioengineering graduate programs. Dr. Fung then served as one of three distinguished advisory board members in bioengineering for OUBC. Dr. Fung visited OU twice, in 2000 and 2001. He visited my lab, and talked with my students when I was just settling down as the first Bioengineering faculty hired by the University. My students and I will never forget those very precious times Dr. Fung spent in Norman, Oklahoma. He inspired us to continue our endeavors in biomedical engineering.

1.2. *New Research in Hearing*

Since 1999, with Dr. Fung as my mentor, my major research direction shifted from lung biomechanics to hearing and ear biomechanics based on available resources and collaboration opportunities in Oklahoma with OU Medical School and Hough Ear Institute. A new research area, "Biomechanics for Restoration of Hearing" is now established at OU, and my own laboratory is now the only one in the country that combines researches in experimental measurements of sound transmission, material tests of ear tissues, 3D reconstruction and finite element modeling of human ear, and design and evaluation of implantable hearing devices.

As we celebrate Y. C. Fung's 90th birthday, I deeply appreciate Dr. Fung's mentoring of my career development in biomedical engineering, specifically in biomechanics, over the past 30 years. In honor of his great contributions to biomechanics, I will now briefly report on applications of Fung's biomechanics principles in hearing research: mechanical tests of ear tissues and finite element modeling of the human ear.

2. *Biomechanical Tests of Ear Tissues*

The human ear includes complex morphological structures of the external ear canal, middle ear, and inner ear (or cochlea) that transfer sound from the environment into the central auditory system. Sound signals stimulate the air media in the ear canal and induce vibration of the tympanic membrane or eardrum. Movement of the tympanic membrane (TM) initiates the acoustic-mechanical transmission through three ossicular bones (*i.e.*, malleus, incus, and stapes) which are connected by two joints and suspended in an air-filled middle ear cavity by ligaments and muscle tendons. The output portion of the middle ear is the stapes footplate which sets into the oval window and transmits vibrations into cochlear fluid. The goal of our research is to determine how the structure and mechanical properties of human ear affect sound transmission from the ear canal to cochlea.

The soft tissues in the middle ear include the TM, four suspensory ligaments, and two muscle tendons (Fig. 1). Mechanical properties of these tissues affect the energy transmission through the ear or the normal hearing process. However, our knowledge on mechanical properties of these tissues is poor because of the extremely small size and irregular geometry of the tissues.

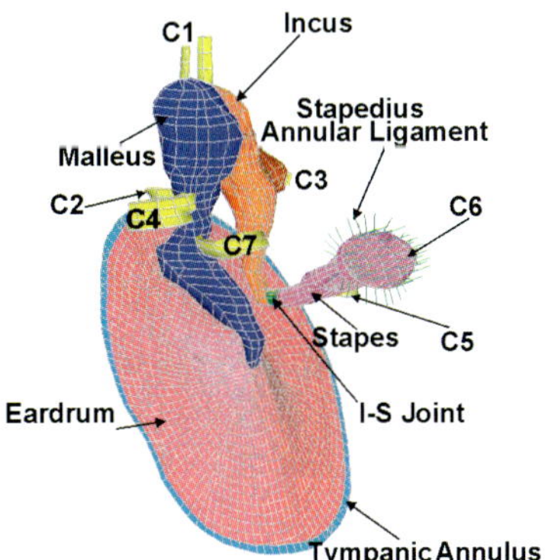

Fig. 1. Anterior-medial view of the human middle ear structure (FE model). C1, C2, C3, C4, C5, and C7 represent the ligaments and muscle tendons. C6 represents cochlear load. C2: lateral malleus ligament; C3: posterior incus ligament; C4: anterior malleolar ligament; C5: stapedial tendon; and C7: tensor tympani tendon [1].

There were no reports on the mechanical properties of the middle ear ligaments and muscle tendons and the viscoelastic behavior of the TM in frequency domain, until our recent studies using Fung's methodology on soft tissue biomechanics [2]. A series of experimental measurements and modeling analysis on four middle ear tissues: the TM, stapedial tendon, tensor tympani tendon, and anterior malleolar ligament (AML) were conducted in our lab. The methods used for the study include the uniaxial tensile tests, digital image correlation analysis, finite element modeling, and constitutive material modeling [3-6].

2.1. *Specimen Preparation*

All specimens were harvested from fresh or freshly-frozen human cadaver ears (*i.e.*, temporal bones) obtained from the University of Oklahoma Health Sciences Center. To conduct the mechanical test of the TM in material testing system (MTS), a strip was cut from the posterior site of the TM (Fig. 2A) and then mounted to a fixture with a ruler fixed on one end and placed in the MTS machine for testing (Fig. 2B) [3].

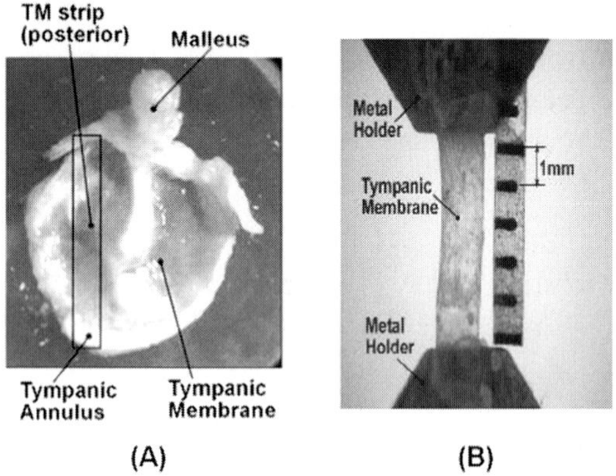

Fig. 2. (A) A left ear TM specimen preparation. (B) The specimen attached to a fixture in the MTS.

The specimen of ligament or tendon was prepared with two bony ends attached. The surgical window of superior view in Fig. 3A shows the AML (C4 in Fig. 1) which was connected between the malleus and middle ear bony wall. Fig. 3B shows the AML sample was mounted in the MTS machine [6].

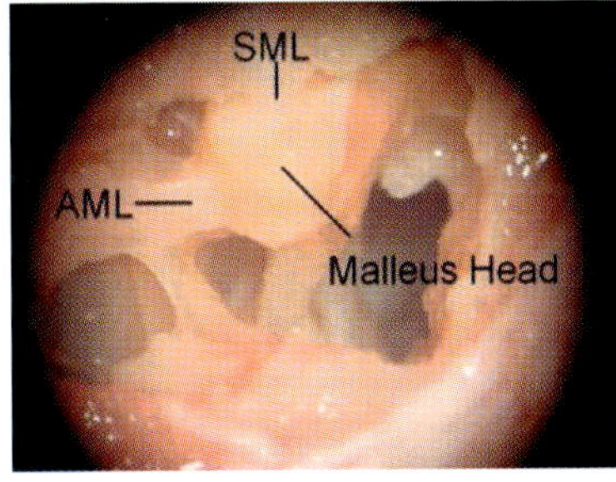

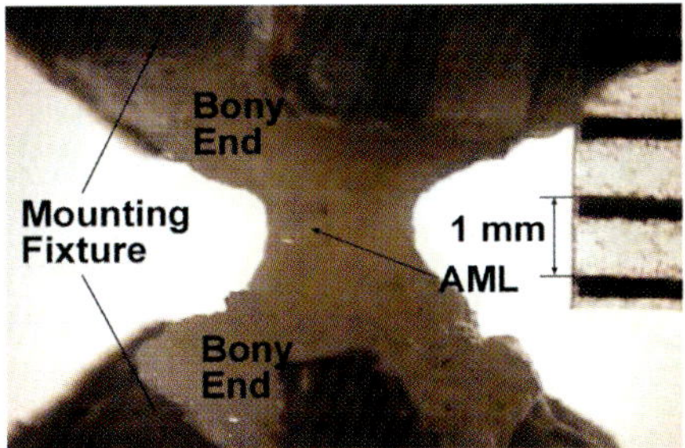

Fig. 3A. Left panel shows surgical window of superior view of the AML in a cadaver ear. Fig. 3B. Right panel is an AML specimen with two bony ends was placed in the MTS.

2.2. *Digital Image Correlation Method*

It is important to verify the boundary effect on mechanical testing of the specimen with such a small length (*e.g.* less than 1 mm for tendon or ligament). The digital image correlation (DIC) method was employed to measure the tensile strain distribution across the specimen. The simultaneously recorded images during the uniaxial tensile test were analyzed based on a gray-scale pattern of pixels in three steps: displacement mapping, bicubic spline interpolation, and least squares correlation [3-6].

As an example, Fig. 4 shows the DIC process with four images of a TM specimen from uniaxal tension testing at the initial time zero (reference image), 0.5, 1, and 1.5 seconds (deformed images). The horizontal and vertical lines on those sequential images were used for strain calculation. A good comparison of the stress-stretch loading curve measured from the MTS and derived from the DIC method indicated that the experimental measurements were reliable without boundary effect. The Ogden model of 1st or higher orders was then used for analysis of the experimental data.

                                    R. Z. Gan

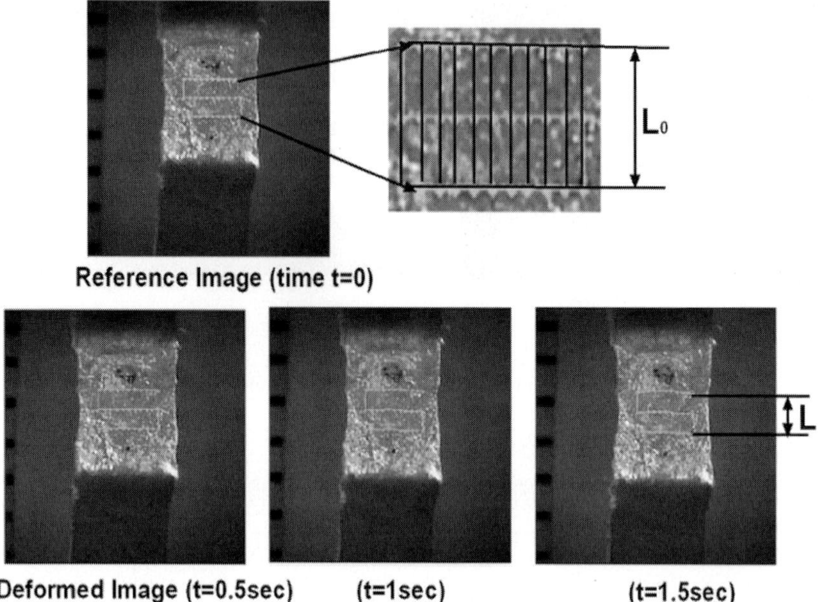

Fig. 4. DIC process with 4 images of a TM specimen obtained from the uniaxial tensile test [3].

2.3. *Summary of Mechanical Properties of Four Ear Tissues*

The stress-strain relationships of the TM, stapedial tendon, tensor tympani tendon and AML based on mean experimental data are compared under the stress range from 0 to 1.0 MPa, or the stretch ratio from 1 to 1.4 in Fig. 5 (N = 11, 12, 10, and 9 for the TM, spapedial tendon, tensor tympani tendon and AML, respectively). The results show a nonlinear stress-strain relationship for all four human middle ear tissues. The stress-strain behavior of the TM is significantly different from the behaviors of middle ear ligaments and tendons. Within the same strain level, the stress induced in the TM raised faster than the stress increase in ligament and tendons, which indicates that the TM has higher elastic modulus than the stapedial tendon, tensor tympani tendon and AML. The stress induced in the ligament or tendons was very small at the beginning of the tensile test compared to stress in the TM. At a stretch ratio of 1.1, the stress in the TM reached 0.4 MPa, while the stresses in the stapedial tendon, tensor tympani tendon and AML were only 0.025, 0.016 and 0.0162 MPa, respectively.

The constitutive equations of these middle ear tissues are summarized below:

TM:
$$\sigma = 0.03(\lambda^{25.76} - \lambda^{-14.38}) \qquad (0 \le \sigma \le 1.0 \text{ MPa}, 1 \le \lambda \le 1.15) \qquad (1)$$

Stapedial Tendon:
$$\sigma = 5.8 \times 10^{-3}(\lambda^{16.40} - \lambda^{-9.70}) \qquad (0 \le \sigma \le 1.45 \text{ MPa}, 1.0 \le \lambda < 1.4) \qquad (2)$$

Tensor Tympani Tendon:
$$\sigma = 8.5 \times 10^{-4}(\lambda^{22.52} - \lambda^{-12.76}) \qquad (0 \le \sigma \le 1.0 \text{ MPa}, 1 \le \lambda < 1.4) \qquad (3)$$

AML:
$$\sigma = 1.04 \times 10^{-3}(\lambda^{12.68} - \lambda^{-7.84}) + 0.01(\lambda^{12.69} - \lambda^{-7.85}) \qquad (0 \le \sigma \le 0.5 \text{ MPa}, 0 \le \lambda \le 1.4) \qquad (4)$$

where σ is stress and λ is stretch ratio.

Fig. 6 shows the Young's modulus ($d\sigma/d\lambda$) and stress σ relationships of the TM, stapedial tendon, tensor tympani tendon and AML. The Young's modulus of the TM was derived over three stress ranges: 0~0.1 MPa, 0.1~0.3 MPa and 0.3~1 MPa, while the modulus of the ligament or tendons was derived in an overall stress range of 0~1.5 MPa. As can be seen in this figure, the TM has higher Young's modulus than the ligament and tendons as predicted from their stress-strain relationships shown in Fig 5. However, two tendons have almost the same Young's modulus, and the ligament (AML) has the lower Young's modulus than other tissues.

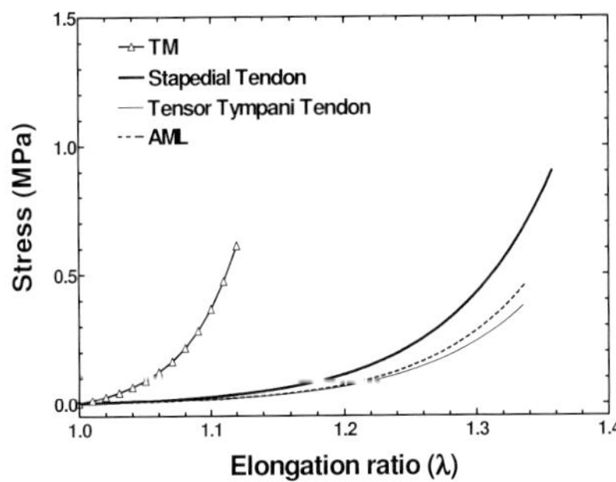

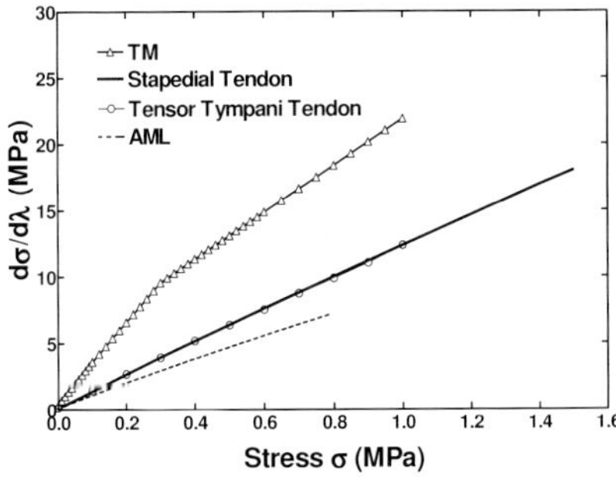

Fig. 5. Mean stress-strain curves measured from four different ear tissues: TM, stapedial tendon, tensor tympani tendon and AML.

Fig. 6. Mean elastic or Young's modulus-stress relationship curves obtained from four ear tissues.

                                    R. Z. Gan

The normalized stress relaxation function G(t) of the TM, stapedial tendon, tensor tympani tendon and AML obtained from stress relaxation tests are reported in Fig. 7. It is noted that all the tissues tested in this study were fully relaxed in about 120 seconds and the stapedial tendon showed largest relaxation among the tissues tested.

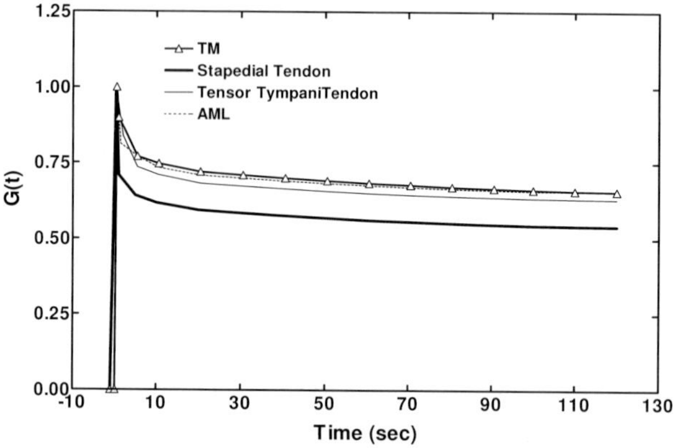

Fig. 7. Comparison of normalized stress relaxation function measured from four ear tissues (mean data).

3. 3D Finite Element Model of the Human Ear

3.1. *Construction of 3D Finite Element Model of Human Ear*

The finite element (FE) method, a general numerical procedure, has the capability to represent the realistic structures and use appropriate constitutive laws for high-fidelity simulations of complex biological systems. Since 1999, with supports from the Whitaker Foundation, NIH, and NSF, we have combined technologies of FE analysis and 3D reconstruction of human ear to develop a 3D FE model of the human ear. To date, one of the most advanced 3D FE model was developed by our group [1, 7-9]. It was constructed on the basis of a complete set of histological section images of the human temporal bone. Initially, the model consisted of the external ear canal, tympanic membrane, ossicular bones, middle ear suspensory ligaments/muscles, and middle ear cavity [7-9]. It has been further developed by including characteristics of the complete structures of the external ear canal, middle ear and cochlea [10]. Apart from the detailed geometric characteristics, a unique feature of our 3D model is the utility for acoustic-structural-fluid coupled FE analysis. Fig. 8 shows the most recently modified FE model #3 which was published by Gan and Wang [11]. By attaching the uncoiled cochlea to the middle ear at the oval window and round window, the model is able to derive the middle ear transfer function and energy transmission through the middle ear and cochlea when fluid is accumulated in the middle ear cavity (*e.g.* otitis media with effusion) as measured in temporal bone experiments [12]. The frequency selectivity of the basilar membrane (BM) motion and intracochlear pressure induced by sound in the ear canal can also be predicted along the length of the basilar membrane from the basal turn to apex.

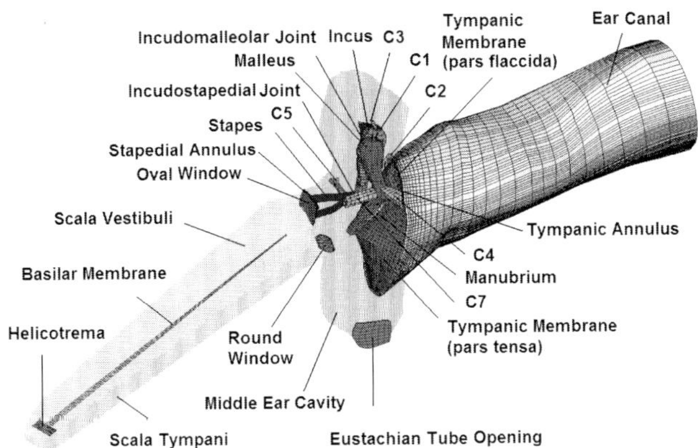

Fig. 8. 3D FE model #3 of human ear with external ear canal, middle ear and simplified (uncoiled) cochlea in anterior-medial view. The middle ear cavity and cochlea were assumed transparent

3.2. *Mechanical Properties of Ear Tissues Employed in FE model*

Material properties of the TM and other middle-ear tissues such as ligaments and muscle tendons measured in our previous studies [3-6] provide the basis for further analysis of material properties under static middle ear pressure. First, a five-parameter hyperelastic Mooney-Rivlin model was utilized to represent the 3D constitutive relations of the TM and ligaments for quasi-static analysis. Five mechanical constants of the Mooney-Rivlin model for all middle ear tissues were determined based on non-linear stress-strain curves obtained from measurements in Fig. 5. Next, the elastic modulus of the tissue for static deformation, $d\sigma/d\lambda$, was expressed as

$$\frac{d\sigma}{d\lambda} = \alpha(\sigma + \beta) \tag{5}$$

                                    R. Z. Gan

where σ is the normal stress, λ is the stretch ratio, α and β are two constants [2]. By combining the Mooney-Rivlin material model and geometry nonlinearity induced by middle ear static pressure, an empirical formula converting the static stress-dependent modulus into the elastic modulus E_d for dynamic analysis is derived by Wang et al. [9] as

$$E_d = kE_{d0}(\frac{E_p}{E_0})^r \tag{6}$$

where k and r are correction factors, E_{d0} is the elastic modulus for dynamic analysis at zero static pressure across the TM, Ep is the elastic modulus corresponding to a specified stress level, which is relative to static pressure p as $d\sigma/d\lambda$, and E_0 is the Young's modulus of the TM or ligaments at a reference strain which was assumed as nominal strain (ε) of 1%. Fig. 9 shows stress-dependent Ed of the TM and ligaments obtained from Eq. (6) when middle ear pressure was varied from -2.0 to 2.0 kPa. Thus, dynamic behavior of the middle ear in response to sound pressure in the ear canal was predicted under various positive and negative middle ear pressures and the FE model-derived tympanogram can be created [9].

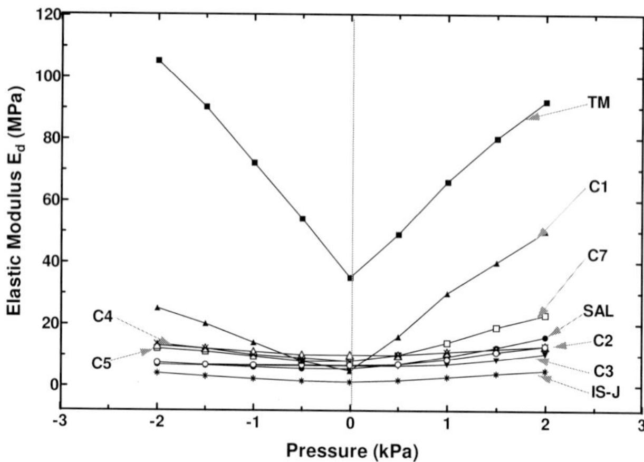

Fig. 9. Variation of elastic modulus with middle ear pressure for dynamic analysis. All Cs are the ligaments and tendons shown in Fig. 1 or Fig. 8.

3.3. *Applications of the FE Model – Simulation of Middle Ear Diseases*

Middle ear diseases such as the otitis media with effusion (OME) and otosclerosis of ossicular chain can be simulated in the FE model to predict the hearing loss induced by those disorders. Fig. 10 displays the model-predicted effect of OME at two fluid levels: up to the umbo (38% of middle cavity volume) and up to the stapes footplate (61% of the cavity volume), on the TM, footplate and basilar membrane (BM) movements. The huge reduction of TM and footplate displacements at over 1 kHz was observed when fluid level increased from the umbo to stapes, but the response of BM displacement as the ratio of BM and stapes footplate (d_{BM}/d_{FP}) had slightly increasing at higher frequencies (e.g., from 3.25 to 3.6 at 6 kHz with the control value of 2.8).

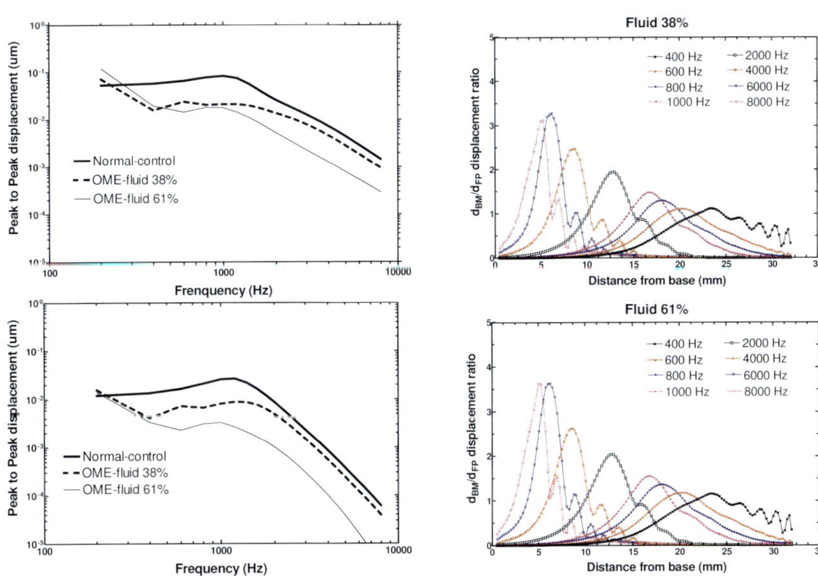

Fig. 10A. FE model-derived TM (top) and footplate (bottom) displacement curves at control, 38% and 61% fluid level in the middle ear cavity.

Fig. 10B. Distribution of the BM/FP ratio along the BM length (from base to apex)

3.4. *Applications of the FE Model – Assisting Design and Function Evaluation of Implantable Hearing Devices*

The design of totally implantable hearing system (TIHS) in our BME lab at University of Oklahoma has incorporated bioengineering approaches based on

                                 R. Z. Gan

3D FE model of the human ear and the FE analysis of electromagnetic coupling of the transducer. The mechanical driver or transducer of the TIHS consists of an implant magnet placed on the ossicles and an implantable coil placed under the ear canal wall. The location, orientation, and dimensions of the implant transducers including the magnet, ossicular attachment, and coil, were determined in the model within the constraints of the middle ear anatomy. Fig. 11 displays two implant transducers: Model I and Model II designed to meet different patient hearing situations [13]. Model I is designed for the ears with normal ossicular chain and the implant is attached to the long process of the incus and the head of stapes. Model II is designed for the ears with disrupted ossicular chain such as missing of the incus. The coil design was completed using optimization process of the electromagnetic coupling to reach the maximum electromagnetic strength. The design of TIHS as well as function evaluation of the transducers has been benefited from the 3D FE model of the human ear we developed.

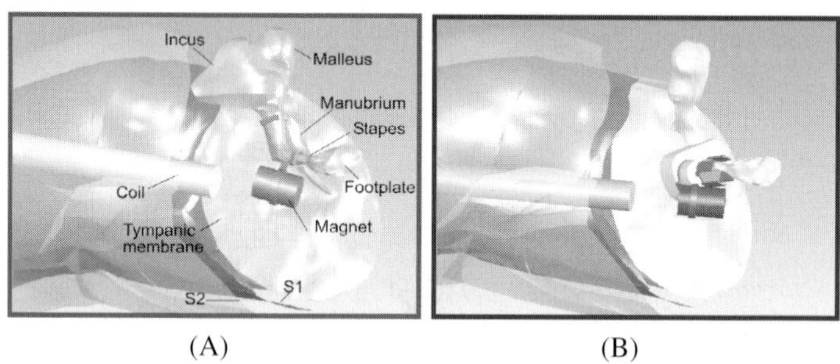

(A)   (B)

Fig. 11. (A) Posterior view of implant Model I and coil placed in the CAD model or FE model #2 of the left ear. (B) Posterior view of implant Model II.

4. Conclusions

As I concluded this chapter, I realize that it is impossible to summarize here all that I learned from Dr. Fung in biomechanics. For example, nanoindentation techniques were used to measure viscoelastic properties of human TM's collagen fiber layer. Measurements with the in-plane (membrane plane) and out-of-plane (through-thickness) setups were conducted on four quadrants of the TM samples [14, 15]. In addition to quasi-static measurement of mechanical properties of the TM, we recently used a self-developed miniature split Hopkinson tension bar for measuring mechanical properties of the TM at high strain rate, corresponding to high frequency range of 100 to 2000 Hz [16, 17]. Characterizing mechanical properties of ear tissues in frequency or time domain will be one of our future studies. To improve the accuracy of the FE model, we will create the spiral cochlea with the organ of Corti for multi-scale modeling. Dr. Fung's influences on my research in hearing biomechanics will continue to be demonstrated through past, present, and future projects.

Acknowledgments

I would like to acknowledge my former students: Drs. Qunli Sun, Tao Cheng, Chenkai Dai, and Bin Feng, and current postdoctoral research fellows Drs. Xuelin Wang and Xiangming Zhang for their contributions to the work presented here. I also want to thank Mark Wood, MD, and Don Nakmali, BSEE, at Hough Ear Institute, and Prof. Hongbing Lu at Oklahoma State University for their expertise and technical assistance. The research work was supported by grants from the Whitaker Foundation, NIH, NSF, and Oklahoma Center for the Advancement of Science & Technology (OCAST).

The author with Dr. and Mrs. Fung in Ottawa, Canada, June 18, 1988.

References

1. Q. Sun, R. Z. Gan, K-H. Chang, and K. J. Dormer. *Biomechanics and Modeling in Mechanobiology*, **1**: 109 (2002).
2. Y. C. Fung. New York, Springer, (1993).
3. T. Cheng, C. Dai, and R. Z. Gan. *Ann Biomed Eng,* **35 (2)**: 305 (2007).
4. T. Cheng and R. Z. Gan. *J Biomech Eng,* **129(6)**: 913 (2007).
5. T. Cheng and R. Z. Gan. *Med Eng Phys,* **30(3)**: 358 (2007).

6. T. Cheng and R. Z. Gan. *Biomechanics and Modeling in Mechanobiology*, **7**(5): 387 (2008).
7. R. Z. Gan, B. Feng, Q. Sun. *Ann Biomed Eng*, **32(6)**: 847 (2004).
8. R. Z. Gan, Q. Sun, B. Feng, and M. W. Wood., *Med Eng Phys*, **28(5)**: 395 (2006).
9. X. Wang, T. Cheng, and R. Z. Gan. *J. Acoustical Society of America.* **122(2)**: 906 (2007).
10. R. Z. Gan, B. P. Reeves, X. Wang. *Ann Biomed Eng*, **35(12)**: 2180 (2007).
11. R. Z. Gan, X. Wang. *J Acoust Soc Am*, **122(6)**: 3527-3538 (2007).
12. R. Z. Gan, Dai C, Wood MW. *J Acoust Soc Am*, **120(6)**: 3799 (2006).
13. R.Z. Gan. *The Hearing Journal*, **61(9)**: 33 (2008).
14. G. Huang, N. Daphalapurkar, R. Z. Gan, and H. Lu. *J Biomech Eng*, **130(1)**: 014501 (2008).
15. N. P. Daphalapurkar, C. Dai, R. Z. Gan, H. Lu. *J. Mechanical Behavior of Biomedical Materials*, **2**: 82 (2009).
16. H. Luo, C. Dai, R. Z. Gan, H. Lu. *J Biomech Eng*, **131(6)**: 064501 (2009).
17. H. Luo, C. Dai, R. Z. Gan, H. Lu. *Int J Experimental and Computational Biomechanics*, **1(1)**: 1 (2009).

Chapter 15

A MODEL FOR A CLASS OF DIFFUSION-BASED INTERCELLULAR COMMUNICATION

SIA NEMAT-NASSER

Center of Excellence for Advanced Materials, Department of Mechanical and Aerospace Engineering, University of California at San Diego, La Jolla, CA 92093-0416, U.S.A.

ALIREZA V. AMIRKHIZI

Center of Excellence for Advanced Materials, Department of Mechanical and Aerospace Engineering, University of California at San Diego., La Jolla, CA 92093-0416, U.S.A.

The diffusion-induced concentration of various molecules (chemo/cytokines or food/oxygen) around a spherical cell is studied based on an exact mathematical evaluation, arriving at closed-form solutions for a cell with various spatial and temporal distributions of secretion or absorption patterns. In addition, a general method for numerical solution and analytical approximation of such phenomena is presented. The asymptotic behavior of the solution is also examined for long times or great distances from the cell. These can be used to estimate the effective communication parameters among cells. Two examples are used to illustrate the application of the model results under physiologically realistic conditions.

1. Introduction

The presence, concentration, and spatial variation of various molecules in the extra cellular environment are some of the main items of information and input to which a cell is exposed. These chemical factors, along with thermal, mechanical (pressure, shear), and radiation (optical, UV, IR) constitute the physical inputs, to which the cell responds and subsequently interacts with the outside world. The chemical factors tend to be more critical (food and oxygen supply) and diverse, both in specificity and concentration. The concentration of various signaling molecules essentially serves as a communication mechanism among cells and it can regulate events such as stem cell differentiation [6, 15, 17, 18]. The main physical process by which chemical signals and supplies operate in the extra cellular environment is diffusion.

In this paper we examine the problem of diffusion-induced chemical concentration around a spherical cell which has a secreting or absorbing membrane. The mathematical basis of such processes is understood and we develop a general method for providing the chemical concentration as a function of time and space. We solve various special cases and look at the associated length and time scales. For example, we examine the possibility of substantial change in communication time and distance if the cell secretion occurs by pulses of various frequencies. Furthermore, we look at the physiological case of oxygen concentration around chondrocytes in the cartilage and specifically intervertebral disc [4, 5, 10-14]. The specific character of chondrocytes as cells that are fixed within the matrix but separated from one another by a substantial distance (compared to the cell size) makes them attractive for such studies. Previous work on the effect of other external factors, such as mechanical transduction [7-9], and also 3D organization of a small number of cells [1], on the activity and function of chondrocytes provide an attractive opportunity for further research.

2. Mathematical Modeling

In this section, we present an exact closed-form solution to the problem of cyto/chemokine secretion by diffusion from a spherical cell, with the distribution of the surface flux-rate being either spatially uniform or directionally variable. The general method of solution for any distribution over the cell surface and any time history of secretion is detailed in the Appendix.

The concentration, v, of a molecule that does not interact with the extracellular medium or decompose within a time scale of interest is dictated by the diffusion equation,

$$\frac{\partial v}{\partial t} = \kappa \nabla^2 v, \tag{1}$$

where t measure time and κ is the diffusion constant. We use a spherical coordinate system (r, θ, ϕ), where r denotes the distance from the origin while θ and ϕ are the zenith and azimuth, respectively, and introduce the non-dimensional variables,

$$x = \frac{r}{\rho}, \tag{2}$$

$$\tau = \frac{\kappa t}{\rho^2}, \tag{3}$$

where ρ is the cell radius. Equation (1) then becomes,

$$\frac{\partial v}{\partial \tau} = \nabla^2 v \quad \text{for} \quad x > 1,$$ (4)

where

$$\nabla^2 \equiv \frac{1}{x^2}\left(\frac{\partial}{\partial x}\left(x^2 \frac{\partial}{\partial x}\right) + \frac{1}{\sin\theta}\frac{\partial}{\partial\theta}\left(\sin\theta\frac{\partial}{\partial\theta}\right) + \frac{1}{\sin^2\theta}\frac{\partial^2}{\partial\phi^2}\right).$$ (5)

The boundary conditions consist of a prescribed secretion rate on the cell surface and that the concentration approaches zero at infinite distance from the cell,

$$\frac{\partial v}{\partial x} = -\frac{\rho F_0(\theta,\phi)}{\kappa} g(\tau) \quad \text{on} \quad x = 1,$$ (6)

$$v \to 0 \quad \text{as} \quad x \to \infty.$$ (7)

Here we have assumed that the time and spatial variation of the secretion from the cell surface are separable functions, represented by $g(\tau)$ and $F_0(\theta,\phi)$, respectively. We give the closed-form results for four different cases. The general method of solution is delineated in the Appendix.

(a) The simplest case is when the secretion rate is uniformly distributed on the surface of the cell and the time-dependence in also steady, represented by a Heaviside step-function,

$$\frac{\rho F_0(\theta,\phi)}{\kappa} = \hat{F}_0,$$ (8)

$$g(\tau) = H(\tau).$$ (9)

The normalized concentration, $\hat{F}_0$, is independent of the direction. The solution is,

$$v_0(\xi,\tau) = \frac{1}{1+\xi}\left(\text{erfc}\left(\frac{\xi}{2\sqrt{\tau}}\right) - e^{\xi+\tau}\text{erfc}\left(\frac{\xi}{2\sqrt{\tau}} + \sqrt{\tau}\right)\right).$$ (10)

Here, for the sake of brevity we have set $\xi = x - 1$, and erfc is the complementary error function,

$$\text{erfc}(z) = 1 - \text{erf}(z) = \frac{2}{\sqrt{\pi}}\int_z^\infty e^{-\varsigma^2} d\varsigma.$$ (11)

It must be noted here that this result is different than the one given by Francis and Palsson [3], which, in our notation, is

$$\psi = \frac{\sqrt{\tau}}{1+\xi}\left(\frac{e^{-\frac{\xi^2}{4\tau}}-e^{-\frac{(\xi+2)^2}{4\tau}}}{\sqrt{\pi}}-\frac{\xi}{2\sqrt{\tau}}erfc\left(\frac{\xi}{2\sqrt{\tau}}\right)+\frac{\xi+2}{2\sqrt{\tau}}erfc\left(\frac{\xi+2}{2\sqrt{\tau}}\right)\right). \quad (12)$$

The reason is that the solution used by Francis and Palsson [3] corresponds to a uniform radial flux from a spherical surface (shell of infinitesimal thickness) instead of from a spherical cell. Therefore, solution (12) does not distinguish between the inside and outside of the sphere. This result can be found in Carslaw and Jaeger [2] page 263, whereas equation (10) above is given on page 248 of this reference. Equation (10) is also given by Yoshida et al. [16]. The difference between the two solutions is depicted in Figure 1. Here the contour lines indicate the time, τ, that takes for a signal to reach a distance ξ from the cell. The normalized concentrations are calculated at 5 intervals between (and including) 0.01 to 0.02, which are chosen based on the physiological estimates in [3]. As expected the difference is small in all cases and expectedly the signal in the correct solution takes a shorter time to travel the same distance.

(b) The next case involves the simplest directional preference in flux:

$$\frac{\rho F_0(\theta,\phi)}{\kappa} = \hat{F}_0\cos\theta, \quad (13)$$

$$g(\tau) = H(\tau). \quad (14)$$

Note that, physically this means that on the north pole $\theta = 0$, the cell releases material at a constant rate, $\kappa\hat{F}_0/\rho$, while on the south pole $\theta = \pi$ it absorbs the material at the same rate. When the considered molecule is abundant in the extracellular fluid, this is a reasonable scenario. However, since the concentration cannot fall below zero, in most cases this boundary condition has to be combined with, for example (8), to represent a physical situation defined by $\rho F_0(\theta,\phi)/\kappa = \hat{F}_0(1+\cos\theta)$ Due to linearity, the solution to this problem is just the superposition of the solutions for (13) and (8). In the case of equation (13), the total secretion rate from the cell is zero.

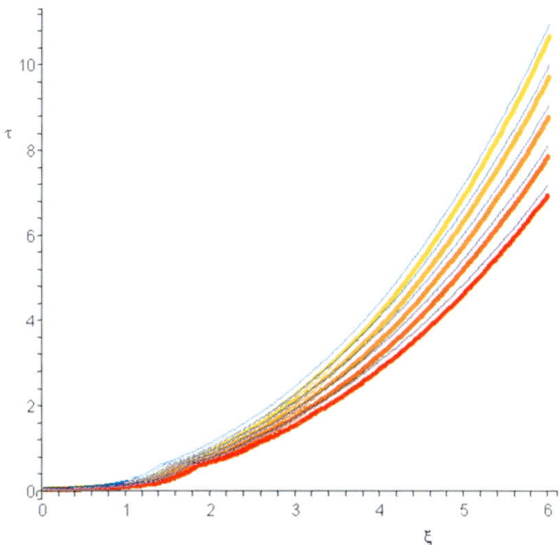

Figure 1. Contour lines for the time it takes a signal of normalized strength 0.01, 0.0125, 0.015, 0.175, and 0.02 to reach a distance ξ from the cell. Both distance and time are normalized as described in the text. The thin lines indicate the solution in [3] while the thick lines are based on the solution (10) above. The curves in each group are ordered top to bottom from highest to the lowest concentration.

The closed-form solution associated with the boundary condition (13) is given by,

$$
v_1(\xi, \theta, \tau) = \frac{\cos\theta}{4(1+\xi)^2}\left(2erfc\left(\frac{\xi}{2\sqrt{\tau}}\right) + \right.
$$

$$
i\left((2\xi+1+i)\,erfc\left(\sqrt{2i\tau}+\frac{\xi}{2\sqrt{\tau}}\right) e^{\xi+i(\xi+2\tau)} \right. \tag{15}
$$

$$
\left.\left. -(2\xi+1-i)\,erfc\left(\sqrt{-2i\tau}+\frac{\xi}{2\sqrt{\tau}}\right) e^{\xi-i(\xi+2\tau)} \right)\right).
$$

While this involves the error function of a complex-valued argument, its value is always real, being equivalent to the short form,

$$v_1(\xi, \theta, \tau) = \frac{\cos\theta}{2(1+\xi)^2}\left(erfc\left(\frac{\xi}{2\sqrt{\tau}}\right) + \right.$$
$$\left. \mathrm{Im}\left((2\xi + 1 - i)\,erfc\left(\sqrt{-2i\tau} + \frac{\xi}{2\sqrt{\tau}}\right)e^{\xi - i(\xi + 2\tau)}\right)\right). \tag{16}$$

In what follows we only report the equivalent short form when applicable. The directionality of this solution is depicted in Figure 2, where we have compared the travel time for various signal strengths, directly above the north ($\theta = 0$) and south ($\theta = \pi$) poles of the cell for the physical boundary condition $\rho F_0(\theta, \phi)/\kappa = \hat{F}_0(1 + \cos\theta)$, using $v_0 \pm v_1$.

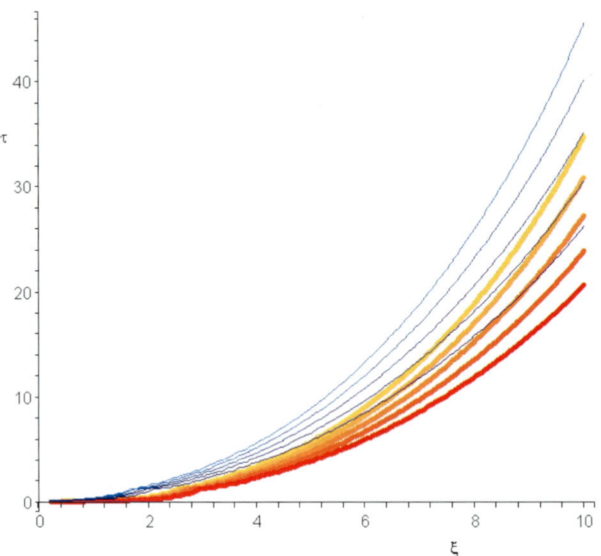

Figure 2. Contours of travel time of signals of normalized strengths 0.01, 0.0125, 0.015, 0.175, and 0.02 (respectively from bottom to top in each group). Thick lines are for location directly above the south pole ($\theta = \pi$) while thin lines are for above the north pole ($\theta = 0$).

(c) In this case we consider a uniform spherical distribution but assume that the secretion rate is decaying exponentially (with non-dimensional time constant p),

$$\frac{\rho F_0(\theta, \phi)}{\kappa} = \hat{F}_0, \tag{17}$$

$$g(\tau) = H(\tau)e^{-p\tau}. \tag{18}$$

This case also involves error functions of complex arguments,

$$v_{p,0}(\xi, \tau) = \frac{1}{(1+\xi)(1+p)}\left(-erfc\left(\frac{\xi}{2\sqrt{\tau}} + \sqrt{\tau}\right)e^{\xi+\tau} + \operatorname{Re}\left(\left(1+i\sqrt{p}\right)erfc\left(i\sqrt{p\tau} + \frac{\xi}{2\sqrt{\tau}}\right)e^{-p\tau+i\sqrt{p}\xi}\right)\right).$$
(19)

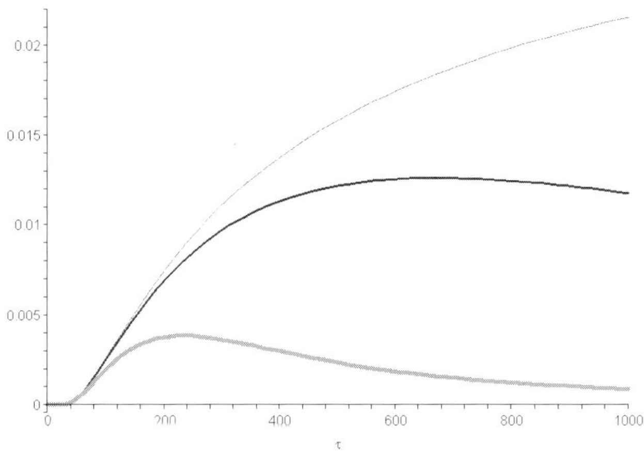

Figure 3. Concentration as a function of time for ξ=25. The thin, medium, and thick curves are v_0, $v_{p,0}$ with $p = 0.01$, 0.001, respectively.

In this case, the maximum traveling distance at various signal strengths are substantially lower than the corresponding case (a); see Figure 3. In the cases that we have considered, the travel times to reach a maximum distance are of the same order of magnitude as in case (a) (Figure 4); however, no general conclusion can be made without further detailed study. In Figure 3, the concentration is plotted as a function of time, τ, at a constant distance, $\xi = 25$, for a constant flux and an exponentially decaying one with $p = 0.01$ and 0.001 (equivalent to decay time $100\ \rho^2/\kappa$ and $1000\ \rho^2/\kappa$, where ρ^2/κ is the time-scale). Figure 4 displays the contours of the concentration, similar to Figures 1 and 2, for the indicated three cases.

(d) The last case is the combination of (c) and (b), that is a preferred direction for secretion as well as an exponentially decaying rate,

174   *S. Nemat-Nasser & A. V. Amirkhizi*

$$\frac{\rho F_0(\theta, \phi)}{\kappa} = \hat{F}_0 \cos\theta, \tag{20}$$

$$g(\tau) = H(\tau)e^{-p\tau}. \tag{21}$$

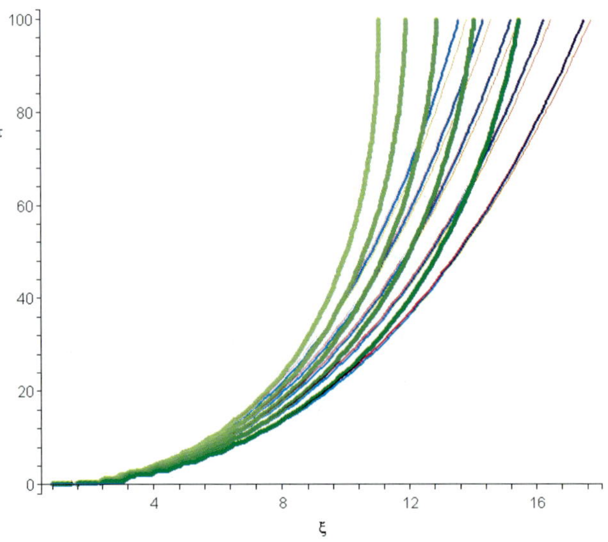

Figure 4. Contours of the travel time, τ, to attain concentration levels 0.01 to 0.02 at a distance ξ. The thin, medium, and thick curves are v_0, $v_{p,0}$ with $p= 0.01$, 0.001, respectively.

The solution is,

$$v_{p,1}(\xi, \theta, \tau) = \frac{\cos\theta}{(1+\xi)^2(4+p^2)} \times$$
$$\left(\mathrm{Re}\left(-(p-2i)(2\xi+1+i)\,erfc\left(\sqrt{2i\tau} + \frac{\xi}{2\sqrt{\tau}} \right) e^{\xi+i(\xi+2\tau)} \right) + \tag{22}$$
$$\left(1-i\sqrt{p}\,(1+\xi)\right)\left(1+\left(1+i\sqrt{p}\right)^2\right)erfc\left(i\sqrt{p}\tau + \frac{\xi}{2\sqrt{\tau}} \right) e^{-p\tau+i\sqrt{p}\xi} \right).$$

3. Limiting and Asymptotic Approximations

Here we present the limiting approximation solutions for long-time and/or great-distance from the cell, for the considered four cases. The simplest form of these results may be useful for repeated evaluations in large-scale and computationally intensive modeling. First, consider the asymptotic solution for the concentration decay, far away from the cell and after a given finite time-period. Since all the arguments in the complementary error functions appearing in the closed-form solutions include the term $\xi/2\sqrt{\tau}$ plus a complex number with non-negative real part, an asymptotic expansion of the complementary error function for large absolute values of the argument is expected to provide a good approximation, as long as $\xi/2\sqrt{\tau} \gg 1$. We thus use only the first term of the following expansion, which is valid for any complex-valued argument, z:

$$erfc\left(z\right) = \frac{e^{-z^2}}{\sqrt{\pi}}\left(\frac{1}{z} - \frac{1}{2z^3} + \frac{1\cdot 3}{2^2 z^5} - \frac{1\cdot 3\cdot 5}{2^3 z^7} + \dots\right). \tag{23}$$

Also, in all terms in the closed-form solution, the surviving exponential term is $e^{-\xi^2/4\tau}$ with the remainder being a rational function. Therefore, if we simply replace all error functions as mentioned-above and simplify the resulting expression, we would arrive at a correct asymptotic formula for the solution to the diffusion problem. The results have a common main term,

$$v_{asym.} = \frac{4\tau^{3/2}e^{-\xi^2/4\tau}}{\sqrt{\pi}(1+\xi)}. \tag{24}$$

The complete solutions are now approximated as follows (at a finite time and a long distance from the cell):

$$\frac{v_0}{v_{asym.}} \approx \frac{1}{\xi(\xi+2\tau)}, \tag{25}$$

$$\frac{v_1}{v_{asym.}} \approx \cos\theta\,\frac{\xi^2 + \xi + 2\tau}{\xi(1+\xi)((\xi+2\tau)^2 + 4\tau^2)}, \tag{26}$$

$$\frac{v_{p,0}}{v_{asym.}} \approx \frac{\xi}{(\xi+2\tau)(\xi^2 + 4p\tau^2)}, \tag{27}$$

$$\frac{v_{p,1}}{v_{asym.}} \approx \cos\theta \frac{\xi(\xi^2 + \xi + 2\tau)}{(1+\xi)((\xi+2\tau)^2 + 4\tau^2)(\xi^2 + 4p\tau^2)}. \tag{28}$$

Note that all these approximations can be simplified to $1/\xi^2$ for $\xi/2\tau \gg 1$ and $\xi/2p\tau \gg 1$ (stronger conditions than $\xi/2\sqrt{\tau} \gg 1$).

  Next consider the asymptotic concentration-decay in time, at a given finite distance from the cell (or for cases (c) and (d), under less restricted conditions as discussed below). For cases (a) and (b), where there must exist steady state solutions, formulas (23) and (24) cannot be used, as they have a $\tau^{1/2}$ time-dependence. This is due to the fact that the first error function in either (10) or (15), $erfc(\xi/2\sqrt{\tau})$, cannot be replaced by its asymptotic approximation for finite distance and long time. In fact, as $\tau \to \infty$, the solutions for these cases are dominated by $erfc(\xi/2\sqrt{\tau})$, leading to

$$v_0 \approx \frac{1}{1+\xi}, \tag{29}$$

$$v_1 \approx \cos\theta \frac{1}{2(1+\xi)^2}. \tag{30}$$

One can slightly generalize these formulas. For bounded $C \equiv \xi/2\sqrt{\tau}$, as $\tau \to \infty$,

$$v_0 \approx \frac{erfc(C)}{1+\xi}, \tag{31}$$

$$v_1 \approx \cos\theta \frac{erfc(C)}{2(1+\xi)^2}. \tag{32}$$

  Under similar conditions (bounded $C \equiv \xi/2\sqrt{\tau}$ as $\tau \to \infty$ and finite p) for cases (c) and (d), equations (27) and (28) are valid. After some algebraic manipulation we obtain,

$$v_{p,0} \approx \frac{Ce^{-C^2}}{\sqrt{\pi}(1+\xi)} \frac{1}{p\tau}, \tag{33}$$

$$v_{p,1} \approx \cos\theta \frac{C(1+2C^2)e^{-C^2}}{2\sqrt{\pi}(1+\xi)^2} \frac{1}{p\tau}. \tag{34}$$

Perhaps a more illuminating form of equations (33) and (34) is to write the ratio of these to (31) and (32) to eliminate the ξ-dependence,

$$\frac{v_{p,0}}{v_0} \approx \frac{Ce^{-C^2}}{\sqrt{\pi}\,erfc(C)}\frac{1}{p\tau}, \tag{35}$$

$$\frac{v_{p,1}}{v_1} \approx \frac{C(1+2C^2)e^{-C^2}}{\sqrt{\pi}\,erfc(C)}\frac{1}{p\tau}. \tag{36}$$

4. Examples

4.1. *Consumption of Oxygen by Chondrocytes*

Urban and coworkers [4, 5, 10-14] have studied extensively the global problem of diffusion of oxygen in the intervertebral disc. Since the intervertebral discs lack blood circulation network, this problem can be modeled as a continuous sink distribution throughout the disc and boundaries with given prescribed (external) concentration derived from blood concentration levels. The mathematical modeling of this process becomes a valuable tool in the assessment of the activity and the mode of energy production (aerobic versus anaerobic) in the chondrocytes under various conditions (including injuries that limit blood circulation to the surface of the discs).

Here we ask whether local distribution of oxygen around a chondrocyte may be affected by presence, density, and distribution of other chondrocytes. We consider a case where the input of oxygen through the boundaries has been disrupted through some mechanism. Chondrocytes are modeled as spherical cells of diameter 15μm. At a concentration of 4.3×10^6 cell/ml (disc nucleus human, [11]) the average center-to-center separation will be around 61.5μm.

This limits the application of our method since the cells will not be acting independent of the effects of one another. On the other hand one can determine the order of magnitude of time it takes for one chondrocyte to *feel* the presence of neighboring cells in the event of the disruption of oxygen inflow. Except at very low level of oxygen, the rate of consumption is independent of the concentration. We use a value of 2.2×10^6 molecules/cell/sec (~ 0.063 μmoles/g(wet tissue)/hour) for disc nucleus in human, form [4]) for the oxygen consumption rate. The diffusion coefficient of oxygen in nucleus tissue is $1400 \mu m^2$/sec ($\sim 0.05 cm^2$/hr, from [4]). We use a value of 39000 molecules/μm^3 for initial oxygen concentration (equivalent to 5kPa partial pressure (based on Henry's law) at around 0.5mm from endplate; see [10]).

Noting that the concentration of oxygen can be written as $v = v_0 - v'$ where v_0 is the equilibrium concentration (here 39000 molecules/μm^3) one can write the equation for v' which is the same as equation (1) for v, with the boundary conditions at the cell surfaces to be secretion for v' replacing consumption for v. Therefore one can use the similar asymptotic solution and look for the magnitude of the time it takes for the concentration to drop to half its equilibrium level. It must be noted here that the asymptotic solution, in [3] (equations (7-11) in that reference), are correct despite the fact that the exact solution needs the modifications discussed here, as can be easily seen from comparison with equations (29) and (31) of the present paper. Putting the quoted parameters in the formula for $\alpha = F_0 \rho / \kappa K_m$ and assuming that the cells will "sense" a state of low oxygen concentration at half the steady state partial pressure (i.e. $K_m = 39000/2$ molecules/μm^3) gives $\alpha = 8.55 \times 10^{-4}$. In other words, for this specific situation, the diffusion time constant $t_{diff.} = \rho^2 / \kappa \approx 0.04$s is substantially shorter than the secretion time scale $t_{sec.} = K_m \rho / F_0 \approx 47$s and the concentration approaches the equilibrium so fast that the local presence of single cells becomes inconsequential. This effect, however, is due to the very high concentration of oxygen as well as its very small molecular size. For rarer and larger molecules, such as proteoglycans and other proteins, even in the intervertebral discs the secretion rates may be regulated by the presence and three dimensional arrangements of the neighboring chondrocytes [7-9]. In other words, the range of the secretion/diffusion time scales may be very wide and therefore an accurate representation of the boundary conditions and their time dependences becomes necessary. Our presentation here for the general case (see Appendix) can be used numerically or analytically for this purpose. Note that our definitions for these time scales differ slightly from [3], and there is a minor mistake in equation (15) of that reference.

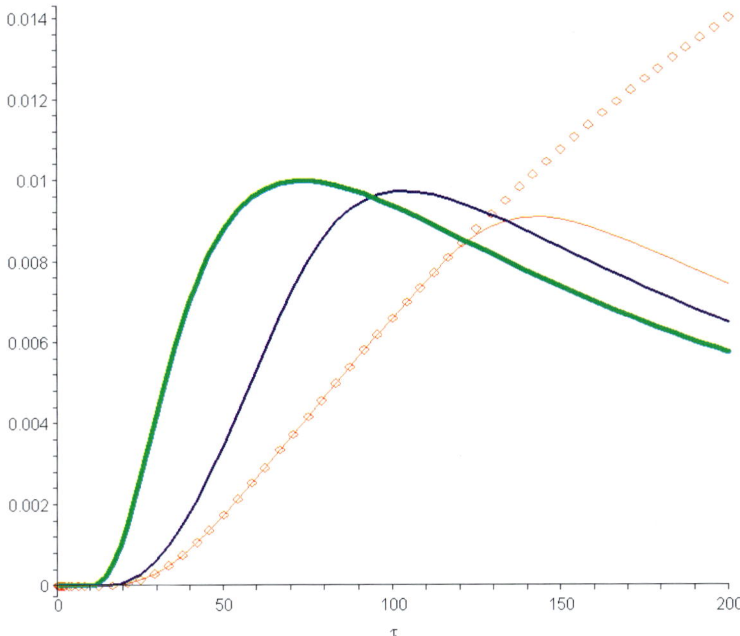

Figure 5. Comparison of the unlimited secretion (hollow symbols, $F_0 = 1$) with limited secretion times ($G_0 = 1$), $\delta = 100$ (thin line), $\delta = 50$ (medium line), and $\delta = 1$ (thick line) at a given distance $\xi = 25$, as a function of time, τ. As seen here, all three cases with limited secretion reach relatively same maximum concentration at this distance, however, the fastest release, $\delta = 1$ (thick line), reaches this maximum in less than half the time it takes for the unlimited secretion (hollow symbols).

4.2. *Time of Release for Limited Amount of Cyto/Chemokine*

When the amount of the cyto/chemokine molecules is limited, it becomes interesting to study the effect of the rate of secretion more closely. To this end, we compare a few cases, where a constant amount of material is released in progressively faster times. The comparison must be made among the solutions of the diffusion equation with step function time dependence of boundary condition:

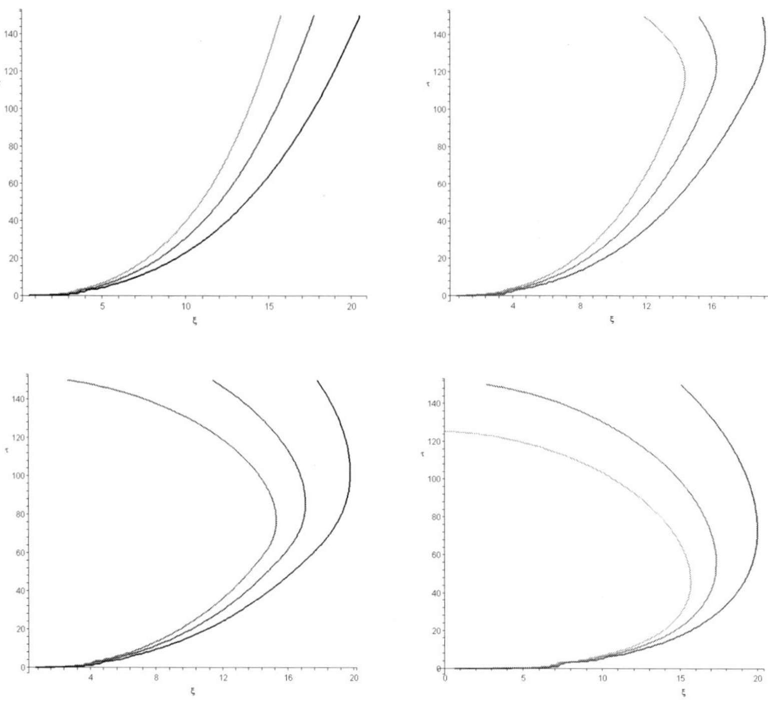

Figure 6. Contour lines for concentration levels 0.01, 0.015, and 0.02. The graphs depict the time, τ, it takes for a point at a distance ξ to reach such a normalized concentration level for infinite release (top left, $F_0 = 1$), and limited secretion times ($G_0 = 1$), $\delta = 100$ (top right), $\delta = 50$ (bottom left), and $\delta = 1$ (bottom right). A close inspection shows that the smaller the release time, δ, the further the signal travels for the same total amount of secretion.

$$v_{s,0}(\xi, \tau, \delta) = \frac{G_0}{\delta}\left(v_0(\xi, \tau) - H(\tau - \delta)v_0(\xi, \tau - \delta)\right) \qquad (37)$$

Here G_0 represents the total amount of secretion, δ is the time of release, and H is the Heaviside step function. In Figure 5, three cases ($\delta = 100, 50, 1, G_0 = 100$) are compared with constant secretion ($F_0 = 1$, i.e. the same rate as $\delta = 100$). The graph represents the concentration at a given distance $\xi = 25$, as a function of time, τ. As seen here, all three cases with limited secretion reach relatively same maximum concentration at this distance, however, the fastest release $\delta = 1$ (dashed line), reaches this maximum in less than half the time it takes for the unlimited secretion (hollow symbols). This shows that once a desired communication distance and signal level are established, it is much more efficient to release a limited amount of material as fast as possible for the chemical signal to reach its target in the shortest time. In Figure 6, a more detailed comparison of the time of travel for three physiologically relevant signal strengths (0.01, 0.015, and 0.02) as a function of the distance from cell is shown. Note that a strict maximum traveling distance exists here which in this case is less than an order of magnitude smaller than that of infinite release case $\xi_{max} = 1/C$ (as $\tau \to \infty$). Also close inspection shows that the smaller the release time, δ, the further the signal travels for the same total amount of secretion.

5. Summary

In this paper we presented an exact mathematical method for the calculation of the diffusion-induced concentration of various molecules around a spherical secreting or absorbing cell. The formulation is general and applicable to any spatial and/or temporal variation of the cell's surface flux. Closed-form analytical solutions are obtained for four special cases. In addition, we have presented asymptotic expressions for long times and great distances from the cell. These expressions may be utilized as simple tools to develop an understanding of the effective communication times and distances among cells. Finally, we used two examples to illustrate the applicability of our mathematical results to physiologically realistic cases.

Acknowledgments

The authors are indebted to Professor Bernhard Palsson for valuable discussions.

Appendix: General Solution

To derive the general solution of (4) with arbitrary boundary conditions (6) and (7), take the Laplace transform, $\overline{f}(s) = \int_0^\infty f(\tau)e^{-s\tau}d\tau$, of these equations,

$$\nabla^2 \overline{v} - s\overline{v} = 0 \quad \text{for} \quad x > 1, \tag{A.1}$$

$$\frac{\partial \overline{v}}{\partial x} = -\hat{F}_0 \overline{g}(s) \quad \text{on} \quad x = 1, \tag{A.2}$$

$$\overline{v} \to 0 \quad \text{as} \quad x \to \infty. \tag{A.3}$$

The separated function

$$\bar{v}_{nm}(x,\theta,\phi,s) = A_{nm}(s)R_n(x,s)P_n^m(\cos\theta)\begin{Bmatrix} \cos m\phi \\ \sin m\phi \end{Bmatrix} \tag{A.4}$$

is a solution of the partial differential equation (A.1) for non-negative integers n and m, when $R_n(x,s)$ satisfies the ordinary differential equation,

$$\frac{d^2R_n}{dx^2} + \frac{2}{x}\frac{dR_n}{dx} - (s + \frac{n(n+1)}{x^2})R_n = 0. \tag{A.5}$$

The solution of (A.5) that vanishes as $x \to \infty$ is

$$R_n(x,s) = K_{n+1/2}(\sqrt{s}x)\frac{1}{\sqrt{x}}, \tag{A.6}$$

where $K_\lambda(z)$ is the modified Bessel function of the second kind of order λ, taking the following forms for $n = 0$, and 1, respectively:

$$K_{1/2}(z) = \sqrt{\frac{\pi}{2z}}e^{-z}, \tag{A.7}$$

$$K_{3/2}(z) = \sqrt{\frac{\pi}{2z}}e^{-z}\left(1 + \frac{1}{z}\right). \tag{A.8}$$

In fact, one can show that for any integer n,

$$K_{n+1/2}(z) = \sqrt{\frac{\pi}{2z}}e^{-z}\sum_{i=0}^{n}\alpha_i z^{-i}. \tag{A.8}$$

The functions $P_n^m(\mu)$ are modified Legendre polynomials,

$$P_n^m(\mu) = (1 - \mu^2)^{m/2}\frac{d^m P_n(\mu)}{d\mu^m}, \tag{A.9}$$

where $P_n(\mu)$ is the Legendre polynomial of order n, being the regular solution of the Legendre differential equation,

$$\frac{d}{d\mu}\left((1 - \mu^2)\frac{dP_n}{d\mu}\right) + n(n+1)P_n = 0, \tag{A.10}$$

which may be expressed using Rodrigues' formula,

$$P_n(\mu) = \frac{1}{2^n n!} \frac{d^n}{d\mu^n}\left(\left(\mu^2 - 1\right)^n\right) \quad . \tag{A.11}$$

Specifically, we have

$$P_0(\mu) = 1, \tag{A.12}$$

$$P_1(\mu) = \mu. \tag{A.13}$$

The functions $A_{nm}(s)$ are calculated by substituting the general series solution

$$\overline{v}(x,\theta,\phi,s) = \sum_{n,m=0}^{\infty} A_{nm}(s) R_n(x,s) P_n^m(\cos\theta)\begin{Bmatrix}\cos m\phi\\ \sin m\phi\end{Bmatrix} \tag{A.14}$$

into the boundary condition (A.2) and using the orthogonality of the eigenfunctions, $P_n^m(\cos\theta)\{\cos m\phi, \sin m\phi\}$. In special cases discussed in this paper, the considered boundary conditions are such that only terms $(n, m) = (0, 0)$ or $(1, 0)$ survive. The time dependence of the boundary conditions enter the solution through the Laplace transform of $g(\tau)$. In the special cases discussed in this paper, we have $g(\tau) = H(\tau)e^{-p\tau}$ and therefore

$$\overline{g}(s) = \frac{1}{s + p}. \tag{A.15}$$

References

1. D. R. Albrecht, G. H. Underhill, T. B. Wassermann, R. L. Sah, S. N. Bhatia. *Nature Methods* **3**, 369 (2006).
2. H. S. Carslaw, J. C. Jaeger. *Conduction of Heat in Solids, second edition.* Oxford University Press, (1959).
3. K. Francis, B. O. Palsson. *Proc. Natl. Accad. Sci. USA* **94**, 12258 (1997).
4. S. Holm, A. Maroudas, J. P. G. Urban, G. Selstam, A. Nahemson. *Connective Tissue Research* **8**, 101 (1981).
5. H. Ishihara, J. P. G. Urban. *Journal of Orthopaedic Research* **17**, 829 (1999).
6. W. Prudomme, G. Q. Daley, P. Zandstra, D. A. Lauttenburger. *Proc. Natl. Accad. Sci. USA* **101**, 2900 (2004).
7. T. M. Quinn, A. J. Grodzinsky, M. D. Buschmann, Y.-J. Kim, E. B. Hunziker. *Journal of Cell Science* **111**, 573 (1998).
8. T. M. Quinn, A. J. Grodzinsky, E. B. Hunziker, J. D. Sandy. *Journal of Orthopaedic Research* **16**, 490 (1998).

9. T. M. Quinn, A. A. Maung, A. J. Grodzinsky, E. B. Hunziker, J. D. Sandy, *Annals of the New York Academy of Sciences* **878**, 420 (1999).
10. J. W. Stairmand, S. Holm, J. P. G. *Spine* **16**, 444 (1991).
11. J. P. G. Urban. In: D. W. L. Hukins (ed.) *Connective Tissue Matrix*, **2**, CRC Press, Boca Raton, FL, 44 (1990).
12. J. P. G. Urban, S. Holm, A. Maroudas. *Biorheology* **15**, 203 (1978).
13. J. P. G. Urban, S. Holm, A. Maroudas, A. Nachemson. *Clinical Orthopaedics and Related Research* **170**, 296 (1982).
14. J. P. G. Urban, S. Smith, J. C. T. Fairband. *Spine* **29**, 2700 (2004).
15. S. Viswanathan, T. Benatar, S. Rose-John, D. A. Lauffenburger, P. W. Zandstra. *Stem Cells* **20**, 119 (2002).
16. F. Yoshida, K. Horike, H. ShiPing. *J. Phys. Soc. Japan* **69**, 3736 (2000).
17. P. W. Zandstra, D. A. Lauffenburger, C. J. Eaves. *Blood* **96**, 1215 (2000).
18. P. W. Zandstra, H.-V. Le, G. Q. Daley, L. G. Griffith, D. A. Lauffenburger. *Biotechnology and Bioengineering* **69**, 607 (2000).

Chapter 16

STEM CELLS, BIOMECHANICS, AND Y. C. FUNG

TABY AHSAN

Department of Biomedical Engineering, Tulane University
Lindy Boggs Center Suite 500, New Orleans, LA 70118, U.S.A.

ADELE M. DOYLE

Institute for Bioengineering and Bioscience, Georgia Institute of Technology
315 Ferst Drive, Atlanta, GA 30332, U.S.A.

ROBERT M. NEREM

Institute for Bioengineering and Bioscience, Georgia Institute of Technology
315 Ferst Drive, Atlanta, GA 30332, U.S.A.

Professor Yuan-Cheng Fung's pioneering work provides a foundation for our understanding of cellular mechanics. Early studies in cellular biomechanics include extensive study of differentiated vascular cells. The more recent proposal to use stem cells for regenerative medicine has prompted study of the role of biomechanics on stem cells. In this brief chapter, we report on work in the Nerem Laboratory evaluating the effects of physiologically-relevant applied physical forces on embryonic stem cells and bone marrow-derived mesenchymal stem cells for vascular applications. Embryonic stem cell-derived endothelial cells (ESC-ECs) respond to applied fluid shear stress similar to vascular endothelial cells. During early differentiation of ESC-ECs, application of fluid shear stress promotes an endothelial phenotype. Mesenchymal stem cells (MSCs), a potential smooth muscle cell-substitute, respond to equibiaxial cyclic strain with cellular rearrangements in a substrate-dependent manner. Gene expression comparison of MSCs with aortic smooth muscle cells reveals that the two cell types have different initial cell signaling profiles and unique responses to applied strain. Taken together, these results demonstrate that the application of physical forces to stem cells can be used to promote differentiation, assess cell phenotype, and discriminate between cell types. Professor Fung's legacy has thus impacted the field of stem cell biology and regenerative medicine.

1. Introduction

In the world of biomechanics, Professor Yuan-Cheng Fung has influenced so many, both his own students and also their students as well as many, many others including the authors of this chapter. Y.C.'s influence has been pervasive. Thus, although he has never worked in the area of stem cells, he in fact also has had an influence here. For the senior author (RMN) this influence goes back 40 years as will be discussed later. For the first author (TA) it starts when she was a

185

Ph.D. student at the University of California, San Diego. And for the middle author (AMD) the influence comes through working with biomechanics collaborators who have been influenced by Y.C. This will be all discussed in the final section of this chapter.

The study of the role of biomechanics in understanding cellular function goes back to the early days of what might be called the modern era of biomechanics. One of the cell types that has been extensively studied over the past thirty years is the vascular endothelial cell [1,2]. It is this cell that has probably more than any benefited from the in vitro study of its function. The success of such studies is perhaps due the fact that in vitro the vascular endothelial cell is studied as a monolayer and in vivo the vascular endothelium also exists as a monolayer.

In the last few years stem cells of various types have been studied extensively as part of the surging interest in regenerative medicine. As with other cell types, stem cells also are very much influenced by their microenvironment [3]. Our collective efforts have been investigating both embryonic stem cells and mesenchymal stem cells. This brief article thus summarizes our results to date with each of these two cell types, illustrating the role of biomechanics in orchestrating the functional characteristics of stem cells. In the next section we start with the embryonic stem cell.

2. The Embryonic Stem Cell and Biomechanics

Every cell in the body can be traced back to a fertilized egg, the result of the union of an egg and sperm. This gives rise to over 200 different cells that carry out the specialized functions required by the body. For studies in the laboratory, an embryonic stem cell is derived from a blastocyst, i.e. a pre-implantation embryo created in the in vitro fertilization process used to help couples conceive. In the interior of the blastocyst is the inner cell mass which is composed of a few cells characterized as pluripotent in that they can ultimately differentiate into all the cell types of the body [4]. Embryonic stem cells are of interest for many reasons. These include the seeking of a better understanding of developmental biology, deriving cells for use in drug discovery, and in the emerging area of regenerative medicine.

As with all cell types, important to our understanding is the influence of a cell's microenvironment on cellular activities and on its function. In the case of an embryonic stem cell, what controls its fate, i.e. whether it replicates itself or differentiates into a specialized cell? If the latter, what controls the type of cell into which it will differentiate? This microenvironment includes the soluble molecules to which a cell is exposed, the substrate/matrix to which it might be in contact, cell-cell contact, and the mechanical environment in which it resides. It is the last of these that is of interest here.

In entering into the world of embryonic stem cell research, our first objective was to differentiate such cells down the mesoderm pathway and into vascular endothelial cells (ECs). This was successfully done with mouse ESCs [5], although others have shown this also for human ESCs [6]. In order to fully demonstrate that what had been achieved was a vascular EC, the cells resulting from the differentiation process were then studied in a laminar flow, parallel plate flow chamber [2], this to demonstrate that their response to flow was like that of an EC [7]. These studies showed that the response to flow and the associated shear stress was as one would expect from a vascular EC with there being both cell elongation and an alignment with the direction of flow.

Having successfully differentiated ESCs into vascular ECs, there then was the following question. In the early stage of the differentiation of an ESC to an EC, how does exposure of a differentiating ESC to flow and the associated shear stress modulate the process? Does it possibly accelerate the process of an ESC differentiating into an EC? These questions became clear when the published protocols for differentiating ESCs into vascular ECs were framed as a flow chart consisting of a series of stages including labor intensive selection techniques (Figure 1). Using this approach, it was found that the application of a steady laminar shear stress of 15 dynes/cm^2 to embryonic stem cells during an early stage of the differentiation process promoted an endothelial phenotype.

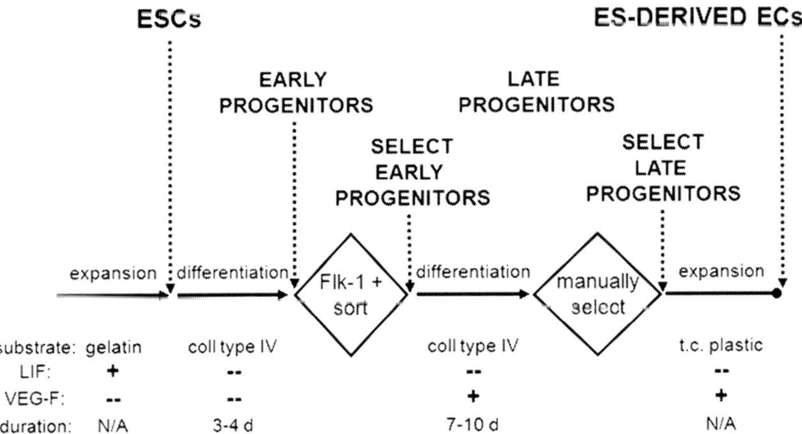

Figure 1. A flowchart of the procedure required to derive vascular endothelial cells (ECs) from embryonic stem cells (ESCs).

Application of quantitative engineering principles to qualitative biological assessments allowed for determining statistically significant changes in protein expression through flow cytometry. The results of those changes are represented in Table 1, and what is observed is an upregulation of markers associated with the EC lineage.

Table 1. Effect of fluid shear stress on protein expression.

STAGE	PROTEIN MARKER	SHEAR
ES cell	Oct4 ⟷ SSEA-1 (mouse) SSEA-3 or 4 (human)	⟷
Mesoderm	vimentin brachyury	↑↑
Endothelial Progenitors & Endothelial Cells	Flk-1	↑↑
	Flt-1	
	PECAM	↑↑
	VE-Cad	↑
	SMC-actin	⟷

This research indicates that the physical force of shear stress can help direct the differentiation of an embryonic stem cell towards a vascular endothelial cell. It also suggests, however, that this process of differentiation can be accelerated, and recent preliminary data indicates that the application of shear stress will result in higher percentage of cells being Flk-1+.

3. Mesenchymal Stem Cells and Biomechanics

Mesenchymal stem cells (MSCs) are multipotent stem cells capable of differentiating along osteogenic, chondrogenic, and adipogenic lineages, in addition to potentially other lineages. These cells are isolated from the adherent fraction of mononuclear cells taken from adult bone marrow. MSCs are characterized using a wide range of positive and negative cell surface protein markers, although MSC-specific markers have not been established. Isolation of multipotent stromal cells with MSC-like characteristics using alternative methods, e.g., immunoselection, or alternative tissue sources, e.g., adipose tissue, has expanded the range of potential derivation options for a therapeutic cell source. MSCs are clinically promising in part due to their low immunogenicity and immunomodulatory capabilities and potential to be used in autologous therapy.

Vascular therapies may benefit from MSCs. Work by others has shown that in response to biochemical or mechanical cues MSCs may adopt traits characteristic of differentiated vascular cells, namely endothelial or smooth muscle cells [8-10]. Our focus has been to determine the effects on MSCs of vascular-relevant levels of equibiaxial cyclic strain, a more complex mechanical cue than uniaxial strain. Human MSCs rearranged to form multicellular structures on gelatin- or collagen type I-coated collagen within hours of applied strain (10% area change, 1 Hz). These mechanosensitive cellular rearrangements were dependent on the protein type used to promote cell-substrate adhesion, as shown in Figure 2 [11].

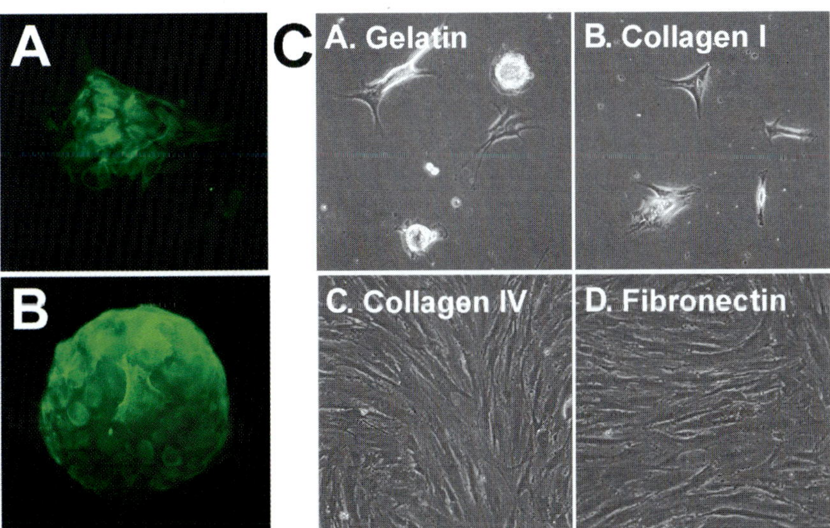

Figure 2. Multicellular rearrangements of human adult bone marrow-derived mesenchymal stem cells in response to cyclic strain. MSCs on gelatin-coated silicone form clusters (A) and knobs (B) in response to applied strain (10%, 1 Hz for 48 hours). Formation of multicellular clusters varies with the underlying adherent protein substrate (C). Images shown are vimentin-stained for intermediate filaments (A, B) or phase images (C).

Cell signaling assessments of MSCs exposed to cyclic strain have shown that these cells alter gene expression differently than vascular smooth muscle cells exposed to the same mechanical cue. Combined, these studies demonstrate that cellular response to mechanical strain is affected by biological parameters including cell-matrix connections and cell type.

4. Concluding Comments

There thus are clear influences of a stem cell's mechanical environment on its fate and on its functional characteristics. This has been demonstrated here for both ESCs and MSCs. Furthermore, this should not be surprising. This is because for every cell type for which the influence of the mechanical environment on cell function has been studied there has been a significant effect observed.

Taking into account the results presented in this chapter, studies that link to the role of biomechanics in stem cell biology, how is this related to Professor Y.C. Fung? Of course there is the fact that Y.C. Fung's work, his teachings, and his mentorship has provided a broad foundation that is applicable to a variety of different tissues and to organs. It also is applicable to the molecular and cellular levels, and thus it is applicable to stem cells.

One can, however, relate stem cells, biomechanics, and Y.C. Fung at a personal level. For the first author (TA) this goes back to her graduate studies at University of California, San Diego, where Y.C. Fung was instrumental in establishing the bioengineering program. Walking past then-emeritus Professor Fung in the hallways, it was humbling to realize the tremendous impact he had in laying down the foundation of biomechanics and bioengineering. More importantly, it was inspiring to witness his ongoing passion for science and engineering, as he continued to pursue his research well into retirement.

For the second author (AMD), Y. C. Fung's pioneering work in bioengineering provides a foundation for study of cellular mechanics. The quality and breadth of Y. C Fung's contributions to the field of bioengineering are motivating for younger generations of students and researchers.

Finally, for the senior author (RMN) this begins in the early 1960s when he was a Ph.D. student at Ohio State University in aerospace engineering and where he studied out of the book on aeroelasticity written by Professor Fung. Then, when spending the better part of a year at Imperial College London in 1970, he first met Professor Fung. A very special experience was to be part of a delegation led by Professor Fung that went to China in 1983 to participate in the first China/Japan/U.S. Conference on Biomechanics. This meeting was held in

Wuhan and afterwards the group did some touring. This included the Huang Shan mountains, i.e., the yellow mountains, and in Figure 3 is shown a photo of Professor Fung and his wife Luna during the climb to the hotel at the top. Certainly, the Huang Shan mountains were a very special highlight of the trip. Since then their paths have crossed many times, and on many occasions they have been in each other's home. Through this life long association, Y.C. Fung has been a friend, a teacher, and a mentor.

Figure 3. Professor Y. C. Fung and his wife Luna in the Huang Shan Mountains in 1983 during the visit of the U.S. delegation to China to participate in the first China/Japan/U.S. Conference on Biomechanics.

Professor Fung has in so many ways been a pioneer. This is true not only in terms of his own research, but also because of the foundation that he has provided. This equally is a foundation for research on stem cells as we hope our work has demonstrated. Thus, this chapter has been written as a tribute to Professor Y.C. Fung and in thanks for all he has done in pioneering the modern era of biomechanics.

References

1. S. Chien. *Am J Physiol Heart Circ Physiol.* **292,** H1209 (2007).
2. M. J. Levesque and R. M. Nerem. *J Biomech Eng.* **107**, 341 (1985).
3. T. Ahsan, A. M. Doyle, and R. M. Nerem. *Principles of Regenerative Medicine.* 28 (2008).
4. J. A. Thomson, J. Itskovitz-Eldor, S. S. Shapiro, M. A. Waknitz, J. J. Swiergiel, V. S. Marshall, and J. M. Jones. *Science.* **282**, 1145 (1998).
5. K. E. McCloskey, S. L. Stice, and R. M. Nerem. **330**, 287 (2006).
6. S. Levenberg, J. S. Golub, M. Amit, J. Itskovitz-Eldor, and R. Langer. *Proc Natl Acad Sci U.S.A..* **99**, 4391. (2002).
7. K. E. McCloskey, D. A. Smith, H. Jo, and R. M. Nerem. *J Vasc Res.* **43**, 411 (2006).
8. S. G. Ball, A. C. Shuttleworth, and C. M. Kielty. *Int J Biochem Cell Biol.* **36**, 714 (2004).
9. Z. Gong and L. E. Niklason. *FASEB J.* **22,** 1635 (2008).
10. J. Oswald, S. Boxberger, B. Jørgensen, S. Feldmann, G. Ehninger, M. Bornhäuser, C. Werner. *Stem Cells.* **22**, 377 (2004).
11. A. M. Doyle, R. M. Nerem, and T. Ahsan. *Ann Biomed Eng.* **37**, 783 (2009).

Chapter 17

OF MICE AND MEN.....AND A CHINA CONNECTION

DON P. GIDDENS, PH.D.[1], JIN SUO, PH.D.[1], W. ROBERT TAYLOR, M.D., PH.D.[1,2], HABIB SAMADY, M.D.[2] AND JOHN OSHINSKI, PH.D.[1,3]

[1]*Wallace H. Coulter Department of Biomedical Engineering at Georgia Tech and Emory University School of Medicine*

[2]*Division of Cardiology, Emory University School of Medicine*

[3]*Department of Radiology, Emory University School of Medicine*

Corresponding author: Don P. Giddens, Ph.D.
Tech Tower, 225 North Avenue
Georgia Institute of Technology
Atlanta, GA 30332-0360

Often, it can be a chance encounter, a seemingly insignificant conversation, a word of encouragement, or a simple gesture from a respected person that has a tremendous impact on another person's life; and this impact can then be amplified through a series of circumstances, dedicated work, and an inspired passion so as to yield an impact on many, many people that could not have been predicted at the onset. Such was the case when Professor Y.C. Fung visited Taiyuan University at the invitation of Dr. Yang, its president, in 1995 just prior to the 4th China-Japan-US-Singapore Conference on Biomechanics. During that visit, Professor Fung went to the laboratory of Mr. Jin Suo and spent perhaps 10-15 minutes with him, discussing experimental fluid dynamics. This encounter, and subsequent events, changed the career path of Mr. Suo and had an impact on our understanding of hemodynamics and atherosclerosis.....

1. Introduction

The importance of hemodynamics, especially the local fluid dynamic wall shear stress (WSS), to localization of atherosclerosis in human arteries has been long appreciated (e.g., [1-3]). This observed connection between biomechanics and biology has led to numerous studies aimed at uncovering mechanisms for atherogenesis, plaque progression and plaque rupture; and models for such studies include endothelial cell cultures and various animals, ranging from mice to non-human primates. Early correlations between hemodynamic factors and atherosclerosis in humans were limited to relationships between hemodynamic variables obtained from laboratory models of vessels and plaque characteristics obtained from autopsy. However, more recently, use of subject-specific computational fluid dynamics (CFD) enabled by various imaging modalities is beginning to permit investigation of atherogenesis and plaque progression [4-6] in human subjects and will hopefully lead to a better understanding of the disease process and, importantly, to better methods of treatment and prevention.

The general methodology we employ is to (i) obtain vessel geometry and flow conditions in individual subjects, whether animal or human,; (ii) compute detailed pulsatile hemodynamic flow fields using CFD; (iii) obtain information on various biomarkers associated with atherosclerosis from that subject; and (iv) analyze the CFD and biomarker data for possible relationships that suggest mechanisms. The hemodynamic variables are primarily associated with WSS, although particle trajectories and residence times may also provide useful information. The biomarkers employed depend upon the subject (animal or human), the specific vessel (e.g., carotid or coronary artery), and the imaging modality (e.g., CT, MRI, IVUS, OCT, nanoparticles). For example, carotid arteries in humans can be imaged noninvasively with much better spatial resolution than can coronary arteries. On the other hand, invasive coronary angiography for evaluation of acute or chronic coronary syndromes allows concomitant IVUS imaging with excellent spatial resolution of plaque, as well as indications on plaque composition. With animal models, it is possible to obtain data at the molecular level, such as the expression of adhesion molecules on endothelial cell surfaces or gene expression, since animals can be sacrificed to obtain tissue.

The interplay between studies in "mice and men" is a key to understanding atherosclerosis across its disease spectrum, from early initiating mechanisms to therapies and interventions.

2. Hemodynamics and Markers of Atherosclerosis in Mice

Because mice are a popular model for studying atherosclerosis, we have developed methodology to image mouse aortas, including arch vessels and major branches, and to use the imaging data to develop CFD models of individual mice [7]. Our research colleagues have developed quantum dot methods that can be employed to target various biomarkers of atherosclerosis, such as Vascular Cell Adhesion Molecule (VCAM-1) and Intra-cellular Adhesion Molecule (ICAM-1) [8], thus allowing investigation of relationships between hemodynamics and known precursors of atherosclerosis.

Mice from a group of twelve-week old wild-type mice (C57 Bl/6, Jackson Laboratory) were employed in a study to investigate hemodynamics in their aortas. Animals were housed at Emory University and cared for according to National Institutes of Health guidelines for the care and use of experimental animals. Mice were euthanized with CO_2, the abdominal cavity was dissected to cannulate the abdominal aorta, and the thoracic cavity was left unopened to avoid distorting the thoracic aorta. The aorta was perfusion fixed at 100 mmHg with normal saline, followed by buffered formalin. The aorta was then filled with a casting material (Microfil, Flow Tech, Inc.) and the preparation left in a formalin solution for two days and then decalcified (Cal-Ex II, Fisher Scientific). The thorax was dissected and micro-CT imaging was performed (viva CT40, Scanco Medical). Three dimensional images of the aorta were obtained through image segmentation and reconstruction methods [7,9], and these were employed to develop a CFD model of the individual mouse. An example is shown in Figure 1, where it can be seen that very detailed anatomic information can be obtained.

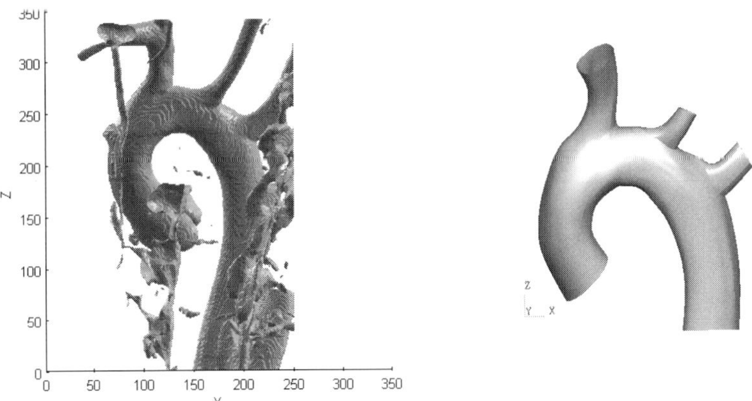

Figure 1. (a) Micro-CT image rendering of the in situ cast of a mouse thoracic aorta. (b) Model of the thoracic aorta after removing smaller branches and smoothing.

The imaging data were used to develop a grid for CFD calculations. Velocity waveforms were measured in mice using a 14 MHz Doppler ultrasound probe, and outflows into the arch vessels were estimated from data in the literature. The Navier-Stokes equations for laminar flow of a Newtonian fluid were used as the governing equations, and computations were performed with CFD-ACE commercial software (CFD Research Corporation).

We are interested in relationships between WSS and the expression of adhesion molecules on endothelial cells lining the arterial surface. In another set of mice, animals were euthanized and the aortas were perfusion fixed, as described previously. The aortas were dissected in situ and opened, and the preparations were subjected to a two-step immunohistochemical protocol using quantum dot-biconjugated secondary antibodies (Invitrogen). The measurement technique involved confocal imaging of the unique quantum dot fluorescence spectra to detect intensity and surface distribution. We targeted VCAM-1 and ICAM-1, two adhesion molecules known to mediate monocyte adhesion during atherogenesis. Details of the methodology can be found in [7,8].

Figure 2 presents a comparison of the CFD and biomarker results. After computing the pulsatile flow field, the magnitude of WSS was time-averaged

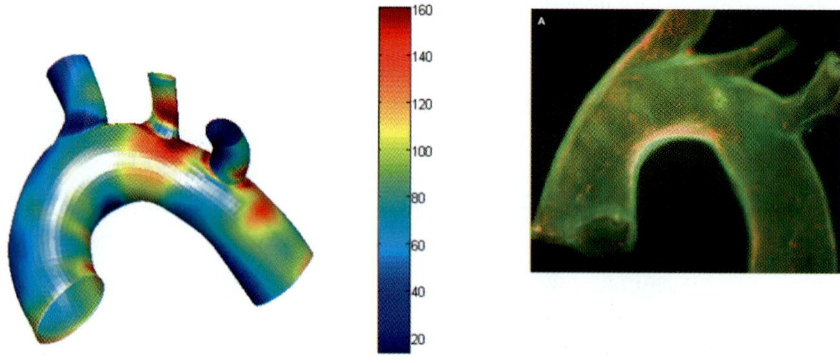

Figure 2. (a) Time averaged values of the magnitude of wall shear stress (WSS). The WSS units are dynes/cm^2 and are scaled by colors as shown. These values are considerably higher than values found in humans. (b) Fluorescence image showing distribution of ICAM-1 expression in the thoracic aorta of a mouse. This image is from a mouse treated with LPS to achieve fluorescence enhancement. Note similarity of high intensity fluorescence with relative low WSS regions.

over the cycle to compute a mean WSS distribution (Figure 2a). WSS values in the mouse aorta are notably higher than in the human aorta [7,9]. However, as can be seen in Figure 2b where the distribution of ICAM-1 expression is presented, there is a tendency for areas of relatively lower mean WSS to coincide with areas of increased ICAM-1 expression. This correspondence also held for VCAM-1 expression [7].

These studies have shown that the magnitude of WSS in mice is roughly an order of magnitude greater than that in humans [7,9] and well above values found to cause endothelial cell disruption in acute studies in dogs [10]. Apparently, endothelial cells in mice live in a quite different hemodynamic environment than in humans, but they also clearly adapt well to these higher WSS values. Further, the locations of adhesion molecule expression in the mouse aorta are similar to sites where atherosclerotic plaques develop in humans and in mouse models of atherosclerosis, primarily along the inner curvature of the arch and at lower WSS regions in branch ostia. Interestingly, the Reynolds numbers (~ 250) and Womersley parameters (~2) for the mouse aorta are in a range similar to those for human coronary arteries, suggesting that the magnitude of WSS may be less important than changes in WSS either temporally or spatially.

2.1. *Hemodynamics in Coronary Arteries of Humans*

Acute myocardial infarction and sudden cardiac death remain a major cause of mortality in the United States. Current therapy for coronary artery disease involves systemic anti-atherosclerotic therapy and revascularization of severe epicardial coronary lesions. Despite this dual approach, patients with coronary artery disease continue to experience significant 1-year rates of myocardial infarction (6.9%) and death (3.9%) following presentation for coronary angiography [11]. Recent data suggest that revascularization of severe coronary lesions improves symptoms, but does not prevent myocardial infarction or reduce mortality [12]. Indeed, myocardial infarctions frequently result from rupture of thin-cap fibroatheromas that are bulky, non-flow limiting atherosclerotic lesions that reside largely in the vessel wall and therefore cannot be identified with coronary angiography. Consequently, these lesions are not often revascularized [13]. Developing imaging modalities adjunctive to angiography that can identify coronary segments "vulnerable" to future rupture may offer opportunities to intervene locally and potentially prevent future myocardial infarction and death.

While atherosclerosis is known to be a diffuse process involving inflammation, oxidative stress, and endothelial dysfunction, the common precursor of plaque rupture, thin cap fibroatheromas, and acute plaque ruptures resulting in myocardial infarction occur in proximal coronary arteries [14]. Accordingly, local triggers of plaque rupture should be targeted to characterize these high-risk coronary locations in patients with clinical and subclinical atherosclerosis. A likely contributor to the vulnerability of coronary segments is wall shear stress (WSS). Alterations of arterial WSS are known to impact endothelial function and atherogenesis. Cell culture studies and experimental models have demonstrated that regions of low and oscillatory WSS induce loss of the physiologic flow-oriented alignment of the endothelial cells, promote the accumulation of lipids, and stimulate expression of leukocyte adhesion molecules, chemotactic factors, and growth factors that cause proliferation of smooth muscle cells and transmigration of macrophages. Low shear stress regions are also associated with a prothrombotic, pro-inflammatory, and pro-oxidant state [15].

In vivo evaluation of coronary WSS in patients with coronary artery disease has thus far been limited. The two major variables required for computation of WSS, namely, high resolution 3-D reconstruction of vessels and blood velocity measurement have been challenging to measure accurately in the coronary circulation with the contemporary non-invasive imaging used to evaluate WSS in the carotid circulation. This is primarily due to the smaller caliber and constant motion of the coronary arteries challenging the contemporary spatial resolution of cardiac computerized tomography (0.5mm) and magnetic resonance imaging (1.0mm) [16]. Invasive coronary angiography and intravascular ultrasound (IVUS) are routinely performed clinical tools that may circumvent these limitations as the spatial resolution of IVUS is 100 micrometers [17]. Accurate flow velocity can be measured using intracoronary Doppler wires [18]. Furthermore, with recent advances in IVUS technology by Volcano Therapeutics, atherosclerotic plaque composition can be assessed with good accuracy when compared to actual histology. This "virtual histology" IVUS (VH™ IVUS) applies spectral analysis of the IVUS backscatter radiofrequency signal to characterize plaque components on the basis of tissue characteristics such as density, compressibility, concentration of various components, and size [19].

Using these imaging tools, we have developed the methodology for performing dynamic WSS analysis of coronary arteries in an effort to identify segments most vulnerable to subsequent plaque rupture. The geometric boundary conditions are constructed through the fusion of IVUS images and biplanar coronary angiograms. The inlet and outlet velocity measurements are performed using the Doppler flow wire, and the computational fluid dynamic (CFD) model is constructed based on the coronary artery geometry and the velocity measurements.

Patients with stable angina or acute coronary syndromes presenting to the cardiac catheterization laboratory are enrolled if they are found to have mild to moderate coronary disease by biplanar coronary angiography (Emory IRB protocol numbers 701 and 2835). The data required for computation of WSS are then acquired. These include Doppler velocity measurement, biplanar angiography of the intracoronary wire path in space (z axis), and plaque volume and morphology using VH™ IVUS.

Figure 3 illustrates the image data obtained from digital biplanar angiography and the IVUS system. The guidewire path is tracked in 4D space/time coordinates using software developed in our lab, and the IVUS slices are reconstructed to form a 3D image of the coronary artery and its wall thickness. The VH software of the IVUS system gives an indication of plaque composition.

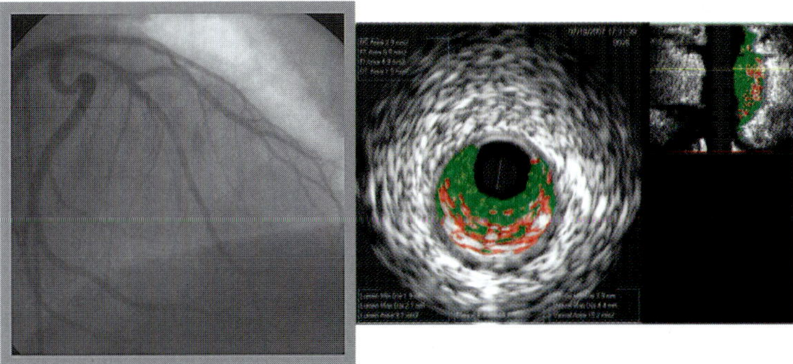

Figure 3. (a) One of the biplanar images from angiography of a patient. (b) Intravascular ultrasound image of a section of the coronary artery from the same patient. The virtual histology rendering of the artery wall is shown, and there is a large plaque evident at this section. Fibrous tissue is shown in green, calcium in white, and necrotic core in red.

Pulsatile flow field computations are performed with the CFD-ACE software as described previously, based on the measured Doppler flow waveform and the vessel geometry as reconstructed from the fused angiographic and IVUS data. Figure 4 illustrates results in a segment of the coronary artery of one of the subjects. The "thickness" of the geometric grid represents wall thickness determined from IVUS data, and the color plots show the spatial distribution of the mean (time-averaged) magnitude of WSS. Note the relatively long plaque in the left anterior descending (LAD) coronary artery. The WSS increases in the stenosis, as expected, and there is a large region of relatively low WSS just distal to the stenosis and in the region of the bifurcation created by a diagonal branch.

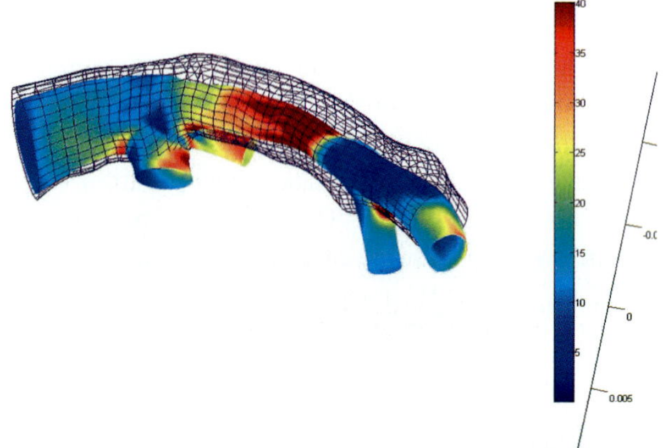

Figure 4. (a) The Doppler-derived flow waveform of a subject. (b) Rendering of the geometry, including wall thickness, and time-averaged WSS as determined by CFD in the same subject. Note the narrowing created by a long plaque in the LAD section. Distribution of WSS is depicted by the color scale in dynes/cm^2. Relatively high values of mean WSS can be observed within the stenosis, while there is a large zone of relatively low WSS in the poststenotic and branching regions.

Our findings in the data analyzed to date are consistent with the following: (i) early plaques localize in areas of low WSS; (ii) there is a relationship between areas of low WSS and co-location of necrotic core in plaques; (iii) plaque progression is associated with areas of low WSS. However, the project is still at a relatively early stage for drawing final conclusions.

And a China Connection.....

The thread that is woven throughout this fabric of research is Mr. Jin Suo, who was inspired to pursue a research career in biomechanics by Professor Fung during that brief visit to his laboratory in China in 1995. Based on his discussions with Professor Fung, Mr. Suo wrote to Professor Giddens seeking a visiting scholar's position. Professor Fung remembered Mr. Suo quite well and gave a strong recommendation. So in 1997, Mr. Suo came to Georgia Tech as a visiting scholar. His passion and his talent for biomechanics showed very early, and he pursued his Ph.D. degree, graduating in 2005 after bringing his wife and daughter to the U.S. Now, Mr. Suo is Dr. Suo, and he is continuing to do creative and pathbreaking work in biomechanics.

So a chance encounter with a great man like Professor Fung, coupled with talent, opportunity and passion, has led to contributions that will affect many people, something that Professor Fung has done over and over. See Figure 5.....

Figure 5. Photograph of Professor Y.C. Fung with Drs. Jin Suo and Don Giddens, taken at the 90th birthday celebration in San Diego.

Acknowledgments

The authors gratefully acknowledge support from the following sources: NIH Grants NHLBI R01 HL70531 and U01 HL080711; The Georgia Tech-Emory Wallace H. Coulter Seed Grant Program; and the Georgia Research Alliance.

References

1. C.G. Caro, J.M. Fitz-Gerald, R.C. Schroter, *Proc R Soc Lond B Biol Sci.* **177,** 109 (1971).
2. C.K. Zarins, D.P. Giddens, B.K. Bharadvaj, V.S. Sottiurai, R.F. Mabon, *Circ Res.* **53**, 502 (1983).
3. D.N. Ku, D.P. Giddens, C.K. Zarins, S. Glagov. *Arteriosclerosis.* **5**, 293 (1985).
4. J. Suo, J. Oshinski, D.P. Giddens, *J Biomech Eng.* **125**, 347 (2003).
5. Q. Long, et al. *Crit Rev Biomed Eng* **26**, 227 (1998).
6. J.A. Moore, et al. *Ann Biomed Eng* **27**, 32 (1999).
7. J. Suo, et al., *Arterioscler. Thromb. Vasc. Biol.* **27**, 346 (2007).
8. D.E. Ferrara, D. Weiss, P.H. Carnell, R.P. Vito, D. Vega, X. Gao, S, Nie, W.R. Taylor. *Am J Physiol Regul Integr Comp Physiol.* **290**, R114 (2006).
9. P.D. Weinberg and C.R. Ethier. *J Biomech.*
10. D L. Fry, *Circ Res.* **22**:165 (1968)

11. G.W. Stone, J.H. Ware, M.E. Bertrand, et al. *JAMA* **298**, 2497 (2007).

12. W.E. Boden, R.A. O'Rourke, K.K. Teo, et al. *N Engl J Med* **356**, 1503 (2007).

13. R. Virmani, A.P. Burke, A. Farb. *Cardiologia* **43**, 267 (1998).

14. J.C. Wang, S.L. Normand, L. Mauri, R.E.Kuntz. *Circulation* **110**, 278 (2004).

15. D. Harrison, K.K. Griendling, U. Landmesser, B. Hornig, H. Drexler, *Am J Cardiol* **91**, 7A (2003).

16. J.P. Laissy, V. Sebban, J.F. Deux, V. Huart, E. Mousseaux. *J Radiol* **85**, 1798 (2004).

17. G.M. Sangiorgi, F. Clementi, C. Cola, G. Biondi-Zoccai, *Catheter Cardiovasc Interv* **70**, 203 (2007).

18. A.L. Gaster, L. Korsholm, P. Thayssen, K.E. Pedersen, T.H. Haghfelt, *Catheter Cardiovasc Interv* **53**, 449 (2001).

19. A. Nair, B.D. Kuban, E.M. Tuzcu, P. Schoenhagen, S.E. Nissen, D.G. Vince, *Circulation* **106**, 2200 (2002).

Chapter 18

MULTI-PATIENT FSI STUDIES FOR ATHEROSCLEROTIC CAROTID PLAQUE PROGRESSION BASED ON SERIAL MAGNETIC RESONANCE IMAGING[1]

DALIN TANG[†]

Mathematical Sciences Department, Worcester Polytechnic Institute, Worcester, MA 01609 U.S.A.

CHUN YANG

School of Mathematics, Beijing Normal University, Beijing, P.R.China

GADOR CANTON, CHUN YUAN

[3]Deparment of Radiology, University of Washington, Seattle, WA 98195 U.S.A.

THOMAS S. HATSUKAMI

Division of Vascular Surgery, VA Puget Sound HCS and University of Washington, Seattle, WA. 98195 U.S.A.

Multi-patient fluid-structure interaction (FSI) studies based on serial magnetic resonance imaging (MRI) data were conducted to quantify correlations between plaque progression and both plaque wall stress (PWS) and flow maximum shear stress (FMSS) and introduce plaque growth functions to predict progression. In vivo serial MRI carotid data from 6 patients (all male; age: 59-73, mean: 67. 3-4 scans/patient; scan time interval: 18 months) were acquired for this study. For each artery, wall thickness (WT), PWS, and FMSS data from 700-900 matched lumen points (100 points per matched slice) were collected for analysis. Point-wise progression was expressed by increase in wall thickness (WTI) at each lumen point. Six growth functions with different combinations of WT, PWS, plaque wall strain (PWSN) and FMSS terms were introduced to predict WT using data from previous scans. 12 cases (out of 15 time-pairs) showed negative correlation between WTI and PWS at current time. FMSS (at current time) showed positive correlation (10 out of 15) with WTI. The growth function including all WT, PWS, PWSN and FMSS factors provided best fit for progression prediction, compared to other 5 growth functions containing fewer factors. More longitudinal case studies are needed before a clinical application can be devised.

[*] This work was supported in part by a National Science Foundation grant DMS-0540684.

[†] Corresponding author: Dalin Tang, Mathematical Sciences Department, Worcester Polytechnic Institute, Worcester, MA 01609, Phone:508-831-5332, fax:508-831-5824, e-mail: dtang@wpi.edu.

1. Introduction: "To Learn Swimming, One Must Get In the Water"

Cardiovascular disease is the leading cause of death in developed countries. Tremendous effort has been made to gather baseline biological data for blood vessels in the circulatory system and to investigate mechanisms governing atherosclerosis initiation, progression and plaque rupture, which often leads to heart attack and stroke [3-8, 11, 16, 20-21, 26]. Research methods used include in vitro, *ex vivo* and *in vivo* studies using laboratories, animal models and patient-specific studies. Various measurement methods have been developed ranging from direct mechanical testing, flow-chambers, to imaging techniques such as ultrasound, magnetic resonance image (MRI), intra-vascular ultrasound (IVUS), computed tomography (CT), micro-CT, optical coherence tomography (OCT), etc. Computational models based on experimental, animal, or human data for atherosclerotic plaques have been introduced to study blood flow and mechanical stress/strain conditions in arteries [2, 9-10, 12, 18-19, 22-3, 8-11, 13-19, 22-34].

In 1980's to early 1990's, models for blood flow in arteries were mostly rigid-walled models. One-dimensional models with compliant or collapsible tubes were introduced by Kamm, Shapiro, Ku and other researchers to study flow limitation in stenotic arteries and other collapsible tubes [13, 17]. With some ideas about developing elastic arterial models, I flew to San Diego in 1992 to seek guidance from Dr. Fung. That was my first meeting with him. Dr. Fung appraised my ideas and suggested me to form collaborations with engineers. Since then he has been giving me advices whenever there was a chance. At the 1996 ASME Bioengineering conference, I had the honor to have a photo taken with him (Fig. 1). He reminded me: "To learn swimming, you must get in the water." Ever since then, I have been in the process of "getting in the water." Starting with collaborations with Ku and Vito at Georgia Institute of Technology, Kobayashi from Shinshu University, to recent collaborations with Woodard, Zheng, and Sicard at Washington University and Yuan, Hatsukami, Canton, and Ferguson at University of Washington in Seattle, models with fluid-structure interactions were developed based on in vitro, *ex vivo* and *in vivo* data. The aim is to investigate blood flow in arteries and try to gain better understanding of mechanisms governing atherosclerosis progression and causing plaque rupture [10, 26-31, 34].

Early arterial models with idealized geometries included three groups of controlling factors: a) geometry (diameter, bifurcation, curvature, aneurysm), b) pressure (inlet/outlet, pressure drop, transmural pressure), and c) material properties. Investigations were focused on flow limitation (collapsible tubes) and flow behaviors which may be linked to atherosclerosis initiation. One important finding that has influenced the blood flow research area in recent years was that intimal thickening and atherosclerosis initiation have positive correlation with low and oscillating wall shear stresses [3, 8, 18]. With the advancement of medical image technology, image-based computational models have been introduced with realistic geometries of arteries and fluid-structure interactions have been taken into consideration. The list of additional controlling factors for image-based models includes: plaque cap thickness, lipid pool size, calcification, ulceration and thrombi, and plaque component material properties. To investigate atherosclerotic plaque progression, patient-tracking information must be obtained to quantify plaque growth functions which may be used for possible predictions. It is an exciting field with great advances made in recent years and many more questions to be answered with future effort.

Figure 1. Photo with Dr. Fung at 1996 ASME Bioengineering Conference. Left: Dr. Y. C. Fung; Right: Dalin Tang.

To join the celebration of Dr. Fung's 90th birthday, we are making a brief report about our recent effort investigating human carotid atherosclerotic plaque progression using MRI-based models with fluid-structure interactions.

2. Data Acquisition, Models and Methods

2.1. *Multi-Contrast MRI Data Acquisition*

3D *in vivo* serial MRI carotid data (3-4 time points, time interval: 18 months) were acquired from 6 patients (all male; age: 59-73, mean: 67) at the University of Washington (UW), after informed consent. The UW institutional review board approved the consent forms and study protocols. MRI scans were conducted on a GE SIGNA 1.5-T whole body scanner using an established protocol outlined in the work of Yuan and Kerwin et al. [14,35]. A computer package CASCADE (Computer-Aided System for Cardiovascular Disease Evaluation) developed by the Vascular Imaging Laboratory (VIL) at the University of Washington (UW) was used to perform image analysis and segmentation [14,35]. Upon completion of a review, an extensive report was generated and segmented contour lines for different plaque components for each slice were sent to Tang's group for model construction and further computational mechanical analysis. Figure 2 shows 20 (selected from 24) MRI slices with 2 different weightings obtained from a human carotid plaque sample, together with segmented contour plots and stacked 3D view of the plaque showing all the slices. Figure 3 gives re-constructed 3D plaque morphology views with components from two patients acquired at three time points showing plaque progression and regression. Serial MRI data were used for model construction, correlation analysis, and quantification of plaque growth functions.

2.2. *3D Reconstruction, Shrink-Stretch Process, and Mesh Generation*

For *in vivo* image-based artery models, an important modeling issue is how to determine zero stress state from *in vivo* plaque geometry. Dr. Fung wrote in his celebrated three-book "Biomechanics" series: "A body in which there is no stress is at the zero stress state. If strain is calculated with respect to the zero stress state, then the strain is zero when the stress is zero, and vice versa. This is an important feature of the constitutive equation. Hence ***the analysis of stress and strain begins with the identification of the zero stress state***" [4]. Under the *in vivo* condition, the artery is axially stretched and pressurized, thus axial and circumferential shrinking was applied a priori to generate the no-load plaque geometry which would be used as the starting shape for the computational simulation. Then an axial stretch and lumen pressure would be applied to recover the *in vivo* loaded geometry. This shrink-stretch process is to partially recover the zero-stress state of the artery with limited *in vivo* data. The shrinkage in axial direction was 9% so that the vessel would regain its *in vivo* length with a 10% axial stretch. Circumferential shrinkage for lumen and outer wall was determined so that plaque geometry had the best match with the original *in vivo* geometry under loaded condition [10]. Because advanced plaques have complex geometries, a curve-fitting mesh generation technique was developed to generate a mesh for these models. The 3D plaque domain was divided into hundreds of small "volumes" to curve-fit the irregular plaque geometry. 3D surfaces, volumes and computational mesh were created with ADINA (ADINA R & D, Inc., Watertown, MA, USA) computing environment.

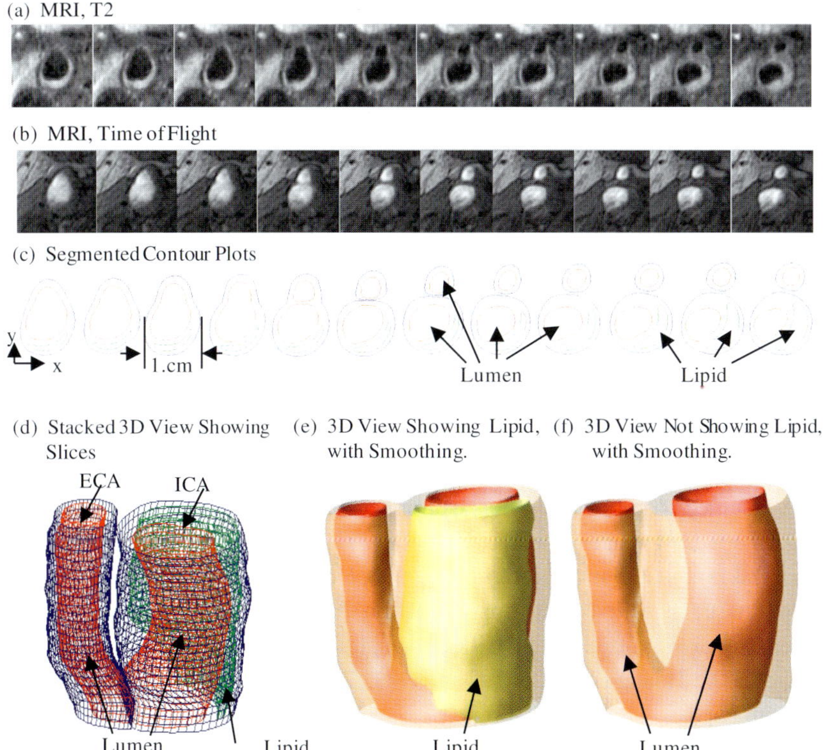

Figure 2. *In vivo* 3D MRI images of a human carotid plaque and re-constructed 3D geometry. (a)-(b) 10 (out of 24) MRI slices (S1-S10); (c) Segmented contour plots showing plaque components; (d)-(f) Three views of the re-constructed 3D geometry.

a) Patient Data Showing Plaque Progression.

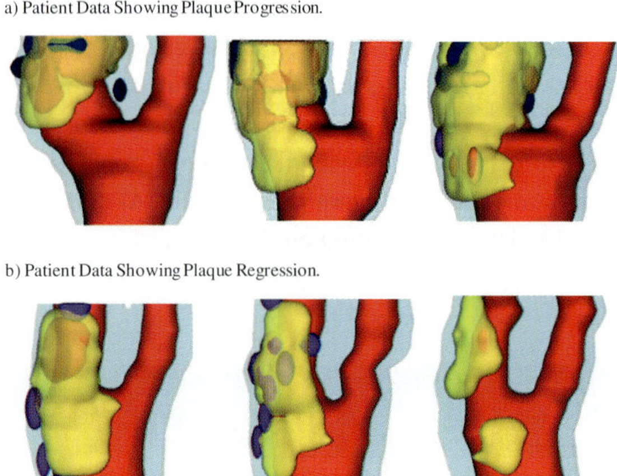

b) Patient Data Showing Plaque Regression.

Figure 3. 3D plaque samples re-constructed from *in vivo* MR images. (a) one patient data at three time points showing plaque growth; (b) one patient taking Statin showing plaque reduction. Time interval: 18 months. Red: lumen; Yellow: lipid; Dark blue: calcification; light blue: outer wall.

2.3. *The 3D FSI Model and Solution Method*

Blood flow was assumed to be laminar, Newtonian, viscous and incompressible. The incompressible Navier-Stokes equations with arbitrary Lagrangian-Eulerian (ALE) formulation were used as the governing equations. Both arterial wall (normal tissue) and plaque components (calcification, lipid core, and others) were assumed to be hyperelastic, isotropic, incompressible and homogeneous. Since multi-component model construction was overly time-consuming (each model takes about 2 months to construct), plaque components were treated as normal tissue as a first-order approximation. The nonlinear modified Mooney-Rivlin model was used to describe the material properties of the vessel wall (Bathe, 2002; Huang et al., 2001; Tang et al., 2004b). The strain energy function is given by [1],

$$W = c_1 (I_1 - 3) + c_2 (I_2 - 3) + D_1 [\exp(D_2 (I_1 - 3)) - 1], \tag{1}$$

where I_1 and I_2 are the first and second strain invariants, c_i and D_i are material parameters chosen to match experimental measurements found in current literature [11,15]. No-slip conditions, natural traction equilibrium boundary conditions and continuity of displacement were assumed on the interface between the vessel wall and fluid. The complete FSI model is given below:

$$\rho(\partial \mathbf{u}/\partial t + ((\mathbf{u} - \mathbf{u}_g) \cdot \nabla)\, \mathbf{u}\,) = -\nabla p + \mu \nabla^2 \mathbf{u}\,, \tag{2}$$

$$\nabla \cdot \mathbf{u} = 0, \tag{3}$$

$$\mathbf{u}\,|_\Gamma = \partial x/\partial t\,, \quad \partial \mathbf{u}/\partial n|_{\text{inlet, outlet}} = 0, \tag{4}$$

$$p|_{\text{inlet}} = p_{\text{in}}(t), \quad p|_{\text{outlet}} = p_{\text{out}}(t), \tag{5}$$

$$\rho\, v_{i,tt} = \sigma_{ij,j}\,, \quad i,j=1,2,3; \text{ sum over } j, \tag{6}$$

$$\varepsilon_{ij} = (\,v_{i,j} + v_{j,i} + v_{\alpha,i}\, v_{\alpha,j})/2, \quad i,j,\ \alpha=1,2,3; \quad \text{sum over } \alpha, \tag{7}$$

$$\sigma_{ij} \cdot n_j\,\,|_{\text{out\_wall}} = 0, \tag{8}$$

$$\sigma^r_{ij} \cdot n_j\,\,|_{\text{interface}} = \sigma^s_{ij} \cdot n_j|_{\text{interface}}\,, \tag{9}$$

where $\mathbf{u}$ and p are fluid velocity and pressure, $\mathbf{u}_g$ is mesh velocity, μ is the dynamic viscosity, ρ is density, Γ stands for vessel inner boundary, $f_{\bullet,j}$ stands for derivative of f with respect to the jth variable, $\boldsymbol{\sigma}$ is stress tensor, $\boldsymbol{\varepsilon}$ is strain tensor, $\mathbf{v}$ is solid displacement vector, superscript letters "r" and "s" were used to indicate different materials. Patient-specific systole and diastole arm pressure values were used to specify a pulsating pressure profile for each patient studied.

The fully coupled 3D FSI models were solved by ADINA. This computational package uses unstructured finite element methods for both fluid and solid models. Nonlinear incremental iterative procedures are used to handle fluid-structure interactions. The governing finite element equations for both solid and fluid models were solved by Newton-Raphson iteration method. More details of the computational models and solution methods can be found in our previous publications [10, 29-31] and Bathe [1].

2.4. *Plaque Progression and Stress/Strain Data Collection*

For each carotid artery, slices from each time point were matched using carotid bifurcation as the registration point (see Fig. 4). Vessel wall thickness (WT) was selected as the measure for plaque progression. In our previous paper, the "shortest distance" method was used to determine vessel thickness, i.e., for a selected nodal point on the inner boundary (lumen), the shortest distance between that point and the outer boundary was defined as the vessel thickness at that luminal point. That led to uneven selection of nodal points from the outer boundary as shown by Fig. 5(a), or unreasonable WT definition when plaque morphology became very irregular. A piecewise equal-step method is introduced to fix the problem. A slice is divided into several sections according to its geometry (4 in Fig. 5). For each section, equal step is used for inner and outer boundaries respectively to choose equal number of nodal points. The corresponding points on the inner and outer boundaries are paired and the distance between the paired points are defined as vessel wall thickness at the given luminal point. This method is sufficient for the cases covered in this paper. Vessel wall thickness, computational plaque wall maximum principal stress (PWS), maximum principal strain (PWSN), and flow maximum principal stress (FMSS) at each lumen point from matched cross sectional slices (with 100 matched data points per slice) were collected for each time (t_1, t_2, and t_3, and t_4 when available). Point-wise plaque progression at each luminal point is expressed by increase in wall thickness (WTI) between two scan time points. Correlations between plaque progression as measured by WTI and PWS, PWSN, and FMSS were quantified for each t_i / t_{i+1} pair. Using all time points (MRI scans) from the 6 patients, 15 time pairs were formed for correlation analysis.

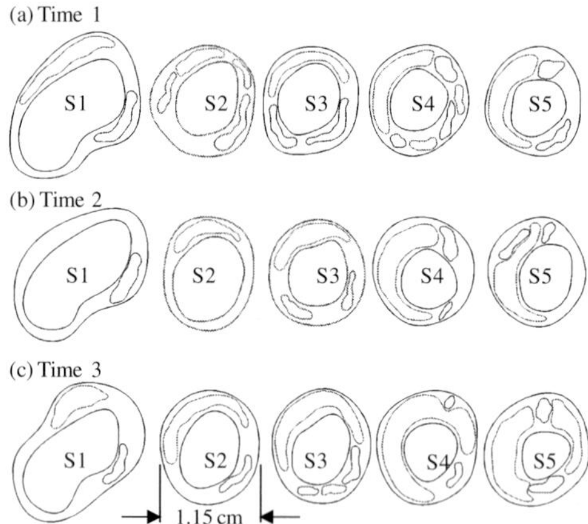

Figure 4. Segmented contour plots from three time points showing the registration process. 5 slices were selected with the bifurcation serving as the registration point.

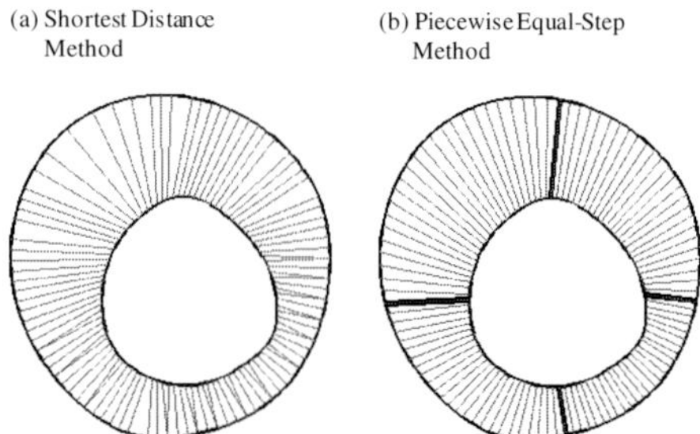

Figure 5. Piecewise equal-step method for determination of vessel wall thickness. (a) Shortest distance method; (b) Piecewise equal-step method. For this selected slice, the vessel wall was divided into 4 sections. 25 points were equally distributed on each section.

2.5. *Predicting Plaque Progression*

Six growth functions (F0-F5) were used to predict wall thickness (WT) at t_{i+2} using data at t_i and t_{2i+1}, i=1,2. F5 contains all the terms:

$$F5=a_1 \times WT(i,t_1)+a_2 \times WT(i,t_2)+a_3 \times PWS(i,t_1)+a_4 \times PWS(i,t_2)+a_5 \times PWSN(i,t_1)+$$
$$a_6 \times WSN(i,t_2)+ a_7 \times FMSS(i,t_1)+a_8 \times FMSS(i,t_2). \tag{10}$$

where a_i values were chosen to reach best fit with WT at t_3. Other functions have some omissions of the WT, PWS, PWSN, and FMSS terms. The same formulas were also used for t_2-t_4 data set. R^2 values of the fitting results from F5 for all patients were compared.

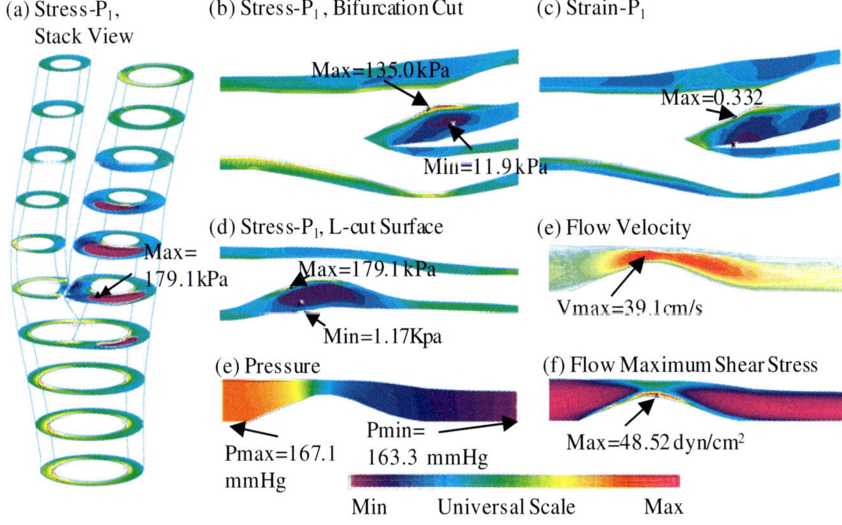

Figure 6. Plots of plaque wall stress (maximum principal stress, Stress-P₁) , strain (maximum principal strain, Strain-P₁), flow velocity, shear stress and pressure from a plaque sample showing solution behaviors.

3. Results

Figure 6 presents baseline results obtained from our FSI model using a sample plaque. Plots for maximum principal stress (Stress-P_1), maximum principal strain (Strain-P_1), flow velocity and flow maximum shear stress (FMSS) on two different cut-surfaces are shown: an L-cut surface showing the bifurcation, a Y-cut showing the lipid pool position and cap thickness.

3.1. *Negative Correlation between Plaque Progression and Plaque Wall Stress (PWS) at Current Time*

Using PWS values from the current time, results from Table 1 shows that 12 out of 15 cases had **negative** correlation between plaque progression (WTI) and PWS, 1 case had positive correlation, and 2 had no significance. However, using PWS values from the previous scan, 11 cases had positive correlation, 1 had negative correlation, and 3 had no significance.

3.2. *Positive Correlation between Plaque Progression and Flo Maximum Shear Stress (FMSS) at Current Time*

Using values from the current time, Table 1 shows that FMSS from 10 out of 15 cases had **positive** correlation with plaque progression, 4 had negative correlation, and 1 case had no significance. When using FMSS values from the previous scan, 4 cases had negative correlation, 8 had negative correlation, and 3 had no significance.

3.3. *Predicting Plaque Progression Using Growth Functions*

All data were standardized and then used in the six growth functions to fit wall thickness at next scan. Using one patient data, Table 2 shows that F5 provided best fitting results. Table 3 presents the fitting results using F5 for 9 three-time-point data sets from 6 patients. Coefficients varied considerably from case to case, even with the same patient, indicating noticeable change in progression patterns over time.

Table 1. Plaque progression (WTI) has a) negative correlation with plaque wall stress (PWS) from current scan (12/15); b) positive correlation with flow maximum shear stress (FMSS) from current scan (10/15); c) positive correlation with plaque stress from previous scan (11/15); d) negative correlation with FSS from previous scan (8/15 negative, 4/15 positive, 3 no significance). r is the Pearson correlation coefficient.

Patient	Time	# of Data Pts	PWS Last Scan		PWS Current Scan		FMSS Last Scan		FMSS Current Scan	
			r	p	r	p	r	p	r	p
P1	t_1-t_2	800	0.03	0.397	-0.368	0	0.08	0.024	0.199	0
	t_2-t_3	800	-0.017	0.633	-0.228	0	-0.070	0.047	0.261	0
P2	t_1-t_2	800	0.492	0	0.253	0	-0.576	0	-0.455	0
	t_2-t_3	900	0.097	0.004	-0.080	0.016	0.130	0.000	0.122	0.000
	t_3-t_4	900	-0.135	0	-0.241	0	-0.035	0.301	0.081	0.016
P3	t_1-t_2	700	0.408	0	-0.144	0.000	-0.234	0	0.325	0
	t_2-t_3	800	0.242	0	-0.402	0	-0.022	0.539	0.163	0
	t_3-t_4	800	0.226	0	-0.374	0	-0.153	0	0.236	0
P4	t_1-t_2	700	0.107	0.005	-0.221	0	0.056	0.138	0.162	0
	t_2-t_3	700	0.024	0.531	-0.416	0	-0.413	0	-0.062	0.102
P5	t_1-t_2	900	0.470	0	-0.011	0.741	-0.414	0	-0.245	0
	t_2-t_3	900	0.080	0.017	0.013	0.695	0.188	0	0.144	0
P6	t_1-t_2	800	0.110	0.002	-0.052	0.140	0.339	0	0.385	0
	t_2-t_3	800	0.127	0.000	-0.149	0	-0.423	0	-0.361	0
	t_3-t_4	800	0.119	0.001	-0.158	0	-0.527	0	-0.246	0
Positive Correlations		11		1		4		10		
Negative Correlations		1		12		8		4		
No Significance		3		2		3		1		

Table 2. Fitting results for one patient (P1) using F0-F5 showing that combining WT, PWS, PWSN and FMSS leads to more accurate prediction of plaque progression. a_i's are coefficients defined in Eq. (10). WT_2 means WT at t_2. WT_1 means WT at t_1. The same is true for other terms.

	a_1	a_2	a_3	a_4	a_5	a_6	a_7	a_8	R^2
	WT_2	WT_1	PWS_2	PWS_1	$PWSN_2$	$PWSN_1$	$FMSS_2$	$FMSS_1$	
F0	0.498	0.544							0.814
F1	0.410	0.663	-0.161	0.182					0.835
F2	0.455	0.751	1.042	0.193	-1.041	-0.220			0.869
F3	0.436	0.637	-0.237	0.102			-0.360	0.218	0.890
F4	0.587	0.568					-0.304	0.188	0.872
F5	0.449	0.561	1.130	-1.045	-1.169	0.870	-0.208	0.133	0.903

Table 3. Growth function F5 combining WT, PWS, PWSN and FMSS provided good predictions (R^2 >0.9) for plaque progression (9 three-time-point data sets from 6 patients). Coefficients varied considerably from case to case, even with the same patient, indicating noticeable change in progression patterns.

Pt	Time	a_1	a_2	a_3	a_4	a_5	a_6	a_7	a_8	R^2
		WT_2	WT_1	PWS_2	PWS_1	$PWSN_2$	$PWSN_1$	$FMSS_2$	$FMSS_1$	
P1	t_1-t_3	0.449	0.561	1.130	-1.045	-1.169	0.870	-0.208	0.133	0.903
P2	t_1-t_3	-0.647	1.512	-0.745	1.225	0.553	-1.002	0.074	-0.150	0.967
	t_2-t_4	0.988	0.161	0.555	0.210	-0.643	-0.039	0.009	-0.111	0.972
P3	t_1-t_3	0.827	0.232	0.332	0.845	-0.162	-0.962	0.168	-0.139	0.908
	t_2-t_4	0.284	0.657	-0.254	0.249	0.1145	-0.220	0.120	0.009	0.914
P4	t_1-t_3	0.429	0.114	0.119	-0.147	-0.218	-0.044	-0.035	-0.406	0.923
P5	t_1-t_3	-0.413	1.160	-0.597	-1.143	0.249	1.208	0.352	-0.387	0.886
P6	t_1-t_3	1.088	-0.060	1.200	-0.673	-1.145	0.614	0.307	-0.379	0.952
	t_2-t_4	0.217	0.469	-0.120	-0.307	0.047	0.123	-0.629	0.031	0.909

4. Discussion

4.1. *Our Results and Current Literature for Plaque Progression*

It is widely accepted that intimal thickening and atherosclerosis initiation have positive correlation with low and oscillating flow shear stresses. Our results based on multi-year patient-tracking data are not contradicting the current literature. When interpreting our findings, we emphasize the following: **a)** our results are for patients with advanced plaques. Mechanisms governing plaque progression may be different at different stages responding to changes in flow and stress/strain environment; **b)** both structural forces and flow shear stress play important roles in plaque progression and should be taken into consideration in a combined way to better understand the complex process; **c)** Correlations of mechanical conditions (PWS or FMSS) taken from the current time or previous time with plaque progression were roughly opposite. The reversed correlation behavior for PWS could be understood as a consequence from the definition of plaque progression (WTI=WT2-WT1) and the fact that PWS correlates negatively with wall thickness. FMSS is mainly influenced by lumen narrowing which is also closely related to WT.

4.2. *Difficulty in Predicting Plaque Progression*

In order for anything to be predictable, we must know the governing mechanisms which are applicable to both the past and the future. Our results indicated that the growth functions, even derived for the same patient, may differ considerably from one time frame to the next time frame. More localized approach and more controlling factors (such as blood cells, endothelial cells, smooth muscle cells, cholesterol level, medication, growth factors) should be included in our models for better predicting powers.

4.3. *Limitations*

The study was done with many limitations. a) *In vivo* MRI resolution was $0.3{\times}0.3{\times}2.0$ mm^3. Better resolution will improve the accuracy of our predictions; b) patient-specific vessel material properties were not available; c) plaque components were treated as normal tissue to reduce the model construction cost. That only affected stress predictions over very limited number of lumen nodes where the cap was very thin. It would not affect the overall results. FMSS should have almost no effect from this simplification; d) using wall thickness increase as the measure of plaque progression practically limited the definition to 2D. Better definitions should be devised for better 3D descriptions of plaque progression; e) location-specific study should be performed to link local plaque growth to mechanical factors.

5. Conclusion

For the first time, multi-year patient tracking data and models with fluid-structure interactions were used to quantify the relationship between plaque progression and both plaque wall stress and flow shear stress and to determine quantitative human plaque growth functions to predict plaque progression. Models combining plaque morphology, stress, strain and flow shear stress may provide more accurate predictions, compared to models containing fewer factors. The results support the new hypothesis that plaque progression depends on both structural stress and flow shear stress conditions. More longitudinal studies are needed before a clinical application can be devised.

Acknowledgement

This research was supported in part by NSF grant DMS-0540684.

References

1. K. J. Bathe, *Finite Element Procedures*, (Prentice Hall, Inc. New Jersey, 1996).
2. D. Bluestein, Y. Alemu, I. Avrahami, M. Gharib, K. Dumont, J. J. Ricotta, S. Einav, *J Biomech.* **41**(5), 1111 (2008).

3. M. H. Friedman, C. B. Bargeron, O. J. Deters, G. M. Hutchins, F. F. Mark, *Atherosclerosis*, **68**, 27 (1987).

4. Y.C. Fung, *Biomechanics: Mechanical Properties of Living Tissues*, (Springer, New York, 1993).

5. Y.C. Fung, *Biomechanics: Circulation*, (Second Edition, Springer, New York, 1996).

6. Y.C. Fung, *Biomechanics: Motion, Flow, Stress, and Growth*, (Springer, New York, 1990).

7. Y. C. Fung, S. Q. Liu, *J. Appl. Physiol.* **70**, 2455, (.1991).

8. D. P. Giddens, C. K. Zarins, S. Glagov, *J. Biomech. Engng.* **115**,588, (1993).

9. G. A. Holzapfel, M. Stadler, C. A. J. Schulze-Bause, *Ann. Biomed. Eng.* **30**(6), 753 (2002).

10. X. Huang, C. Yang, C. Yuan, F. Liu, G. Canton, J. Zheng, P. K. Woodard, G.A. Sicard, D. Tang, *Mol & Cell Biomechanics*, **6**(2),121 (2009).

11. J. D. Humphrey, *Cardiovascular Solid Mechanics*, (Springer-Verlag, New York, 2002).

12. M. R. Kaazempur-Mofrad, A. G. Isasi, H. F. Younis, R.C. Chan, D. P. Hinton, G. Sukhova, G. M. Lamuraglia, R. T. Lee, R. D. Kamm, *Annals of Biomedical Engineering*, **32** (7), 932 (2004).

13. R.D. Kamm, A.H. Shapiro, *J.Biomech Engng*, **101**, 1-78 (1979).

14. W. Kerwin, A. Hooker, M. Spilker, P. Vicini, M. Ferguson, T. Hatsukami, C. Yuan, *Circulation.* **107**(6), 851 (2003).

15. S. Kobayashi, D. Tsunoda, Y. Fukuzawa, H. Morikawa, D. Tang, D. N. Ku, *Proc of 2003 ASME Summer Bioengineering Conf.*, 497 (2003).

16. D. N. Ku, *Annu. Rev. Fluid Mech.* **29**, 399 (1997).

17. D.N. Ku, M.N. Zeigler, J.M. Downing, *J. Biomech. Engng*, **112**, 444 (1990).

18. D. N. Ku, D. P. Giddens, C.K. Zarins, S. Glagov, *Arteriosclerosis.* **5**, 293 (1985).

19. Q. Long, X.Y. Xu, B. Ariff, S. A. Thom, A. D. Hughes, A. V. Stanton, *J. Magn. Reson. Imaging.* **11**, 299 (2000).

20. R. M. Nerem, *J. Biomech. Eng.*, **114**, 274 (1992).

21. T. J. Pedley, *The Fluid Mechanics of Large Blood Vessels*, (Cambridge University Press, New York, 1980).

22. K. Perktold, M. Resch, *J Biomed Eng.* **12**(2), 111 (1990).

23. K. Perktold, M. Hofer, G. Rappitsch, M. Loew, B. D. Kuban, M. H. Friedman, *J. Biomech.* **31**, 217 (1998).

24. M. Prosi, K. Perktold, Z. Ding, M. H. Friedman, *J. Biomech.* **37**, 1767 (2004).

25. D. A. Steinman, *Ann. Biomed. Eng.,* **30**(4), 483 (2002).

26. D. Tang, "Flow in healthy and stenosed arteries," *Wiley Encyclopedia of Biomedical Engineering,* Article 1525 (New Jersey, John Wiley & Sons, Inc., 2006).

27. D. Tang, C. Yang, S. Kobayashi, D. N. Ku, *J. Biomech. Eng.,* **126**, 363 (2004).

28. D. Tang, C. Yang, S. Mondal, F. Liu, G. Canton, T. S. Hatsukami, C. Yuan, J. *Biomechanics,* 41(4), 727 (2008). Featured article by the Society for Heart Attack Prevention and Eradication (SHAPE).

29. D. Tang, C. Yang, J. Zheng, P. K. Woodard, J. E. Saffitz, G. A. Sicard, T. K. Pilgram, C. Yuan, *J. Biomech. Eng.,* **127**(7), 1185 (2005).

30. D. Tang, C Yang, J. Zheng, P. K. Woodard, J. E. Saffitz, J. D. Petruccelli, G. A. Sicard, C. Yuan, *Ann. Biomed. Eng.,* **33**(12), 1789 (2005).

31. D. Tang, C Yang, J. Zheng, P. K. Woodard, G. A. Sicard, J. E. Saffitz, C. Yuan, *Ann. Biomed. Eng.,* **32**(7), 947 (2004).

32. Y. Vengrenyuk, L. Cardoso, S. Weinbaum, *Mol Cell Biomech.* **5**(1), 37 (2008).

33. D. A. Vorp, J. P. Vande Geest, *Arterioscler Thromb Vasc Biol.,* **25**(8), 1558 (2005).

34. C. Yang, D. Tang, C. Yuan, T. S. Hatsukami, J. Zheng, P. K. Woodard, *CMES: Comput. Model. Eng. Sci.* **19**(3), 233 (2007).

35. C. Yuan, L. M. Mitsumori, K. W. Beach, K. R. Maravilla, *Radiology,* **221**, 285 (2001).

Chapter 19

CURRENT STATUS ON COUNTERMEASURES FOR INTRADIALYTIC HYPOTENSION

J. S. LEE

Global Monitors, Inc. and University of California at San Diego
Department of Bioengineering, La Jolla, CA 92093-0412, U.S.A.

Intradialytic hypotension is one of the most adverse effects of hemodialysis. It has been postulated that the cause of hypotension is hypovolemia as more fluid is extracted by the dialyzer than the fluid restituted from the tissue. Many maneuvers have been developed to counter intradialytic hypotension under the premise of hypovolemia. However, there are indications suggesting intradialytic hypotension is due primarily to pooling of blood to abdominal organs. This new hypothesis calls for a new approach to develop effective countermeasures for intradialytic hypotension. After the review of the methodologies to assess hypovolemia and microvascular pooling, countermeasures for intradialytic hypotension, and the use of anti-pooling effect to counter the development of hypotension; we conclude with the recommendation to develop: (1) a monitoring system capable of identifying the responsible mechanisms, (2) a device and/or drug therapy that can counter the effect of blood pooling and/or hypovolemia, (3) the integration of these two for the delivery of personalized countermeasure, and (4) a hemodialysis process that does not induce pooling of blood to abdominal organs for the avoidance of intradialytic hypotension.

1. Introduction

Some 30% of hemodialysis patients would experience hypotensive symptoms (dizziness, fainting, headache and nausea) at one or more treatments, a problem of hemodialysis mentioned in a 1995 article of Daugirdas [3]. In a 2008 review, Palmer and Henrich stated that a symptomatic reduction in blood pressure during or immediately after dialysis occurs in approximately 20 to 30% of dialysis sessions [12]. Many maneuvers have been developed as countermeasures of intradialytic hypotension over this decade. The current status of treatment for intradialytic hypotension may be summarized by this 2008 clinical commentary of Henrich [6]: "What are badly needed in this area of clinical research are improved methods to reduce the frequency of intradialytic hypotension, thereby avoiding its untoward effects!"

For each hemodialysis treatment, 3 to 4 liters of fluid is ultrafiltrated or extracted from the blood circulating through a hemodialysis machine or dialyzer. In view that the blood volume in the patient's circulation system is about 5 liters, it is a common perception that the development of hypotension in hemodialysis patients may be the result of low blood volume (i.e., hypovolemia) when fluid restitution from the body tissue to the circulation is not sufficient to replenish the volume of fluid extracted by the hemodialysis machine. Using a one-compartment model to simulate a circulation with a uniform hematocrit [18], the observed increase in systemic hematocrit has been taken to support the hypovolemia hypothesis [2, 17].

The well known Fahraeus effect indicates that the hematocrit of blood flowing in many microvessels is less than the systemic hematocrit of blood circulating in large blood vessels [5, 11]. The measurement of red blood cell (RBC) and plasma volume by the dual indicator dilution technique also indicates that the whole body hematocrit is less than the systemic hematocrit [14]. To examine the implication of non-uniform hematocrit, a two-compartment model was developed [7, 9, 10]. In this article, several new examples are presented to highlight the interrelation between volume and hematocrit changes.

The two-compartment analysis on hemodialysis [15] and endotoxin shock [14] and one result on orthostatic hypotension [19] suggest that pooling of blood to abdominal organs could be activated by the hemodialysis process and be the leading cause to induce intradialytic hypotension.

Many countermeasures for intradialytic hypotension are developed under the hypothesis of hypovolemia. They may not be effective, if blood pooling to abdominal organs is the mechanism of hypotension in hemodialysis patients. Literatures on inflatable bands or compressible garments show that their compression of abdominal organs reduces blood pooling to those organs. This anti-pooling effect leads to improved cardiovascular function of patients with hemorrhagic and orthostatic hypotension. Based on the results of anti-pooling devices and two-compartment analysis, this article is concluded with four recommendations on what to develop for more effective alleviation and prevention of intradialytic hypotension.

2. Simulation of the Circulation

2.1. *Formulation of a Two-Compartment Model*

The Fahraeus effect indicates that the blood flowing in microvessels with a diameter less than 250 μm has a tube hematocrit smaller than the systemic hematocrit H_{sys} [5, 11]. Using this demarcation, we can divide the circulation into two compartments as that sketched in Fig. 1.

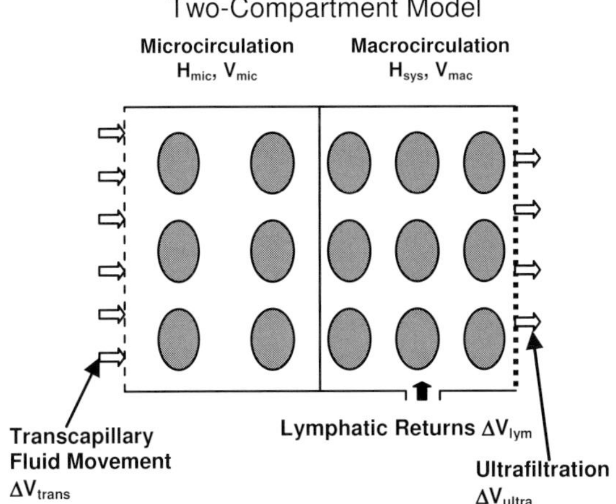

Figure 1. A two-compartment model showing a lower hematocrit for blood flowing in the microcirculation. There are three fluid exchanges passages: the wall of capillaries, the lymphatics and the ultrafiltration site of the hemodialysis machine. The arrow direction corresponds to a net fluid movement at that passage.

The microcirculation compartment has V_{mic} as the blood volume and H_{mic} as the microvascular hematocrit (a volume weighted tube hematocrit of all microvessels) while the macrocirculation has V_{mac} and H_{sys}. The sum of these two volumes is the total blood volume V_b. By determining the RBC and plasma volume for this model and defining one constant α and one ratio β as:

$$\alpha \equiv H_{mic}/H_{sys} \quad \text{and} \quad \beta \equiv V_{mic}/V_b \tag{1}$$

we can derive the whole body hematocrit (H_w = total RBC volume/total blood volume) and the F_{cell} ratio as

$$F_{cell} \equiv H_w/H_{sys} = 1 - \beta + \alpha\beta \tag{2}$$

If one considers that the circulation has similar blood volume distributions as exemplified by the heart, lung and mesentery, then the microcirculation may contain 40% to 50% of the total blood volume [10]. If we take the microvascular volume fraction β as 0.5 and F_{cell} as 5/6, a value close to that obtained for dog by dual indicator dilution technique [14]; we obtain from Eq. 2 an α as 2/3. These values of α and β are employed to construct the two-compartment model depicted in Fig. 1.

The figure also depicts three sites for fluid exchanges: transcapillary fluid movement within the microcirculation, lymphatic returns, and ultrafiltration through the hemodialysis machine. Over a period of time, the volumes of these three fluid exchanges are accumulated to ΔV_{trans}, ΔV_{lym} and ΔV_{ultra} respectively. If a volume is positive, their movement is indicated by the arrow direction shown in the figure. These three form the total blood volume change in the following way:

$$\Delta V_b = \Delta V_{trans} + \Delta V_{lym} - \Delta V_{ultra} \tag{3}$$

In general each of these pathways for fluid exchange may have different value in the protein concentrations C_{trans}, C_{lym} and C_{ultra}. However, we like to simplify the derivation with the following assumptions, where γ is a constant:

$$C_{trans} = C_{lym} = \gamma C_{pl}, \qquad C_{utra} = 0 \tag{4}$$

By adding these three fluid and protein movements into one net fluid volume change which is ΔV_b , then the "average" protein concentration C_f of this net volume change becomes:

$$C_f = (\Delta V_{trans}C_{trans} + \Delta V_{lym}C_{lym})/\Delta V_b = [1 + (\Delta V_{ultra}/V_b)/(\Delta V_b/V_b)]\gamma C_{pl} \tag{5}$$

The conservations of plasma protein mass and RBC volume for this two-compartment model lead to the derivation of the following two formulas relating the change in blood volume to plasma protein concentration C_{pl} (before) and C_{pl}' (after) and the change in microcirculation volume to hematocrit H_{sys} and H_{sys}':

$$\Delta V_b/V_b = (1 - H_w)(C_{pl} - C_{pl}')/(C_{pl}' - C_f) \tag{6}$$

$$\Delta V_{mic}/V_b = [\Delta V_b/V_b - F_{cell}(H_{sys}/H_{sys}' - 1)]/(1 - \alpha) \tag{7}$$

Because C_f and $\Delta V_b/V_b$ appear in Eq. 5 and 6, an iterative procedure is used to determine their values. Then Eq. 7 determines $\Delta V_{mic}/V_b$ from the measured hematocrits and calculated $\Delta V_b/V_b$. Finally the macrovascular volume change is calculated as:

$$\Delta V_{mac}/V_b = \Delta V_b/V_b - \Delta V_{mic}/V_b \tag{8}$$

To complete the calculation, we need the value of F_{cell}, β and γ. The review done by Lee suggests that their range can be 0.8 to 0.9, 0.3 to 0.5 and 0.2 to 0.3 respectively.

The macrovascular volume change in Eq. 8 reflects the combined effect of hypovolemia and microvascular pooling on cardiac filling. It is a more effective index than hypovolemia or pooling alone in characterizing how hemodialysis produces intradialytic hypotension.

2.2. *Effect of Microvascular Pooling and Hypovolemia on Systemic Hematocrit*

Analyzed through the two-compartment model with 0.833 for F_{cell}, 0.5 for β and 0.2 for γ, the redistribution of RBC between the micro and macrocirculation for the following four cases are depicted in Fig. 2:

A. Control
B. An reduction in total blood volume by 5% $(= -\Delta V_b/V_b)$
C. A shift of blood volume from the macrocirculation to microcirculation by 15% $(= -\Delta V_{mic}/V_b)$
D. The sum of case B and C.

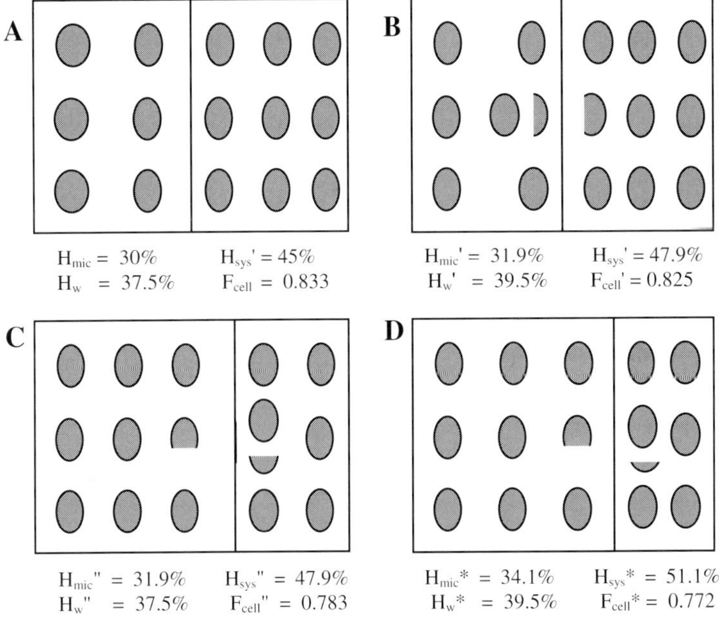

Figure 2. Four distributions of RBC in the micro- and macrocirculation compartments as they experience changes in blood volume. A: Control; B: With the total blood volume decreased by 5% through fluid extraction without change in microcirculation volume; C: No change in the total blood volume but with a dilation of microvascular blood volume by 15% of the total blood volume. As one notices that the systemic hematocrits H_{sys} for both B and C are the same. D: the sum of B and C. A decrease in total volume is illustrated as a decrease in the width of the compartments and a shift of blood volume from the micro- to macrocirculation is identified by shifting the midline in between the two compartments. All four conditions have 15 disks to show no change in total RBC volume.

If one were to use the one-compartment model [18] to calculate the systemic hematocrit from case A to B, the new hematocrit will be 47.4%, which is a slight underestimate of that of the two-compartment model. However the most striking difference between these two models is that the increases in systemic hematocrit observed from A to B and from A to C are identical for two-compartment model while no hematocrit change from A to C is predicted by the one-compartment model.

2.3. *Hypovolemia vs. Microvascular Pooling*

The implication contains in the similar hematocrit changes between A to B and A to C of Fig. 2 is central to this review article. The former reflects the changes due to hypovolemia in that the fluid loss at 5% of total blood volume becomes the volume reduction for the macrocirculation. On the other hand, the latter describes a three-time change in microvascular pooling (15% reduction in the macrocirculation blood volume) for the same change in systemic hematocrit.

One may also notice the changes in F_{cell} for these four cases. These changes reflect primarily the interrelations between the two compartments in terms of their individual hematocrit and volume change.

3. Causes for the Development of Hypotensive Symptoms

3.1. *Volumes Changes for Hemodialysis*

The hemodialysis experiments done by Schneditz et al. [15] have the necessary data for Lee to carry out the two-compartment analysis on the likely mechanism for the development of hypotension [10]. The experiment was done at an ultrafiltration rate higher than normal for 20 minutes. Over this period, it yields a value of 14% for $\Delta V_{ultra}/V_b$. The measured hematocrit and plasma protein concentration before and after are listed in row 2 and 3 of Table 1 and the analyzed results in the last five rows. The values presented in column A are for $F_{cell} = 0.8333$, $\beta = 0.5$ and $\gamma = 0.2$. The fluid volume restituted from the tissue and collected from the lymph of the patient combine with the dialysis or ultrafiltrated volume to produce a reduction in the total blood volume by 4.6%. However the calculated results indicate that the microcirculation is activated by the hemodialysis to dilate by 20.9% of the total blood volume. Its combination with the reduction in total blood volume, the estimated volume decrease for the macrocirculation comes to 25.5% of the total blood volume over this 20-minutes period, a value large enough to induce hypotension.

Table 1. Volume changes and redistribution in hemodialysis (experimental data from Schenditz et al. [15]

Measurements	t = 20 min		t = 40 min.	
Plasma Protein Concentration	68.9 g/l		75.1 g/l	
Hematocrit	27.9%		30.3%	
Volume changes and redistributions	A		B	C
Dialysis (UF) Volume $\Delta V_{ultra}/V_b$	14%		14%	14%
Fluid restitution from Tissue and Lymph Returns $(\Delta V_{trans} + \Delta V_{lym})/V_b$	9.4%		11.1%	8.9%
Change in Blood Volume $\Delta V_b/V_b$	-4.6%		-3.4%	-4.6%
Microvascular Pooling $\Delta V_{mic}/V_b$	20.9%		20.6%	12.5%
Change in Macrovascular Volume $\Delta V_{mac}/V_b$	-25.5%		-24.0%	-17.1%

Column A: $F_{cell} = 0.8333$, $\alpha = 2/3$, $\beta = 0.5$, $\gamma = 0.2$
B: $F_{cell} = 0.8333$, $\alpha = 2/3$, $\beta = 0.5$, $\gamma = 0.3$
C: $F_{cell} = 0.8333$, $\alpha = 4/9$, $\beta = 0.3$, $\gamma = 0.2$.
(Case C has a smaller microcirculation compartment than A and B. Case A, B and C all have the same F_{cell}, i.e. 0.833)

3.2. *Mechanisms Causing Intradialytic Hypotension*

The analytic results on hemodialysis and other evidences summarized next suggest that there are two mechanisms contributing to the development of hypotensive symptoms [10]

- Hypovolemia (low blood volume) as induced by the ultrafiltration process of hemodialysis or hemorrhage,
- Blood pooling to the abdominal organs as induced by the hemodialysis process or toxics in blood.

Hypovolemia and/or blood pooling lead to poor venous return and subsequently low cardiac filling. According to the Starling's Principle of the Heart, the heart pumps whatever it receives. A low cardiac filling thus, produces a low cardiac output, leading to poor blood flow to the brain. Subsequently, the patient experiences hypotensive symptoms.

Unfortunately, the lack of clinical methods to accurately assess changes in blood volume and blood pooling during hemodialysis make it impossible to quantify how each of these two mechanisms contributes to the development of intradialytic hypotension. The analytic result on hemodialysis provides one indirect support that blood pooling may be the dominant mechanism leading to hypotension. This hypothesis is further supported by the result of endotoxin shock and orthostatic hypotension that are examined next.

3.3. *Other Supporting Results on Blood Pooling*

Endotoxin can induce severe shock in dogs with a drop of arterial blood pressure to 34% of the control pressure [14]. However, the measurements of plasma volume by the dual indicator dilution technique indicate that the total blood volume might only be reduced by 3.6%. Using an equation similar to Eq. 7 but including the effect of sequestered RBC, Lee estimated that the microcirculation dilates by 14.2% (= ΔVmic/Vb,control) and the macrocirculation reduces its volume by -17.8% (= ΔVmac/Vb,control) of the control blood volume [9]. This large volume reduction may be the key factor to induce the observed shock. Chien et al. reported that the splanchnic circulation had markedly increase blood volume following endotoxin shock [1]. The results of Roth et al. and Chien et al. suggest that abdominal organs may be the site for hemodialysis induced blood pooling.

Yamamoto et al. showed that the use inflatable abdominal band to compress the abdominal organs could significantly reduce the reduction in systolic blood pressure (ΔSBP) as the patient stands up from supine position post-hemodialysis [19]. They interpreted that the band compression reduces the blood pooling to the abdominal organs so that the resulted improvement in cardiac filling produces higher cardiac output for the alleviation of orthostatic hypotension. They found that ΔSBP measured before hemodialysis is much lower than that after. This difference may be the result that more blood volume is pooled away from the central circulation to the abdominal organs during hemodialysis.

3.4. *Accuracy and Implications of One and Two-Compartment Model*

The microcirculation plays an important role in blood volume control. If blood is withdrawn from the circulation, what fraction of it comes from the microcirculation? By using a cyclic hemorrhage-reinfusion protocol, LaForte et al. [7] found that 60% of hemorrhaged blood volume comes from the microcirculation, 33% from the macrocirculation and 7% from the tissue via fluid restitution.

The one-compartment model and hematocrit monitoring [18] have been employed widely to assess the reduction in blood volume [2, 17]. The validity of this methodology will be examined here. When the one-compartment model is applied to the hematocrit measurements of hemodialysis study of Schneditz et al. [15], one finds that the relative volume reduction is 8.6%. Taking the volume distribution results of LaForte et al. into account, one estimates the reduction in macrocirculation volume is 2.8%. The application of similar analysis to the endotoxin data [14] yields a reduction in total blood volume as 5.9%, 33% of it is the macrovascular reduction of 1.9%. In contrast to the significant reduction in macrovascular volume predicted by the two-compartment model, one can only conclude that the use of one-compartment model would likely fail in predicting intradialytic hypotension and endotoxin shock.

The analytic results shown in column B and C of Table 1 are calculated to cover the likely ranges in the protein concentration of the transcapillary fluid movement and lymphatic returns and the range in microvascular fraction. Even though the volume changes are altered with these value changes, the conclusions of a significant reduction in macrovascular volume and a minimal change in total blood volume remain similar.

4. Countermeasures for Intradialytic Hypotension

4.1. *Current Maneuvers*

The standard maneuver employed to counter intradialytic hypotension is either to reduce the ultrafiltration rate or to infuse saline with the goal to minimize the impact of hypovolemia to blood pressure. The main critique is that their effectiveness is limited, especially if the pooling is the cause of hypotension. They also introduce some complications. For example, the saline infused to counter hypotension needs eventually to be ultrafiltrated out from the patient to meet the requirement set for the hemodialysis.

In the study done by Dheenan and Henrich [4], they measured the reduction in hypotensive episodes per treatment for a number of maneuvers. They found that sodium modeling, cooler temperature and higher sodium dialysate were effective in reducing hypotensive episodes. Whether the improvement came from the alleviation of hypovolemia or blood pooling was not examined in this study. In conjunction with the episode reduction, they also found that many patients could not tolerate the protocol of sodium modeling routinely. Excessive thirst occurred in some patients with high sodium dialysate. Cooler temperature dialysis was tolerated by most patients, and shivering and cramping occurred in some patients.

4.2. *Use of Anti-Pooling to Counter the Development of Hypotension*

Developed by NASA as a civilian use of anti-G suit, medical anti-shock trousers (MAST) have been used as a device to counter the development of low blood pressure in trauma patients. The increase in venous return in hemorrhagic shock by MAST was studied by HR Lee et al. [8] in dogs. Their data indicate that the blood volume being squeezed out by the MAST from the abdominal organs and legs (i.e. the auto-infusion effect) was found to induce an increase in the stroke volume, cardiac output and arterial blood pressure. Modified anti-G suits have been used daily for more than one year with beneficial results for patients with severe orthostatic hypotension [13].

More recently, inflatable abdominal band, a simplified version of MAST and anti-G suit, are introduced and shown to significantly reduce the precipitous decrease in systolic blood pressure as patients with orthostatic hypotension stand up [16, 19]. The band produces an auto-infusion from abdominal organs to enhance stroke volume and ejection fraction and subsequently the easement of orthostatic hypotension. Being easily worn by the patient, the use of the band for its auto-infusion and anti-pooling effect should be explored for use by hemodialysis patients.

5. Summary

The analyzed results on hemodialysis, endotoxin shock and post-dialytic orthostatic hypotension summarized here provide indirect support that the key mechanism for intradialytic hypotension is blood pooling to abdominal organs. Under this mechanism, it is appropriate to use anti-pooling device like abdominal band to counter the development of intradialytic hypotension. In order to meet the needs for the avoidance of intradialytic hypotension, we must go beyond the entrenched hypothesis of hypovolemia to pursue the following development efforts:

1. A micro-invasive monitoring technology that can identify the underpinning mechanisms leading to the development of intradialytic hypotension for a particular patient. One potential technique is to measure both the changes in hematocrit and plasma protein concentration and then use the two-compartment model to examine whether intradialytic hypotension results from hypovolemia or blood pooling. The macrovascular volume change so obtained can serve as a more effective index than blood volume reduction or blood pooling alone in characterizing how hemodialysis induces intradialytic hypotension.

2. A device and/or drug therapy that are specifically design to deal with the mechanism of blood pooling and/or hypovolemia.

3. The integration of these two into a countermeasure for personalized relief of the untoward effects of intradialytic hypotension.
4. An improved hemodialysis that does not induce intradialytic hypotension. The monitoring technology may allow researchers to identify the factors causing the abdominal microcirculation to dilate and regulating blood pooling to the microcirculation. The better understanding of these factors may lead to means to make the hemodialysis process to impose less blood pooling to abdominal organs.

Acknowledgement

The author would like to express his deep appreciation for the teaching, guidance and visions offered by Professor Y. C. Fung throughout his career development in biomedical engineering and dedicate this work to honor his 90th birthday. This research is supported by grants from National Heart, Lung and Blood Institute.

References

1. S. Chien, R.J. Dellenback, S. Usami, K. Treitel, C Chang and MI Gregersen, *Am J Physiol* **210**, 1411 (1966).
2. J.J. Dasselaar, R.M. Huisman, P.E. De Jong and C.F.M. Franssen *Hemodialysis Inter* 11, 448 (2007).
3. J.T. Daugirdas, *Kidney Int.* **39**, 233–246 (1995).
4. S. Dheenan and W.L. Henrich:, *Kidney Inter.* **59**, 1175(2001).
5. Y.C. Fung, *Biodynamics Circulation* (Springer-Verlag, NY. Chapter 5. Microcirculation, 1984) pp. 224.
6. W.L. Henrich, *Am J Kidney Dis* **52**, 209 (2008).
7. A.J. LaForte, L.P. Lee, G.F. Rich, T.C. Skalak and J.S. Lee, *Am J Physiol* **266**, H2268 (1994).
8. H.R. Lee, W.F. Blank, W.H. Massion, P. Downs and R.J. Wilder, *Am Emerg Med* **1**, 7 (1983).
9. J.S. Lee, *Am J Physiol* **267**, H1142 (1994).
10. J.S. Lee, *Ann Biomed Eng* **28**, 1 (2000).
11. H.H. Lipowsky and B.W. Zweifach, *Microvasc Res* **7**, 73 (1974).
12. B.F. Palmer and W.L. Henrich, *J Am Soc Nehphrol* **19**, 8 (2008).
13. G. Rosenhamer, C. Thorstrand, *Acta Med Scand* **193**, 277 (1973).
14. C.J. Rothe, R.H. Murray and T.D. Bennett, *Am J Physiol* **236**(33), H291 (1979).
15. D. Schneditz, J. Roob, M. Oswald, H. Pogglitsch, M. Moser, T. Kenner and U. Binswanger, *Kidney Int* **42**, 1425 (1992).
16. A.A. Smit, W. Wieling, J. Fujimura, J.C. Deng, T.L. Opfer-Gehrking, M. Akarriou, J.M. Karemaker and P.A. Low, *Clin Auton Res* **14**, 167 (2004).

17. R.R. Steuer, J.K. Leypoldt, A.K. Cheung, D.H. Harris, J.M. Conis. *ASAIO J* **40**, M691 (1994).
18. W. van Beaumont, J.E. Greenleaf, and L. Juhos, *J Appl Physiol* **33**, 55 (1972).
19. N. Yamamoto, E. Sasaki, K. Goda., K. Nagata, H. Tanaka, J. Terasaki, H. Yasuda, A. Imagawa and T. Hanafusa, *Kidney Inter* **70**, 1793 (2006).

Chapter 20

PRESSURE ULCER, PRESSURE AND FLOW MOTION

ZENGYONG LI

School of Mechanical Engineering, Shandong University
Jinan, 250061, P.R. China

ERIC W. C. TAM

Dept. of Health Technology and Informatics, The Hong Kong Polytechnic University
Hong Kong SAR, P.R. China

ARTHUR F. T. MAK

Dept. of Health Technology and Informatics, The Hong Kong Polytechnic University
Hong Kong SAR, P.R. China

This article summarized some of our recently published works on the effects of epidermal pressure on skin flowmotion and discussed the implications to the development of pressure ulcer. Studies were conducted in 5 normal subjects and 5 persons with spinal cord injury (SCI). A series of animal studies were conducted using an established rat model. In the human subject study, external pressure of 16.0kPa was applied to the ischial tuberosity for 30 minutes. In anaesthetized rats, external pressure of 13.3kPa was applied to the trochanter area for 6 hours/day for 4 consecutive days. The normalized amplitude of the metabolic component (0.01-0.02 Hz) and neurogenic component (0.02-0.06 Hz) for persons with SCI was found to be significantly lower during the resting (F=5.26, p=0.032) and post loading conditions (F=5.44, p=0.029) respectively as compared with normal individuals. In anaesthetized rats, prolonged tissue compression induced significant decrease in the normalized amplitude in the frequency interval of 0.01-0.05 Hz in the trochanter area (p<0.001). These findings suggest that the contributions of endothelial related metabolic and neurogenic activities to the blood perfusion regulation became relatively less for persons with SCI during the resting and post loading periods respectively. Also prolonged compression might induce endothelial damage and affect the endothelial related metabolic activities.

1. Introduction

Pressure ulcers are common complications of tissue damage developed in spinal cord injured (SCI) subjects and frail individuals with diminished pain sensation, diminished mobility and/or prolonged unrelieved pressure applied to body surface. Pressure ulcer (PU) causation is a multifactorial process associated with a number of extrinsic factors (pressure, shear, duration of loading, temperature, moisture, hygiene) [12] and intrinsic factors (general physical conditions, morphology of bone and tissues, muscle tissue mechanical properties) [5, 19]. Among these factors, prolonged compression of vascularized soft tissues is well accepted to be the most important mechanical cause for the onset of tissue breakdown at the skin, subcutaneous tissues and underlying striated muscles [6].

Although there are many factors contributing to the formation of pressure ulcers, a common pathway is associated with disturbed flowmotion within the affected tissues [23, 16]. Flowmotion is the rhythmic oscillations in the vascular network due to contraction and dilation of the local smooth muscles [4]. SCI patients are often suffered from disturbed flowmotion. Schubert *et al* [23] reported that there was a disturbed flowmotion in the sacrum in some SCI patients and elderly subjects during resting conditions. However, the specific origins of such disturbance have not been recognized and it is unclear which frequency band(s) manifest the effect of prolonged compression on the flowmotion.

Spectral analysis of the Laser Doppler flowmtery (LDF) signal deals with changes in the dynamics of blood flow and has been introduced as an approach for the evaluation of microvascular control mechanisms [4, 2]. However, the LDF signals consist of notably different features in both time and frequency. To capture these features, methods in the time-frequency domain are required. The method of wavelet analysis was applied to study those cardiovascular signals [3, 25]. Five characteristic frequencies of blood flow signals have been identified in the human cutaneous circulation as recorded by LDF signals [3, 25, 14]. These oscillations reflect the influence of heart beat, respiration, intrinsic myogenic activity of the vascular smooth muscle, neurogenic activity on the vessel wall and endothelial related metabolic activity with frequencies around 1, 0.3, 0.1, 0.04 Hz and 0.01Hz, respectively [2 , 3, 25, 24].

In anesthetized rats, characteristic frequencies in the range 0.01–5 Hz have also been demonstrated in skin blood flow [11, 17]. In rats, the dynamics of the cardiovascular-respiratory system possesses features similar to those observed in humans. However, cardiac and respiratory rhythms in rats were reported to be approximately three times faster than in humans in these studies (with frequency around 3 Hz corresponding to heart beats, and that around 1 Hz corresponding to respiration) [17]. This factor corresponds to the ratio of total blood volume to the cardiac output.

Since wavelet analysis of blood flow reveals various spectral peaks corresponding to specific origins, in the present study the peripheral vascular oscillations of tissue overlying the ischial tuberosity (IT) (high risk area for pressure ulcer) was assessed in normal and persons with SCI using spectral analysis based on wavelet transform. Also in order to recognize the specific origins of disturbance in flowmotion, the effect of prolonged compression on the flowmotion was assessed using a rat model. This study may help to shed new insight on the possible causes of tissue breakdown, especially due to prolonged unrelieved pressure.

2. Materials and Methods

2.1. *Subjects*

Ten male subjects were recruited from a local Medical Rehabilitation Centre to participate in this study. Of which, five were normal subjects (average age =31.2±3.3 y) and five were persons with spinal cord injury and complete paraplegia (average age=37.2 ±7.3 y). Three experimental trials were collected from each subject.

All subjects were asked for absence of the following conditions: 1) known case of vascular disease; 2) vasomotor drugs in use; 3) active pressure ulcer of any grade; 4) severe hypertonicity or remarkable contracture; 5) previous myocutaneous flap(s) or skin transplant surgery; and 6) heavy smoking or drinking. Heavy smoking was defined as the use of more than two cigarettes per day. No alcohol drinks was permitted before the experiment.

Before the experiment, basic parameters including the age, weight, height and brief medical history about the subject were taken and a consent form was signed. The experimental protocol was reviewed and approved by the Hong Kong Polytechnic University Human Ethic Committee.

2.2. *Animals*

Twenty one adult Sprague-Dawley rats weighing 400-500g were used to assess the effect of pressure on flowmotion. These rats were fed with a standard nutrient diet, with no supplement of any extra vitamins. Each rat was anesthetized by intraperitoneal delivery of xylazine (20mg/ml) and ketamine (100mg/ml). Additional small doses were administered at appropriate intervals during the study to maintain the proper level of anesthesia and the insensate state of the animal. The sufficiency of the level of anesthesia was determined by a mustache dithering test.

2.3. *Experimental procedure_1*

All subjects were lying with both hips flexed to 90°. A laser Doppler flow sensor was attached onto skin over ischial tuberosity (IT) by an adhesive ring. External pressure was applied to the skin by a pneumatic indentor (diameter =25 mm) aligned perpendicular to the loading site. A loading pressure of 16.0kPa (120mmHg) was chosen to simulate the interface pressure experienced by wheelchair users. The total duration of loading was 30 minutes.

Blood flow measurements were performed using a Laser Doppler flow monitor (Moor Instruments DTR4, software version 4.1, Axminster, UK) with a contact probe (DP1T/7-V2) with a power of 1.0 mW at a wavelength of 780 nm. The LDF signal was sampled at 40 Hz and the measurements were expressed in arbitrary units in the tissue sample volume (flux) for comparative study. Skin blood flow was monitored for 10 minutes prior to loading and two 10 minute consecutive intervals (T1&T2) after the release of loading.

2.4. *Experimental procedure_2*

The rats were randomly divided two groups: experimental group (EXP) (n=14) and control group (CNL) (n=7). In the experimental group, a precalibrated average pressure of 13.3kPa (100 mmHg) was applied to a risk area for pressures ulcers, the trochanter area [20] via specifically designed pneumatic indentor (diameter 14mm). The loading duration was 6 hours/day for 4 consecutive days. Such an in vivo rat model has been established to study the etiology of pressure ulcer [15]. The hair on the loading site was carefully shaved without damage to the skin at least on one day before the experiment. All animals were under anesthesia throughout the loading period. In the control group, no pressure was applied but the rats were under the same anesthesia protocols. LDF measurement was monitored for 20 mins prior to and after the prescribed compression period.

The experimental protocol was reviewed and approved by the Hong Kong Polytechnic University Animal Ethic Committee.

2.5. *Spectral analysis*

The idea of the continuous wavelet transform is to project a signal s on a family of zero-mean functions, the wavelets, deduced from an elementary function, called the mother wavelet $\Psi(u)$, by translations and dilations. Continuous wavelet transform of a signal s is defined as

$$\tilde{g}(s,t) = \int_{-\infty}^{+\infty} \Psi_{s,t}(u)g(u)du$$

$$(1)$$

where $\tilde{g}(s,t)$ is a wavelet coefficient and $\Psi_{s,t}(u)$ is a wavelet function and was defined as

$$\Psi_{s,t} = \frac{1}{\sqrt{s}} \Psi(\frac{u-t}{s})$$

(2)

where t is time, s is scale related to the frequency f as $f = f_0 / s$, and f_0 determines the current frequency resolution. By choosing $f_0 = 1$, we obtain the simple relation $f = 1/s$.

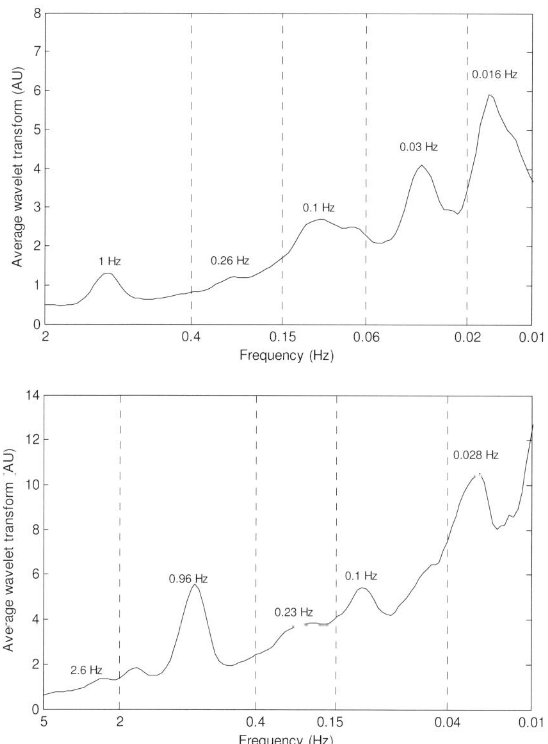

Fig.1. The wavelet transform of the laser Doppler perfusion signal in the resting skin for human (a) and rats (b), which were shown on a log scale. The vertical lines indicate the outer limits of each frequency interval from I to V, which correspond to endothelial related metabolic, neurogenic, myogenic, respiratory and cardiac activities respectively.

The continuous wavelet transform is a mapping of the function $g(u)$ onto the time-frequency plane. By adjusting the window used in wavelet transform, slower and faster events can be categorized accordingly [13]. In this study, Morlet wavelet was chosen for the wavelet transform analysis.

The wavelet transform was calculated in the frequency interval from 0.01 to 2 Hz for subjects and 0.01 to 5 Hz for rats. Typical wavelet transforms of LDF signal for subjects (a) and for rats (b), are presented in figure 1. The wavelet transform is calculated with a logarithmic resolution, and the frequency axes are therefore presented logarithmically. This is particularly appropriate for the estimation of the low frequency component. Periodic oscillations with five characteristic frequencies (centered around 0.015, 0.03, 0.1, 0.3 and 1Hz respectively) were observed for human (figure1a). The outer limits of each spectral interval were determined: (*I*) from 0.01 to 0.02 Hz, (*II*) from 0.02 to 0.06 Hz, (*III*) from 0.06 to 0.15 Hz, (*IV*) from 0.15 to 0.4 Hz, and (*V*) from 0.4 to 2 Hz.

The wavelet transform was calculated in the frequency interval from 0.01 to 5 Hz for rats. The upper limit was set to include the heart rate frequency, while the lower limit was chosen to include all three regulatory mechanisms of the blood flow, the myogenic, neurogenic and metabolic. Periodic oscillations with five different characteristic frequencies (around 2.6, 0.96, 0.23, 0.1, and 0.028 Hz respectively) were observed for rats (figure 1b). The average amplitude within each interval was used to characterize the spectral components [16, 17].

2.6. *Statistical analysis*

All values were expressed as means and standard deviations. T-test was used to test the significant difference in age and body mass index (BMI) between normal and persons with SCI. Two-way repeated measures ANOVA was used to study the significant difference in the normalized amplitude of blood oscillations for each frequency band between pre-loading and post loading periods in persons with SCI and normal subjects (SPSS version 11.5.1). To determine whether prolonged compression causes significant differences in each dependent variable for consecutive 4 days, one way ANOVA repeated measures was used. Post-hoc analyses within the groups were done with Bonferroni multiple comparison tests. A difference with $P<0.05$ was considered statistically significant.

3. **Results**

3.1. *Subject characteristics*

The body mass index (BMI) was 25.2±3.0kgm-2 and 22.5±5.8kgm-2 respectively in normal group and SCI group. No significant differences was found in age (p=0.13) and BMI (p=0.38) in the two groups.

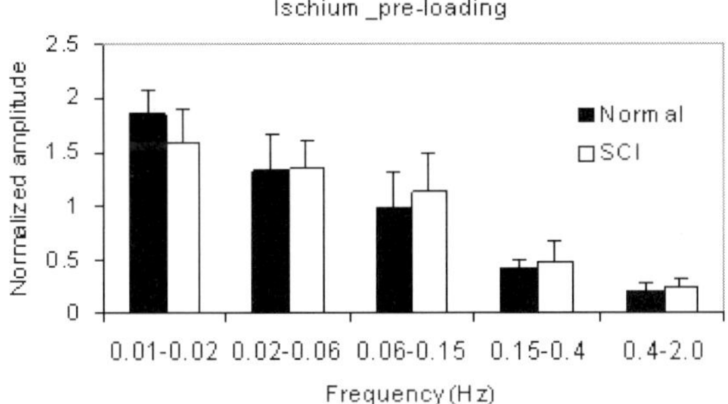

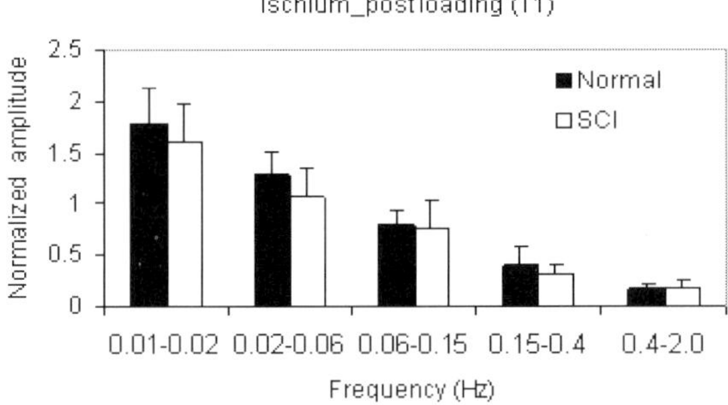

Fig.2. Comparison in normalized amplitude for the five frequency intervals for the resting skin over the ischial tuberosity between the normal subjects and persons with SCI during the pre-loading (a) and post loading periods (b). Significant differences are marked between persons with SCI and normal subjects with *p<0.05. [16]

Comparison of normalized oscillatory activities between the normal subjects and persons with SCI: The normalized amplitude of the metabolic component (I, 0.01-0.02) for the subjects with SCI was found to be significantly lower by 20% during the pre-loading period when compared to normal subjects (figure 2a) (F=5.26, p=0.03). Also the normalized amplitude in the interval II (0.02-0.06 Hz) was significantly lower by 27% for subjects with SCI during the post loading period (T1) when compared to normal subjects (figure 2b) (F=5.44, p=0.03).

Normalized spectral amplitudes within each frequency interval in rats: The results showed that prolonged compression induced significant decrease in the normalized amplitudes in the frequency interval *I* during the pre-occlusion period on day 4 compared to that on day 1 in the compressed tissue at the trochanter area (p=0.010 and p<0.001 respectively) (figure 3a). During the post-occlusive reactive hyperemia (PRH), the normalized amplitude in the interval *II* was significantly higher on day 3 (p=0.027) and day 4 (p=0.06) compared to that on day 1(figure 3b). However, the normalized amplitude in the interval *III* was significantly lower on day 4 (p=0.023). No significant difference in the normalized amplitude was found during the resting period in the four days in the control group (CNL).

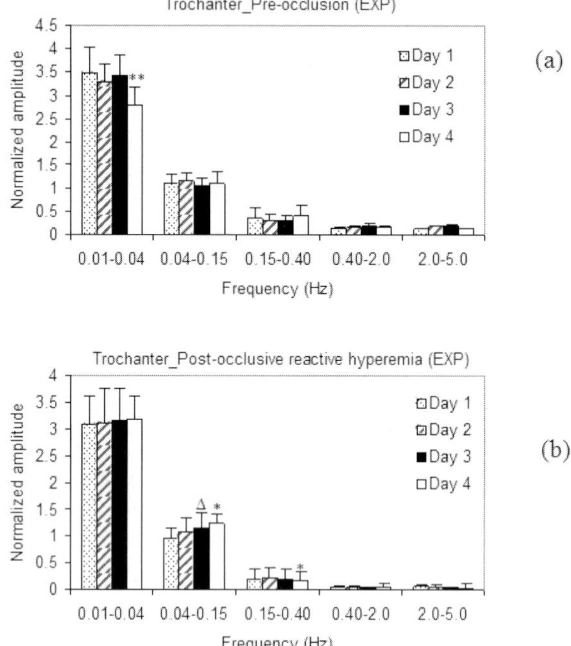

Fig.3. The normalized amplitude for the five frequency intervals for the resting skin at the trochanter area in rats for the experimental group (EXP) during the pre-occlusion period (a) and during the post-occlusive reactive hyperemia (PRH) period (b) in the four consecutive days. Significant differences are marked between Day 4 and Day 1 with *p<0.05 and **p<0.01, Day 3 and Day 1 with [Δ]p<0.05 and [ΔΔ]p<0.01. [17]

4. Discussion

The first difference in the oscillatory activities between persons with SCI and normal subjects was that the endothelial related relative metabolic activity (I, 0.01-0.02 Hz) was significantly lower during the pre-loading period for persons with SCI compared to that for the normal subjects. The most important role of such metabolic regulation is to adjust the blood flow to satisfy the need of cells for oxygen [11]. Such metabolic activity apparently originates from the endothelial layer [13]. Herrman *et al* [9] reported that the average spectral ower in the low-frequency range (0-1Hz) significantly decreased after ischemic stress in the rat. This present study confirms that prolonged compression would lead to a decrease in the flowmotion power in the compressed tissue. In addition, it is also noted that the decreased power was in the frequency interval *I* (0.01-0.05 Hz), which corresponds to the endothelial related metabolic activities [13, 11, 17]. Therefore the lower metabolic activity in the skin over the IT area might be attributed to prolonged unrelieved pressure for persons with SCI. It was also shown that prolonged compression tissue compression would induce a significant decrease in the hemoglobin oxygenation in the frequency interval 0.01-0.05 Hz [18]. This may result in a reduced oxygen delivery due to an increase in vascular resistance.

Another significant difference in the oscillatory activities between persons with SCI and normal subjects was also evident in the response of the oscillatory activities following the release of loading in the skin over the IT area. After pressure loading for 30 minutes, decreased normalized amplitudes in the frequency intervals II (0.02-0.06Hz) (corresponding to neurogenic activity) was found in persons with SCI during the post loading period (T1).

Reactive hyperemia is a consequence of vasodilation following occlusion of arteriole. Vasodilation mechanism involves the stimuli resulting from metabolic changes subjected to hypoxia and accumulation of metabolites [10, 21, 22], the release of mediators from the endothelium or from afferent nerves [1, 7, 8]. Lower relative amplitude of the neurogenic component (0.02-0.06 Hz) found in persons with SCI indicated that the loss of neurogenic control might be a main factor contributing to the disturbance in the post pressure regulation of the oscillatory flow. The neurogenic regulation is based on the global integration of the neural signals. It permits the body to react to its environment and to make adjustments based upon internal and external factors. It is clear from the present study that the ability of the tissue to provide the large reactive flowmotion response in the skin over the IT area was significantly impaired in persons with SCI. The hyperemic response following the release of occlusive pressure represents a protective adaptation to the damaging effects of pressure on the skin. Failure of this protective mechanism may lead to insufficient skin recovery and the tissue toxic following the ischemia.

In summary, the present study demonstrates that in the absence of neurogenic control for persons with SCI, the lower relative metabolic activity in the skin over the IT area at rest might be the result of prolonged unrelieved pressure applied to the body/skin and underlying tissues compared with the normal subjects. However, the reduced response of oscillatory activities following the release of occlusive pressure might be mainly attributed to the loss of neurogenic control. These findings suggest that the contributions of endothelial related metabolic and neurogenic activities to the blood perfusion regulation became relatively less for persons with SCI during the resting and post loading periods respectively. This may result in reduced oxygen delivery and insufficient skin recovery.

Acknowledgments

This project has been supported by the Central Research Grants of the Hong Kong Polytechnic University (G-YD72 and G-YX35).

References

1. C.H. Baker, E.T. Sutton. *Int J Microcirc: Clin Exp* **12,** 275 (1993).
2. S. Bertuglia, A. Colantuoni, M. Arnold, H. Witte. *Microvascular research* **52**(3), 235 (1996).
3. M. Bracic, A. Stefanovska. *Bull. Math. Biol.* **60**, 919 (1998).
4. A. Bollinger, U. Hoffman, U. K. Franzeck *Blood Vessels* **28,** 21 (1991).
5. C.V.C. Bouten, D.L. Bader. ASME Summer Bioengineering Conference, *Snowbird, Utah*, p161 (2001).
6. C.V.C. Bouten, R.G.M. Breuls, E.A.G. Peeters, C.WJ.Oomens, F.P.T. Baaijens *Biorheology* **40**,383 (2003).
7. B. Fromy, P. Abraham, J.L. Saumet, *Brain Res* **811,** 166 (1998).
8. B. Fromy, S. Merzeau, P. Abraham, J.L. Saumet, *British Journal of pharmacology* **131 (6)**, 1161 (2000).
9. E.C. Herrman, C.F. Knapp, J.C. Donofrio, R Salcido, *Journal of Rehabilitation Research and Development* **36**(2), 109 (1999).
10. S.M. Hilton, *J Physiol (London)* **149**, 93 (1959).
11. A. Humeau, A. Koitka A, P. Abraham, J.L. Saumet, J.P. L'Huillier, *Physics in Medicine and Biology* **49** (5), 843 (2004).
12. S.L. Knight, R.P. Taylor, A.A. Polliack, D.L. Bader, *Journal of Applied Physiology* **90**, 2231 (2001).
13. H.D. Kvernmo, A. Stefanovska, K.A. Kirkebøen, K. Kvernebo, *Microvasc Res* **57**, 298 (1999)
14. H.D. Kvernmo, A. Stefanovska, K.A. Kirkebøen, *European Journal of Applied Physiology* **90**(1-2), 16 (2003).
15. M.P.C. Kwan, E.W.C. Tam, S.C.L. Lo, M.C.P. Leung, R.Y.C. Lau, *Experimental Biology and Medicine* **232**, 481 (2007)
16. Z.Y. Li, E.W.C. Tam, J.Y.S. Leung, A.F.T. Mak, Arch. Phys. *Med. Rehabil.* **87,** 1207 (2006).

17. Z.Y. Li, E.W.C. Tam, M.C.P. Leung, A.F.T. Mak, S.C.L. Lo, Leung. *Phys. Med. Biol.* **51**(10), 2681 (2006).
18. Z.Y. Li, E.W.C. Tam, A.F.T. Mak, R.Y.C. Lau, *Phys. Med. Biol.* **51**(21), 5707 (2006).
19. E. Linder-Ganz, A. Gefen. *Journal of Applied Physiology* **96**(6), 2034 (2004).
20. R. Salcido, S.B. Fisher, J.C. Donofrio, M. Bieschke, C. Knapp, R.Z. Liang, E.K. Legrand, J.M. Carney, *J Rehabil Res Dev* **32**(2), 149 (1995).
21. S.S. Segal, B.R. Duling, *Science* **234,** 868 (1986).
22. S.S. Segal, B.R. Duling, *Circ Res* **59**,283 (1986).
23. V. Schubert, P.A. Schubert, G. Breit, M. Intaglietta. *Paraplegia* **33**(7), 387 (1995).
24. T. Soderstrom, A. Stefanovska, M. Veber, H. Svensson, *Am J Physiol Heart Circ Physiol* **284**, H1638 (2003).
25. A. Stefanovska, M. Bracic, H.D. Kvernmo, *IEEE Trans Biomed Eng* **46**, 1230 (1999).

Chapter 21

CORRELATION OF WHOLE BLOOD VISCOSITY WITH REAL-TIME MICROVASCULAR ABNORMALITIES IN TYPE-1 DIABETES MELLITUS (T1DM) PATIENTS

ANTHONY TZE-WAI CHEUNG
Department of Pathology and Laboratory Medicine
University of California at Davis, School of Medicine
Sacramento, CA 95817, U.S.A.

In a recently completed large cohort study, this laboratory has identified a significant presence of real-time microvascular abnormalities (vasculopathy) in type-1 diabetes mellitus (T1DM) patients [1, 2]. In addition, a significant correlation between the severity of the vasculopathy with elevated levels of endothelial dysfunction/inflammation biomarkers in the same patients has also been established [1, 2]. Though it is an accepted concept that endothelial dysfunction is normally caused by changes in intraluminal vessel wall shear stress [3-5], the exact pathogenetic mechanism has not been identified. Based on preliminary studies in this laboratory, a mechanism for the pathogenesis of endothelial dysfunction (which could eventually lead to microvascular changes in T1DM) was proposed. We hypothesized that abnormalities in whole blood viscosity (WBV), an important hemorheologic parameter involved in changes in intraluminal vessel wall shear stress, would be significantly elevated in T1DM and should correlate with real-time *in vivo* microvascular changes in T1DM patients. To test this hypothesis, we have randomly measured WBV in T1DM patients (n=10) to correlate with real-time microvascular abnormalities in the same patients. Using the Rheolog™, WBV in T1DM patients (n=10) and healthy non-diabetic control subjects (n=10) was measured at a shear rate of $300s^{-1}$. The presence of microvascular abnormalities in the same patients and control subjects was analyzed using the real-time videotape sequences on the conjunctival microcirculation obtained non-invasively via intravital microscopy. A Severity Index (SI), scale 1-15, was computed to reflect degree/severity of vasculopathy in the analysis. T1DM patients compared to control subjects had significantly higher WBV (4.0±0.3cp vs 3.3±0.1cp; P<0.05) and SI (7.1±1.8 vs 0.6±0.7; P<0.05). The results strongly support the independent work from the laboratories of Chien and Garcia-Cardena/Grimbone that the biomechanical forces of blood flow – in particular intra-luminal vessel wall shear stress – play a critical role in regulating endothelial cell function and causing endothelial dysfunction and related microvascular complications [3-5]. It is likely that the biomechanical forces of blood flow play a regulatory role in the homeostasis of the endothelial lining. Therefore, the abnormalities in hemorheologic parameters and real-time microvascular abnormalities (vasculopathic characteristics) identified in the microcirculation in T1DM patients may have cascading effects and thereby contribute to the further worsening of endothelial dysfunction, inflammation and microvascular injury.

                                   A. T.-W. Cheung

1. Introduction

Type-1 diabetes mellitus (T1DM) is associated with an increased risk of vascular complications, which accounts for much of its morbidity and mortality. In unrelated studies in the literature and related preliminary studies from this group, evidence has been found to correlate severity of microvascular abnormalities (vasculopathy) with endothelial dysfunction via increased monocyte-mediated inflammation and upregulation of adhesion molecules [1, 2, 6]. Since it is an accepted hemorheologic concept that endothelial dysfunction induced by a change in intraluminal vessel wall shear stress – whole blood viscosity (WBV), vessel diameter and/or vessel flow velocity – is mechanistically responsible for the pathogenesis of vasculopathy, a study on the effect of changes in shear stress and its correlation with real-time microvascular abnormalities in T1DM is warranted.

The following equations provide a relationship of the key variables (hemorheologic parameters) which contribute to changes in shear stress:

Wall Shear Rate (WSR) = 8Vm/D

$$\text{Shear Stress} = \text{WSR} \times \eta \tag{1}$$

where V_m is mean velocity, D is vessel diameter, and η is WBV. Findings of significant associations between perturbations in hemorheologic parameters and coronary artery disease, cerebrovascular disease, hypertension, type-1 and type-2 diabetes mellitus, and metabolic syndrome further suggest that altered hemorheology may play an important role in the development of vascular diseases [7, 8].

Despite the paucity of information on real-time vasculopathy in T1DM, investigations on the nailfold capillary bed and retinal microcirculation have been previously reported, e.g., [9-13]. This study represents a further extension of such real-time investigations to include the conjunctival microcirculation and to quantify microvascular abnormalities which were significantly present in T1DM patients. In fact, this report describes the results of a hemorheologic-microvascular study in which a significant change in shear stress has been directly linked to the existence of real-time microvascular abnormalities in a specific vascular disease.

2. Research Subjects and Methods

2.1. *Research subjects / patients and controls*

T1DM patients (n=87) and healthy non-diabetic control subjects (n=41) were recruited for a large cohort study in an NIH-sponsored Road Map (bench top to bedside) research project. T1DM patients were defined as subjects with onset of diabetes at <20 years of age and on insulin therapy since diagnosis. Detailed patient inclusion and exclusion criteria have been previously reported and the results of this large cohort multi-parameter study have been published [1, 2]. All subjects to be studied were given the Patient's Bill of Rights and informed consent was obtained from all participants. All experimental subject recruitment and experimental protocols were approved by the IRB for human protection at UC Davis Medical Center and were in accordance with the Declaration of Helsinski human use guidelines.

In this hemorheologic-microvascular study, 10 T1DM patients and 10 healthy non-diabetic control subjects were randomly selected from the cohort study patient (n=87) and control (n=41) groups.

2.2. *Study design — Flow chart of the study*

The research was conducted in the pathology research laboratories at the University of California, Davis Medical Center. *On the morning of each patient or control study,* fasting anti-coagulated blood (~5 ml) was obtained via venipuncture to be measured for WBV.   After venipuncture, each subject (T1DM or control) was non-invasively studied using computer-assisted intravital microscopy (CAIM) to quantify the severity of real-time microvascular abnormalities for correlation with WBV.

2.3. *Rheolog™ (rheometer): Measurement of whole blood viscosity (WBV) [14]*

Each venous blood sample was measured for WBV at a shear rate of $300s^{-1}$ using a computerized rheometer (Rheolog™; Rheologics, Inc., Exton, PA; www.rheologics.com). WBV values generated at a shear rate of $300s^{-1}$ were obtained for comparison between T1DM patients and healthy non-diabetic control subjects. The details of this procedure and the Rheolog™ system have been previously reported [14] and are briefly described below.

The Rheolog™ consists of two components: a disposable U-tube unit and a height detection system. The disposable U-tube unit has two riser columns (3.25

mm diameter), which are connected via a 10-cm glass capillary tube (0.8 mm diameter). Each capillary tube is calibrated using a specially built air-calibration system. Venipuncture blood can be shunted directly into the tube or anti-coagulated (K_2EDTA) blood can be transferred from a vacutainer to the tube via an air-syringe compression pump and a 21-gauge butterfly needle. A four-way stopcock controlled by a built-in computer is used to introduce blood from the vacutainer to the U-tube unit and allows blood to move from one riser column to the other. The temperature of the riser columns and the capillary tube unit is maintained at body temperature. Measurement of height variations in the riser columns commences after the temperature of the system has stabilized at 37°C. The height detection system uses a charge coupled device - light emitting diode (CCD-LED) arrangement. Since the transmission of light from the LED to the CCD is blocked by the blood, the changes in the height of blood in the riser column can be detected by the CCD sensor. Automated height measurements are normally accomplished in three minutes. Following the procedural guideline from Rheologics, Inc. (Exton, PA), each venous sample was fed to the U-tube unit for measurement. The analog signal on height displacement from each venous blood sample was digitized and stored as a function of time in an on-line laptop computer, and used for computation of WBV in this study [14]. Data read-out and print-out are produced within five minutes after completion of measurement. Though a WBV profile is generated by the system, only WBV values sheared at a rate of $300s^{-1}$ were used in this study to correlate with microvascular abnormalities and SI.

2.4. *Computer-assisted intravital microscopy (CAIM): Quantification of microvascular abnormalities [2, 14]*

Each subject was asked to sit down and relax for 10 minutes, and was cautioned not to touch or rub the eye (typically the left) before the study. If the subject had any eye irritation, two drops of non-medicated saline were applied. Excess saline was lightly blotted off at the corner of the eye. The subject was then asked to place his/her chin on a chin rest (similar to those used in ophthalmologic examinations). The horizontal height of the CAIM system was then adjusted to align with the perilimbal region of the bulbar conjunctiva. The location of the front lens element was adjusted to provide the flattest surface for focusing on the bulbar conjunctival surface. A CCD video camera (COHU Model CCD-6415-3000) was used for on-screen visualization and videotaping of the microcirculation, using a half-inch video-recorder (Panasonic ProLine Model AG3200). A fiber-optics light source (Fiber-Lite Model 3100) with a Kodak #58 Wratten (anti-red) filter was focused on the perilimbal vessels of the left bulbar conjunctiva for epi-illumination. Once in focus, the conjunctival vessels appear as sharp black tubes or lines against a light grey or white background on-screen. At least five different fields in the perilimbal region were non-invasively videotaped for each subject for a total of at least five minutes.

All video sequences were studied in their entirety to identify individual vasculopathy parameters. Selected video sequences were coded and analyzed independently by two investigators. Rare discrepancies were adjudicated by consensus between these two investigators. Well-resolved video frames were selected from each video sequence and were captured for detailed analysis. Several vessels of interest were selected for quantitative analysis using customized imaging software VASCAN and VASVEL, and public domain software SCION. A Severity Index (SI) has been developed to quantify severity of vasculopathy in the conjunctival microcirculation. Its computation is based on the presence of 15 microvascular abnormalities commonly found in vascular diseases. These abnormalities include abnormal vessel diameter, abnormal arteriole to venule ratio (A/V ratio), abnormal vessel morphology, vessel (flow) sludging, box-car flow pattern (trickled flow), microaneurysm(s) or micropool(s), abnormal flow velocity, beaded vessel(s), abnormal vessel distribution, distended vessel(s), damaged vessel(s), tortuous vessel(s), ischemia or lack of vessels, comma sign(s), and hemosiderin deposits. The SI ranges from a score of "0" (no abnormality present) to "15" (all abnormalities present) and is an arithmetic summation of the presence of all abnormalities found in each T1DM patient or control subject. This technique has been validated in this laboratory by three co-investigators and has an intra-assay variation coefficient of <5%.

2.5. *Statistical analysis*

Patient characteristics are reported as the mean + standard deviation. ANOVA was used to compare T1DM and control SI. Spearman rank correlation analysis was used to analyze the correlation between WBV, microvascular abnormalities and SI in T1DM patients. Statistical significance was set at P-value ≤0.05.

3. Results

3.1. *Whole Blood Viscosity (WBV)*

WBV values generated at a shear rate of 300 s^{-1} were significantly higher in T1DM when compared with non-diabetic control subjects (4.0±0.3cP versus 3.3±0.1cP respectively; P<0.05).

3.2. *Microvascular abnormalities and Severity Index (SI)*

Figures 1-3 show the microvasculature in the bulbar conjunctiva (conjunctival microcirculation) under real-time CAIM. The optical design of the CAIM system utilized macro-optics as an operating platform – sharp focusing of the visual field was achieved via movement of the optical elements toward-or-away from the surface of the bulbar conjunctiva and not by the utilization of a focusing ring. This process assured the same magnification in all imaging sessions and all images can be directly compared without concern for magnification variability (Optical magnification = 4.5x; on-screen magnification = 125x).

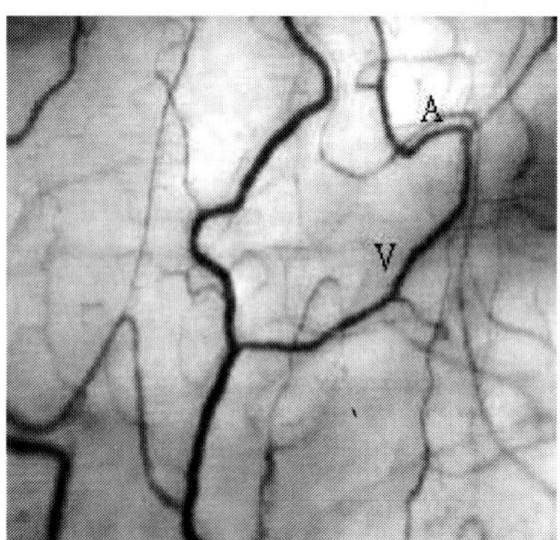

Figure 1. Captured image of the conjunctival microcirculation in a non-diabetic control subject [2, 14]. Optical magnification = 4.5X; onscreen magnification = 125X. "A" indicates arteriole; "V" indicates venule. This image illustrates a typical view of the conjunctival microcirculation in a healthy non-diabetic control subject to serve as a baseline reference for vasculopathy. Note the even and orderly distribution of normal-sized arterioles and venules and the normal presence of capillaries.

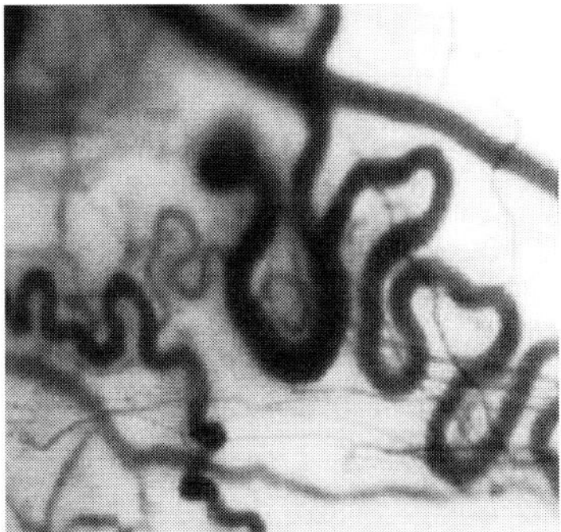

Figure 2. Captured image of the conjunctival microcirculation in a T1DM subject [2]. Optical magnification = 4.5X; onscreen magnification = 125X. This image shows substantial microvascular changes (abnormalities) compared with the non-diabetic control image in Figure 1. Note the presence of vessels with abnormally wide vessel diameter, abnormal vessel morphometry, extensive vessel tortuosity, and presence of a microaneurysm which has ruptured (resulting in the accumulation of hemosiderin particulate deposits). In addition, note the extremely uneven and disorderly distribution of arterioles, venules and arterioles which differ significantly from the non-diabetic control image in Figure 1. This image is consistent with images documented in T1DM patients from previous studies which have been peer-reviewed and published [15, 16].

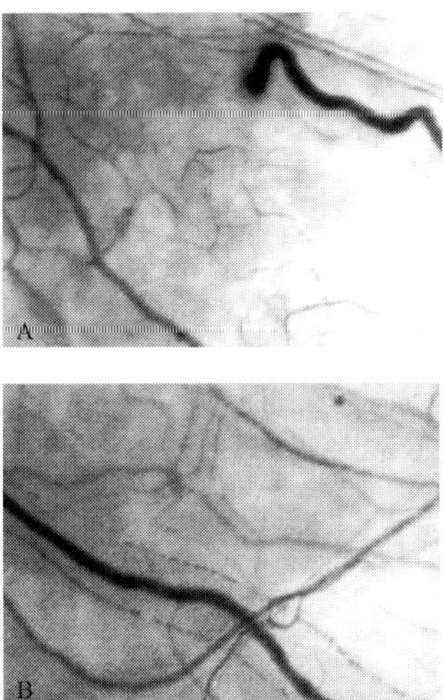

Figure 3A-3B. Captured images of the conjunctival microcirculation in two additional T1DM subjects [2]. Optical magnification = 4.5X; onscreen magnification = 125X. These images demonstrate the microvascular abnormalities also commonly found in T1DM subjects. The even and orderly distribution of normal-sized arterioles, venules and capillaries does not exist. (A) Note the presence of a damaged vessel which also displays an abnormally wide diameter. (B) Note the abnormally wide vessel diameter, vessel sludging and the unique box-car flow (trickled flow) pattern.

The conjunctival microcirculation in healthy non-diabetic control subjects presented a well-organized and evenly distributed network of capillaries, arterioles and venules. The averaged arteriole to venule (A:V) ratio in all control subjects (n=10) was approximately 1:2 and the arterioles and venules were evenly distributed throughout the bulbar conjunctiva. The overall morphologic features of the vessels were normal and dilations/distensions, beaded vessel configuration, microaneurysms (micropools), wide vessel diameter, and vessel tortuosity were rarely present. In addition, evidence of vessel injury or damage (broken microvessels and/or hemosiderin deposits) was rare and not observed in control subjects in this study. Vessel diameter ranged from 10-25 µm for arterioles and 30-70 µm for venules. Blood flow velocity ranged from 0.5 – 2.0 mm/s and vessel sludging (flow stagnation for venules) as well as the box-car flow pattern (trickled flow) were rarely present. Overall, microvascular abnormalities were infrequently observed in the healthy non-diabetic control subjects. Figure 1 shows a typical image of the conjunctival microcirculation in a healthy non-diabetic control subject with no history of vascular disease.

The mean SI (computed as the arithmetic sum of the simultaneous presence of any of the 15 microvascular abnormalities which exist in vascular diseases) for the non-diabetic control subjects (n=10) was 0.6±0.7 and was comparable with results from previous studies in this laboratory [2, 14]. With an SI of "0" for a patient or control subject with no microvascular abnormalities and "15" for the presence of all 15 possible abnormalities, an SI of 0.6 ±0.7 was indicative of a paucity of microvascular abnormalities and would serve as an excellent baseline value for comparison with T1DM SI.

The conjunctival microcirculation in T1DM patients (n=10) differed uniquely from healthy non-diabetic control subjects (See Figures 2 and 3 and compare with Figure 1). The mean SI of T1DM patients (7.1±1.8) differed significantly (P<0.05) from healthy non-diabetic control subjects (0.6±0.7). Figures 2 and 3 show typical images of the microvascular abnormalities commonly found in T1DM patients. Figure 4 shows the prevalent appearance of some microvascular abnormalities (vessel sludging, box-car flow pattern, flow velocity and abnormal vessel diameter) in T1DM patients. Figure 5 shows a direct correlation of WBV with SI in T1DM patients. In addition, Spearman rank correlation analysis revealed significant correlations (P<0.05) between WBV, SI and specific prevalent microvascular abnormalities in T1DM patients (P<0.05).

Microvascular Abnormalities	T1DM Patients									
	A	B	C	D	E	F	G	H	I	J
Vessel Sludging	X	X	X	X	X	X	X	X	X	X
Box-car Flow Phenomenon	X	X	X	X		X	X	X	X	X
Damaged Vessel(s)		X	X		X		X		X	X
Abnormal A:V Ratio	X	X	X	X	X			X		X
Vessel Tortuosity		X	X	X		X	X	X		
Hemosiderin Deposits			X	X						
Abnormal Vessel Distribution		X				X	X		X	X
Abnormal Flow Velocity	X	X	X	X	X	X		X	X	X
Abnormal Vessel Morphometry		X			X	X			X	
Distended Vessel(s)					X					
Abnormal Vessel Diameter	X	X	X	X	X	X	X	X	X	
Microaneurysm(s)					X	X				
Beaded Vessel(s)						X				
Ischemia										
Comma sign(s)										
Severity Index (SI)	5	9	8	7	8	9	6	6	7	6

Figure 4. Presence of microvascular abnormalities in T1DM patients. The presence of real-time microvascular abnormalities in the conjunctival microcirculation in each T1DM patient is tabulated for comparison between patients. The presence of all microvascular abnormalities in each patient is arithmetically computed to give a severity index (SI), which can serve to quantify the severity of vasculopathy. Note the prevalent appearance of vessel sludging, box-car flow pattern, abnormal flow velocity and abnormal vessel diameter in T1DM patients.

T1DM Patients	A	B	C	D	E	F	G	H	I	J
Severity Index (SI)	5	9	8	7	8	9	6	6	7	6
Whole Blood Viscosity (WBV)	3.5	4.3	4.0	4.2	4.3	4.3	3.7	3.8	3.9	3.8

Figure 5. Direct correlation of Whole blood viscosity (WBV) with Severity Index (SI) in T1DM patients.

4. Discussion

Microvascular disease plays a critical role in much of the morbidity and mortality associated with T1DM [6]. To effectively treat, manage and ideally prevent the development and progression of microvascular complications in T1DM, a better understanding of the pathogenesis of microvascular disease in T1DM is warranted. Real-time *in vivo* studies of the microcirculation in T1DM have rarely been conducted. This laboratory has described real-time *in vivo* microvascular abnormalities in T1DM – a study on juvenile patients and a post-pancreas-kidney transplantation efficacy study – in two previous reports [15, 16]. In addition, this laboratory has also reported increased SI in T1DM patients and has also established the association between endothelial dysfunction and real-time microvascular abnormalities in T1DM patients [1, 2]. However, the key shear stress parameter and the specific mechanism causing the endothelial dysfunction have not been identified until this study.

The concept that changes in intraluminal vessel wall shear stress could lead to endothelial dysfunction in vascular diseases is not new. This study not only supports the concept but, in addition, also identifies WBV as the key shear stress parameter which is significantly associated with the pathogenesis of microvascular abnormalities in T1DM. This study has also generated *in vivo* data on real-time systemic microvascular abnormalities in T1DM to correlate with shear stress changes. The possibility that abnormalities in WBV might serve as a mechanistic marker of vascular injury has been previously suggested by independent *in vivo* studies in the laboratories of Chien and Garcia-Cardena/Grimbone [3-5]. The results from this study strongly support their suggestion that intraluminal vessel wall shear stress play a critical role in regulating endothelial cell function — causing endothelial dysfunction and related microvascular complications in vascular diseases.

The severity of microvascular abnormalities which exist in T1DM patients has been computed and presented as an arithmetic SI for comparison with non-diabetic controls. However, a careful analysis of the existence of individual abnormalities shows a prevalence of individual abnormalities in T1DM patients (see Figure 4). All 10 T1DM patients show a prevalence of vessel sludging and nine of 10 patients have box-car flow pattern (trickled blood flow), abnormal vessel flow velocity and abnormal vessel diameter. In the same context in identifying WBV as a mechanistic marker for changes in shear stress and endothelial injury, one can now additionally identify abnormal vessel flow dynamics (vessel sludging, box-car flow pattern and abnormal vessel velocity) and morphometry (abnormal vessel diameter) as microvascular markers for vasculopathy in T1DM. It is interesting to note that vessel sludging is indicative of compromised blood flow arising from up-regulation of adhesion molecules in endothelial dysfunction. In addition, box-car (tricked) flow, abnormal blood flow velocity and wide vessel diameter are component parameters which can induce changes in shear stress. It is highly likely that the biomechanical forces of blood flow play a critical role in the hemostatic regulation of the intraluminal endothelial lining. Therefore, the changes in hemorheologic parameters in addition to WBV (i.e., changes in blood flow velocity and vessel diameter) may have cascading effects and thereby contribute to the further worsening of endothelial dysfunction, inflammation and microvascular injury in T1DM.

In this laboratory, the conjunctival microcirculation has been used as a research platform to study a variety of vascular diseases. It is a very efficient platform because of its easy and non-invasive accessibility, excellent quality of image display, absence of magnification variability, and the reliability of relocating specific vessels for longitudinal assessment. For example, each individual vessel in a T1DM subject can serve as its own reference control in a longitudinal medication efficacy study. The dynamic (flow velocity pattern) and morphometric (diameter) characteristics of the conjunctival microcirculation reflect the characteristics of a true microvascular network (e.g. normal presence and distribution of capillaries, arterioles, and venules) and are comparable with those of the microcirculation at the susceptible soft tissue end organ level in vascular diseases. Furthermore, the conjunctival microcirculation has a direct anatomic connection to the systemic circulation, as it is derived from the ophthalmic artery which originates from the internal carotid artery at a point adjacent to origin of vessels in the Circle of Willis. Consequently, the conjunctival microcirculation offers an ideal non-invasive site to study systemic real-time microvascular complications in cardiovascular and cerebrovascular diseases.

Acknowledgement

Dr. Yuan-Cheng Fung is a good friend, role model, teacher, and mentor to me. It was an honor and privilege that I was invited to make an oral presentation in the international symposium in honor of Dr. Fung's 90th birthday. Dr. Fung and my mentor, Dr. Theodore Yiu-Tsou Wu, have been close friends for over six decades. It was with pride and delight that I could represent Dr. Wu's laboratory to make the presentation as a tribute to Dr. Fung's birthday. This report is written as a brief summary of my oral presentation in the **2008 Yuan-Cheng Fung Symposium** to be included as a chapter in the book, entitled ***"Tributes to***

Yuan-Cheng Fung on His 90th Birthday". Dr. Fung is a pivotal figure who is instrumental in my research planning and professional development. I was introduced into the field of biomechanics by Dr. Fung over 30 years ago when I started to get involved with bioengineering and applied mechanics research at his urging. Dr. Fung's influence on my research career and professional development is respectfully acknowledged and greatly appreciated.

The subject materials presented in the lecture and part of the content of this report represent years of research in my laboratory. Portions of this report have been previously presented in conferences and published. These previous appearances have been appropriately referenced in the text [2, 14].

The work reported in this manuscript was partially funded by a National Institutes of Health (NIH) grant (R21DK69801) and a National Center for Research Resources (NCRR) grant (UL1 RR024146). The contributions of my collaborators (Drs. Sridevi Devaraj, Ishwarlal Jialal, Eric Miguelino, Peter Chen, and Meighan Smith), fellows, residents, students, and staff are very much appreciated.

I wish Dr. Fung a very happy 90th birthday with many happy returns. In addition, I look forward to, very much, paying tribute to him again in the celebration for his 100th birthday.

References

1. S. Devaraj, A.T.W. Cheung, I. Jialal, S. Griffen, D. Nguyen, N. Glaser and T.T. Aoki TT, *Diabetes* **56**, 2790-2796 (2007).
2. A.T.W. Cheung, M.M. Tomic (Smith), P.C.Y. Chen, E. Miguelino, C.S. Li, and S. Devaraj, *Clin Hemorheo Microcirc* **in print** (2009).
3. S. Chien, *Am J Physiol Heart Circ Physiol* **292**, H1209-H1224 (2007).
4. G. Garcia-Cardena, J. Comander, K.R. Anderson, B.R. Blackman and M.A. Grimbone Jr., *Proc Natl Acad Sci USA* **98**, 4478-4485 (2001).
5. G. Garcia-Cardena and M.A. Grimbone Jr., *Handb Exp Pharmacol* **176**, 79-95 (2006).
6. P. Libby, D.W. Nathan, K. Abraham, J.D. Brunzell, J.E. Fradkin, S.M. Haffner, W. Hsueh, M. Rewers, B.T. Roberts, P.J. Savage, S. Skarlatos, M. Wassef and C. Rabadan-Diehl, *Circ* **111**, 3489-3493 (2005).
7. B.K. Lee, A. Durairaj, and A. Mehra, *Clin Hemorheo Microcirc* **39**, 43-51 (2008).
8. I. Velcheva, N. Antonova, and E. Titianova, *Clin Hemorheo Microcirc* **39**, 391-396 (2008).
9. J. Koscielny, R. Latza, S. Wolf, H. Kiesewetter and F. Jung, *Clin Hemorheo Microcirc* **19**, 139-150 (1998).
10. B.R. Zaugg-Vesti, U.K. Franzeck, C. von Ziegler, J. Furrer, G. Pfister, A. Yanar and A. Bollinger, *Int J Microcirc Clin Exp* **15**, 193-198, (1995).

11. O. Arend, S. Wolf, F. Jung, B. Bertram, H. Postgens, H. Toonen and M. Reim, *Br J Ophthalmol* **75**, 514-518 (1991).
12. B. Bertram, S. Wolf, K. Schulte, F. Jung, H. Kiesewetter, F.C. Sitzmann and M. Reim, *Graefes Arch Clin Exp Ophthalmol* **229** 336-340 (1991).
13. A.T.W Cheung, K.L Cox, C.E Ahlfors and W.I. Bry WI, *Transpl Proc* **25**, 1310-1313 (1993).
14. M.M Smith, P.C.Y. Chen, C.S. Li, S. Ramanujam and A.T.W. Cheung, *Clin Hemorheo Microcirc* **41**, 229-239 (2009).
15. A.T.W. Cheung, A.R. Price, P.L. Duong, S. Ramanujam, J. Gut, E.C. Larkin, P.C.Y. Chen and D.M. Wilson, *Microvasc Res* **63**, 252-258 (2002).
16. A.T.W. Cheung, R.V. Perez and P.C.Y. Chen, Transpl **68**, 927-932 (1999).

Disclosure Statement

Rheologics, Inc. provided the Rheolog™ rheometer, supplies and technical support for this study.

Chapter 22

Y. C. FUNG AND BIOMECHANICS:
FROM ORGANS-SYSTEMS TO MOLECULES-GENES

SHU CHIEN

Departments of Bioengineering and Medicine, and Institute of Engineering in Medicine
University of California at San Diego, La Jolla, CA 92093-0412, U.S.A.

Professor Y.C. Fung has made tremendous impacts on humanity through his research and its applications, by setting the highest standards of rigor, via the work of his many students and their students, and due to his exemplary leadership. He has applied his profound knowledge and elegant analytical methods to the study of biomedical problems with rigor and excellence. He established the foundations of biomechanics in a variety of living tissues, including the lung, the heart, blood vessels, blood cells, ureter, intestine, skin, as well as other organs and tissues. Through his vision of the power of "making models" to explain and predict biological phenomena, Dr. Fung opened up new horizons for bioengineering, from organs-systems to molecules-genes, and has provided the cornerstones of research activities in many institutions in the United States and the whole world. He has made superb contributions to education in bioengineering, as well as service to the professional organizations and translation to industry and clinical medicine. He is widely recognized as the Father of Biomechanics and the leading Bioengineer in the world. His extraordinary accomplishments and commands in science, engineering and the arts make him a Renaissance Man whom the world is very fortunate to have.

Dr. Fung is the Father of Biomechanics. He created the field by merging his outstanding expertise in engineering mechanics with his innovative work in biomedical sciences and their applications. His contributions are so immense that it is not possible to cover even a small portion in a short article. This article is based on the presentation that I had the great pleasure and privilege to give at the "International Symposium on Genomic Biomechanics: Frontier of the 21st Century" in celebration of the 90th birthday (Chinese Calendar) of Professor Yuan-Cheng "Bert" Fung held in San Diego, CA, on September 14, 2008. Most of Dr. Fung's publications have been collected in a two-volume treatise entitled *"Selected Works on Biomechanics and Aeroelasticity"* [1], which provides a very valuable reference resource for his superb scientific contributions.

                                    S. Chien

Dr. Fung in China

Dr. Fung was born on September 15, 1919, in Yuhong, Wutzin, Kiangsu, China. He received his college education at the Central University of China, which is one of the top universities in China that moved from the capitol in Nanjing to the temporary capitol of Chongqing following the outbreak of the Sino-Japanese war in 1937. Dr. Fung received his B.S. and M.S. in Aeronautics from the Central University in 1941 and 1943, respectively, graduating at the top of his classes. While he was studying for his M.S. degree, Dr. Fung also taught as an Instructor at the Ta-Kung Technical School in Shapingpa, Chongqing. From 1943 to 1945, Dr. Fung worked as a Research Fellow at the Bureau of Aeronautical Research in Chengdu, before coming to the United States to study at California Institute of Technology (CalTech). Although Dr. Fung has been in the United States for over sixty years and is a most patriotic American, he always feels a strong tie to his motherland and has devoted tremendous amounts of time and energy to advance research and education in China. He played a major role in the birth of biomechanics and biomedical engineering in China by training the leaders, giving lectures, and organizing meetings (such as the China-Japan-US Conference on Biomechanics that he initiated) there.

Dr. Fung at CalTech

Dr. Fung came to the United States in 1945 to pursue his Ph.D. study in CalTech on a highly competitive scholarship that he won. Because of miscommunications during the wartime (World War II), however, he learned upon his arrival at CalTech in January 1946 after a long boat trip via India that the scholarship he had earned was no longer available. However, he was able to get support from his advisor, Dr. Ernst Sechler. In less than three years, Dr. Fung completed his Ph.D. study (Summa Cum Laude) in Aeronautics and Mathematics in 1948. After graduation, he stayed in the Department of Aeronautics at CalTech as a Research Fellow, advanced through the ranks of Assistant and Associate Professors, and became a Full Professor in 1959. Dr. Fung's research in Aeronautics focused on the dynamics of airplanes in turbulent weather, and on safety, performance and design of aircraft and spaceship. Because of his outstanding accomplishments, he was extremely well respected in the field of aeronautical and aerospace engineering. He wrote a definitive textbook on *Aeroelasticity* [2].

Beginning in the late 1950s, Dr. Fung became interested in the mechanics of the eye because his mother was suffering from glaucoma. He became increasingly convinced that understanding of the human functions could be improved by analyzing how forces, motion and stress and strain in the human body. In 1966, he published a landmark book on Biomechanics [3]. He advanced the viewpoint that if scientists can determine the structure and mechanical properties of a living organ, then by the principles of physics, they can predict the functions of that organ.

Coming to UCSD

In 1966, Dr. Fung visited Dr. P. F. Scholander at the UCSD Scripps Institution of Oceanography, together with Dr. Benjamin Zweifach, who was on a sabbatical leave at CalTech from New York University. Dr. Sol Penner, then Chair of the Department of AMES (Aeronautical Mechanics and Engineering Science), wanted to initiate a program of Bioengineering at UCSD. At his invitation, Drs. Fung and Zweifach came to UCSD and started an outstanding Bioengineering Program, with biomechanics and microcirculation as the central research themes. In the 1980s, Dr. Fung coined the term "tissue engineering" as a new direction for bioengineering. This not only has developed into a new integrative research theme at UCSD Bioengineering, but also has become the focus of bioengineering efforts in many programs in the world. These two disciplines - biomechanics and tissue engineering - have been major thrusts in bioengineering in the last two decades and will continue to grow for years to come.

In his studies on biomechanics and mechanobiology, Dr. Fung integrates biology and mechanics at different scales, from organs to genomics. He correlates structure and function in terms of remodeling, geometry, and growth, with emphases on temporal and spatial features and active vs. passive responses. He combines experiments (with innovative technology and rigorous execution) and theory (with creative formulation and elegant analysis), and makes effective iteration and feedback between them. Dr. Fung established the foundation of biomechanics in a variety of organs and tissues, including the lungs, heart, blood vessels, skin, as well as other organs and tissues. His groundbreaking studies on tissue remodeling under stress provide the foundation of tissue engineering.

Residual Stress in Biomechanics

Dr. Fung has numerous accomplishments in biomechanics. One of his very fundamental and innovative contributions is his research on residual stress, which is the stress that remains after the removal of external forces. He developed the ingenious approach by determining the residual stress from the opening angle after cutting open a biological structure in two different

                                          S. Chien

directions. An example is shown in Fig. 1 for an aorta that have been removed and hence without pressure load; a ring is first obtained by cross cuts, and this ring is then cut open in the radial direction to result in a zero stress state [4]. The opening angle provides a measure of the residual stress in the vessel wall due to the geometry and wall constituents. It is truly amazing that such an important parameter in biomechanics can be measured with such a simple experimental procedure, requiring only a pair of scissors. Fig. 2 shows that the ileal artery and ileal vein have striking different residual stress as shown by their opening angles [5]. This opening angle measurement can also be applied to the heart ventricles [6], trachea [7], and other organs and tissues.

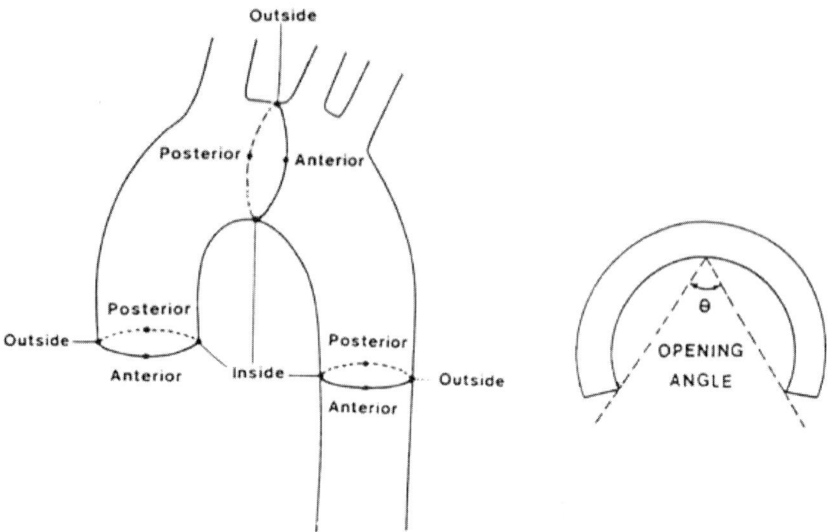

Figure 1. Left: Sketch of an aorta with an indication of the cutting positions. Right: Schematic cross-section of a cut vessel segment at zero-stress, defining the opening angle θ. From [4].

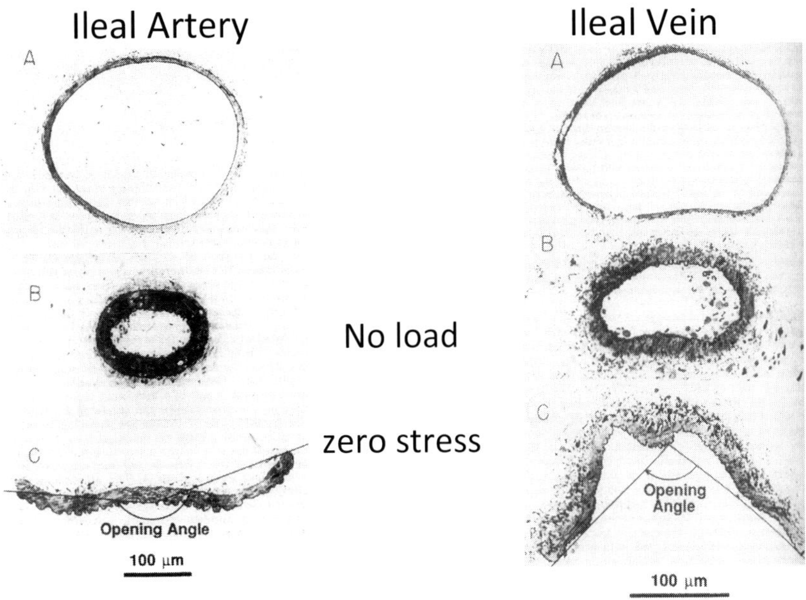

Figure 2. Differences in residual stress of an ileal artery (left) and an ileal vein (right) as reflected by their opening angles [5].

With a combination of experimental and theoretical approaches, Dr. Fung has determined the 3-D stress distribution in an artery [8]. He has also determined the stress-strain relations of collagen and elastin [9,10], which are the interstitial constituents of the vessel wall (Fig. 3), thus elucidating the relationship of the biomechanical behavior of a blood vessel to that of its molecular constituents and establishing an important step in multi-scale and integrative bioengineering.

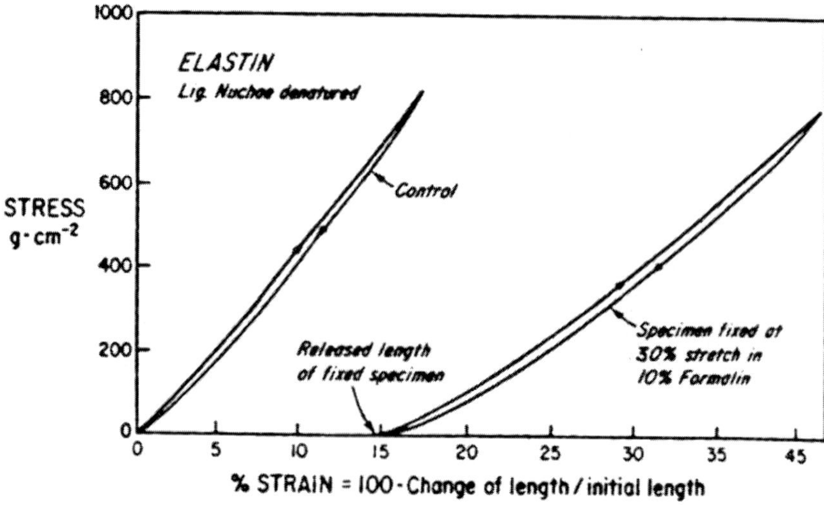

Figure 3. Stress-strain relationship of elastin from a ligament under control state and after fixation at 10% stretch in 10% formalin [9].

Development of Instrumentation for Biomechanics

While Dr. Fung is able to solve complex bioengineering problem with simple procedures, he is also extremely capable in designing sophisticated instrumentation to meet the experimental needs. Examples are the development of the innovative device "Biodyne" [11] for testing the mechanical properties of a variety of biological materials (Fig. 4), and the instrumentation to study the biomechanics of the skin [12], peristaltic transport [13], peeling force in a biological graft [14], as well as other ingenious instruments designed in cooperation with the late Gene Mead and other colleagues. These instruments allowed the performance of innovative experiments on the biomechanical properties of the skin and the adhesive strength of a biological graft; these findings have important bearing on the translational studies on the development of skin graft for clinical application in burns and other disorders. The instrument for the investigation of peristaltic transport led to the elucidation of the biomechanics of this process in the ureter [13], intestine [15] and other organs.

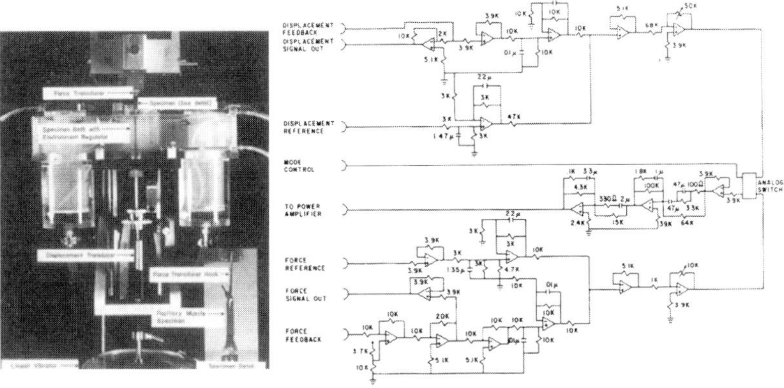

Figure 4. A Device for Testing of Mechanical Properties of Biological Materials: The "Biodyne" [11].

Biomechanics of Blood Cells

Dr. Fung has used elegant engineering analysis to investigate the mechanics and geometry of human red blood cells (RBCs), including stress analysis in RBC swelling [16], determination of RBC 3D geometry by interference microscopy (Fig. 5) [17], and the use of extreme theorem to analyze the extreme value statistics of RBC geometry [18]. At the time of these important studies conducted by Dr. Fung, my main area of interest was the mechanism of RBC deformability in relation to blood rheology, and I benefited immensely from his groundbreaking studies.

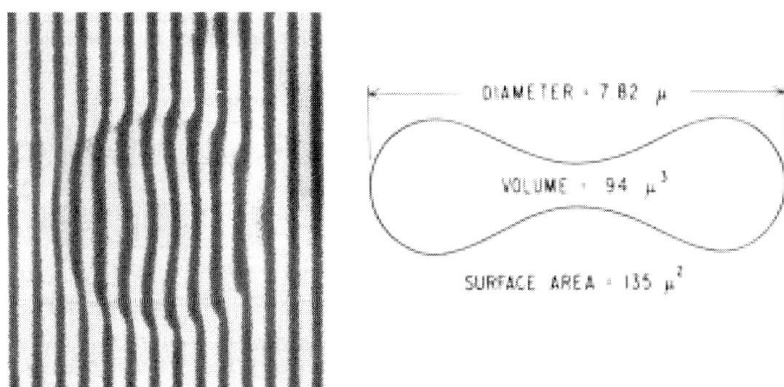

Figure 5. Determination of RBC 3D geometry by interference microscopy. Left: Holographical picture of a human RBC. Right: the geometric dimensions of a biconcave discoid human RBC determined by using the interference microscope [17].

                                           S. Chien

   Dr. Fung extended his studies on blood cells to their behavior in blood vessels, especially at the branch points. He elucidated the mechanism of stochastic nature of flow in capillaries by experimental studies and engineering analysis, showing that a small difference in flow between two daughter branches at a bifurcation can have a dramatic nonlinear effect on blood cell distribution [19] (Fig. 6). He used a lab model system to demonstrate the important role of RBCs in modulating the interaction of white blood cells with the vascular endothelium [20]. These findings showed how high RBC concentration can push the WBCs toward the vessel wall and enhance WBC-endothelial interactions. These findings have important implications in the role of hemodynamics in WBC behavior in the microcirculation in health and disease, including inflammation. Dr. Fung has computed the blood flow pattern in vessels beyond a local constriction [21]. The complex flow behavior in such regions, including flow reversal, reattachment and stagnation (Fig. 7), has been shown to have major effects on endothelial structure and function in the poststenotic region as well as branch points, and may thus play a significant role in atherogenesis [22].

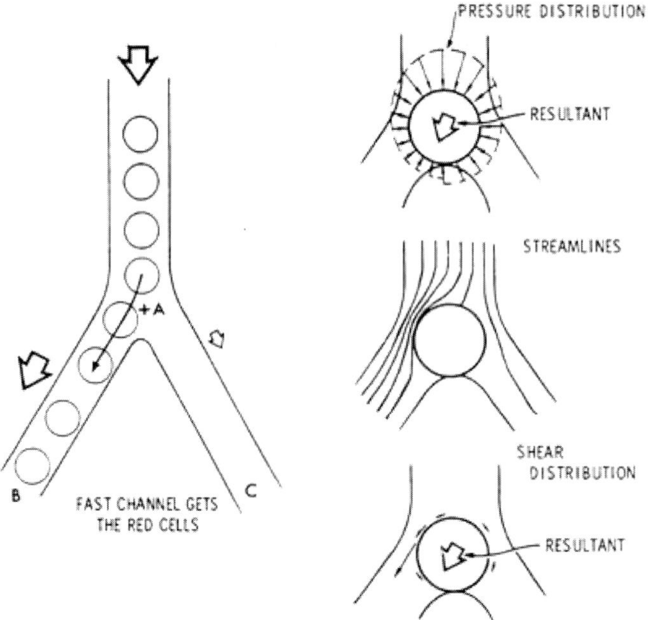

Figure 6. Stochastic nature of flow in a capillary branch point. The distributions of pressure and shear determine the streamlines, which cause the preferential entry of the blood cells into the branch with high flow, leading to unequal distribution of cells in the daughter branches [20].

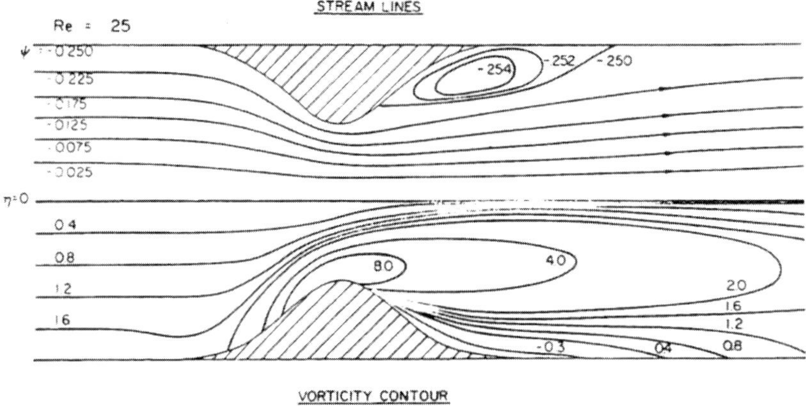

Figure 7. Computation of streamlines in flow beyond a narrow constriction [21]].

Biomechanics of Pulmonary Microcirculation: Coupling of Circulation and Respiration

Dr. Fung has made outstanding contributions on the joining of circulatory and respiratory systems at the microvascular level. In his pioneering work with the late Dr. Sidney Sobin and other colleagues [23, 24], the innovative concept of sheet flow through the pulmonary capillaries between the posts spanning sheets of alveolar wall (Fig. 8) was formulated with rigorous experiments and elegant analysis. The sheet-flow theory has been applied to derive the alveolar sheet thickness and hence the alveolar sheet flow as a function of the capillary-alveolar pressure differential (Fig. 9) [25]. Application of this theory to experimental data provides a quantitative understanding of the interplay among a large number of factors, including alveolar blood flow, blood volume, regional differences, and transit time distributions, thus allowing the representation of the effects on flow of arterial, alveolar and venous pressures, alveolar area, mean a-v path length, alveolar membrane tension [26]. Their findings opened up a new paradigm for research on the pulmonary microcirculation in health and disease. Dr. Fung applied morphometric technique to study the topographical arrangement of arterioles and venules in the cat lung [27], and generated novel findings on the distribution pattern of these two types of pulmonary microvessels (Fig. 10). The results indicate that, on the average, each terminal precapillary arteriole supplies 24.5 alveoli, and that each terminal postcapillary venules drains 17.8 alveoli.

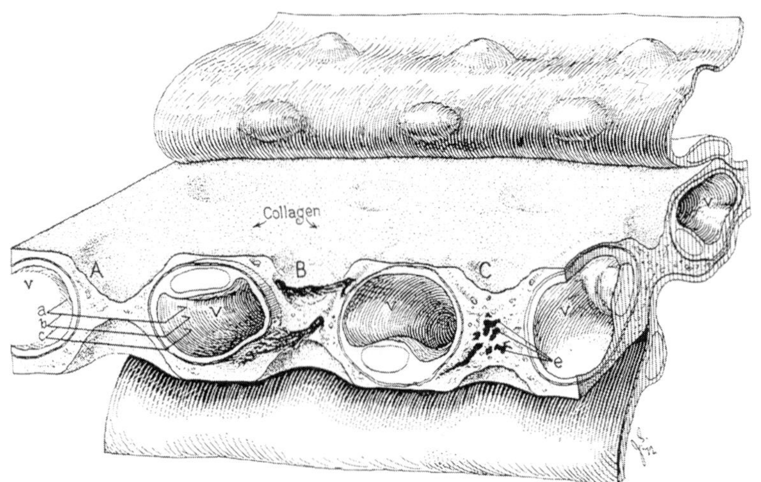

Figure 8. Schematic drawing of a pulmonary alveolar microvascular sheet [24].

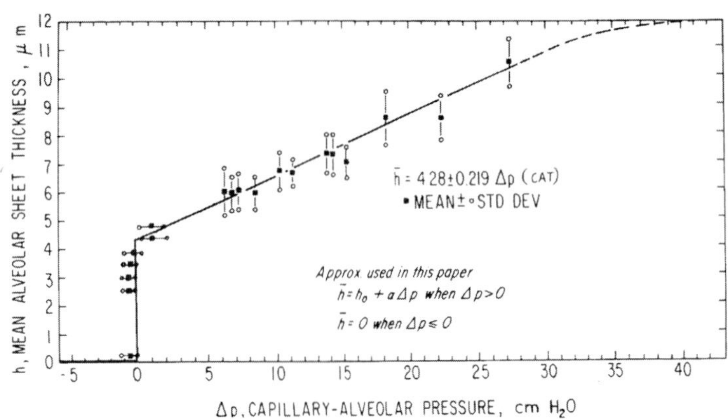

Figure 9. Mean alveolar sheet thickness as a function of the capillary – alveolar pressure difference [24].

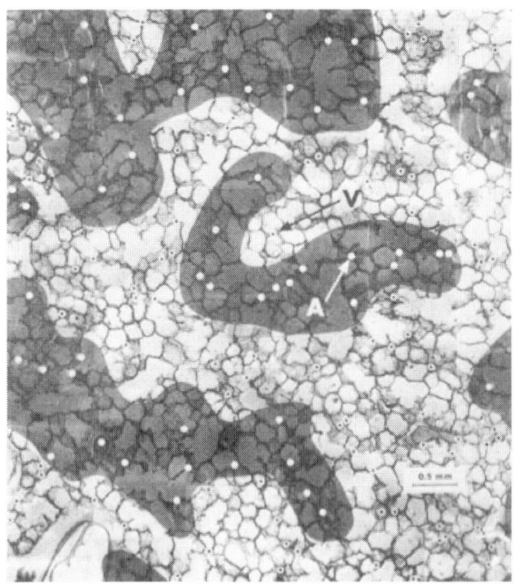

Figure 10. A montage of photomicrographs of a cat lung. Branches of pulmonary blood vessels which are not capillaries and have diameters less than 100 μm are individually identified. Arteries are marked with white dots; veins by black dots. A darker transparency covers the arterial regions [27].

Cardiovascular Remodeling

Dr. Fung applied mophormetric analysis to determine the geometric features of the coronary artery circulation in the right ventricle in the pig under normal condition and in hypertensive state [28]. These results have generated important information on coronary artery remodeling in cardiac hypertrophy.

As a result of his pioneering work on residual stress and remodeling, Dr. Fung formulated the concept that the remodeling of a blood vessel is best described by changes of its zero-stress state and that growth results from changes in cellular and extracellular mass and configuration (Fig. 11) [29]. These concepts have led to his proposal of specific ways to test the biomechanical properties of a tissue-engineered vascular graft in order for it to match the hemodynamic conditions it encounters in vivo.

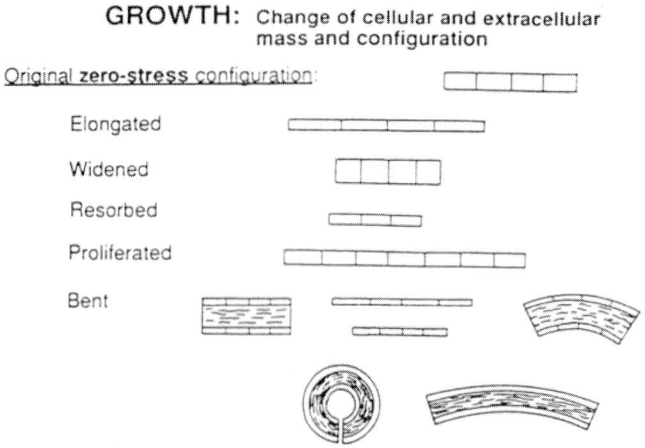

Figure 11. Diagrammatic illustration that the remodeling of a blood vessel is best described by change of its zero-stress state, which results from changes in cellular and extracellular mass and configurations, as depicted in these drawings [29].

Genetic Basis of Pulmonary Artery Remodeling in Hypertension

Dr. Fung has made a systematic study on the time course of the remodeling of the pulmonary artery of normal and hypertensive rats in response to hypoxia (Fig. 12) [30]. With the advent of molecular and genomic approaches to study biological problems, he studied the changes in gene expression under such hypoxia-induced pulmonary hypertension by using the DNA microarray technology [31]. He was able to match the time course of changes in the expression of a variety of genes with the alterations of specific structure and functions in the pulmonary artery, as exemplified in Fig. 13. This important study provides a genetic correlate of functional abnormalities in pulmonary hypertension. Based on these and other findings, Dr. Fung proposed that "DNA Replication and transcription involve molecular motion through a viscous medium to reach the appropriate site, and the lengthening takes place through the balance between the chemical energy of binding and the kinetic energy of molecular motion. Therefore, these processes involve accelerations, strains and strain rates with directions. They need to be treated as force vectors and stress tensor."

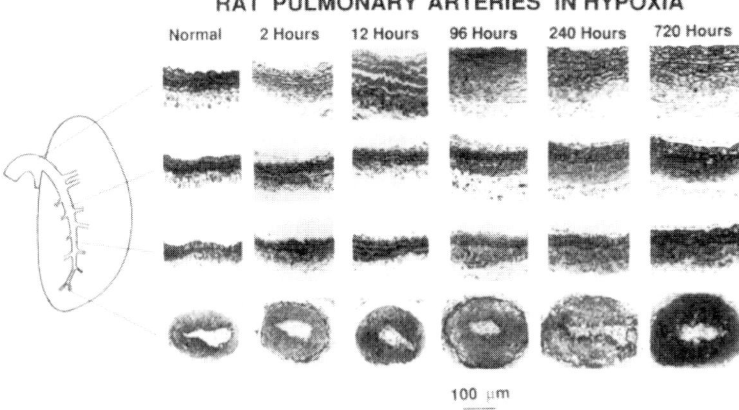

Figure 12. Photographs of histological slides from four regions of the main pulmonary artery of a normal rat and hypertensive rats with different periods of hypoxia from 2 to 720 hours [30].

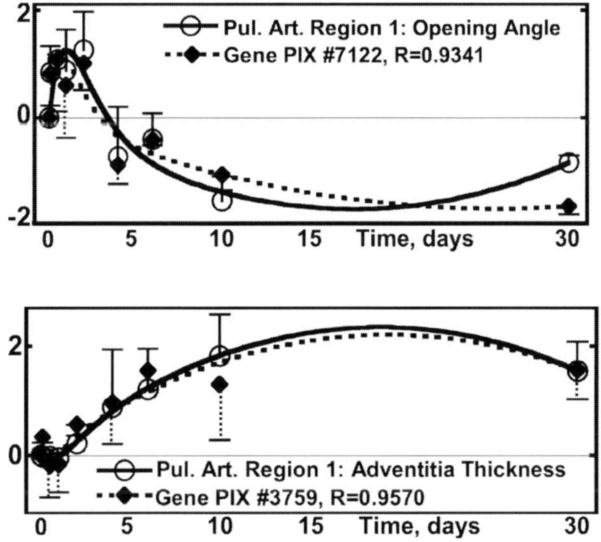

Figure 13. Matching of gene activities, as determined by microarray analysis, and physiological functions and structure of the pulmonary artery obtained after different periods of hypoxia. Note the high levels of correlation between the opening angle of region 1 of pulmonary artery with gene Pix #7122 and between the adventitial thickness of region 1 of pulmonary artery with gene Pix #3759. Similar high levels of correlation were obtained for other functional/structural parameters and other genes [31].

Translation of Biomechanics to Industry and Clinical Medicine

Dr. Fung's superb research accomplishments have major impacts in opening new areas and laying the foundation for bioengineering. As mentioned above, his research also has important translational application to industry and clinical medicine. Dr. Fung was a consultant in many industrial firms, especially when he was in the field of Aeronautics. His research has formed the foundation of industrial applications in a variety of bioengineering fields, including the tissue engineering of cardiovascular, urinary, musculoskeletal, and cutaneous systems. His studies on the biomechanics of the skin contributed importantly to the development of skin substitutes to treat burn patients. He has contributed to the development of tissue-engineered vascular graft with mechanical matching to the native vessels. His research contributes to the advancement of diagnosis and treatment of a variety of important diseases, including pulmonary disorders such as pulmonary hypertension and emphysema, cardiovascular diseases such as myocardial infarction, heart failure, atherosclerosis, and systematic hypertension, dysmotility of the digestive and urinary systems, musculoskeletal disorders, burns, and many others.

Contributions to Education

Dr. Fung has made marvelous contributions to bioengineering education as well as research. He built the superb graduate and undergraduate programs at UCSD, which have educated graduates who have made important contributions to bioengineering endeavors in academic bioengineering, medical schools, hospitals and industry. A large number of his Ph.D. students and postdoctoral fellows hold key faculty positions in universities in the United States and abroad, and many are heads of departments of bioengineering. In the most recent National Academy of Sciences - National Research Council survey of graduate education published in 1995, UCSD Bioengineering was ranked number 1 in graduate education. This is largely attributable to Professor Fung's vision and leadership.

Dr. Fung's educational influence extends beyond his teaching in the classrooms and laboratories at UCSD. He defined the pedagogy of biomechanics by writing several authoritative books that are widely used as textbooks in this country and abroad and are read by virtually everyone in the field. These books have been translated into many foreign languages and updated in several editions, setting the standard for all textbooks in bioengineering. His great contributions to education and training in China were mentioned in the early part of this article.

Services to Professional Organizations

Dr. Fung has devoted himself to the advancement of Bioengineering by serving in many key positions in professional organizations. He was President of the American Academy of Mechanics and Biomedical Engineering Society (BMES), Vice President of the International Society of Biorheology, and Chairman of the Third International Congress of Biorheology, the Second World Congress of Microcirculation, and the First China-Japan-USA Conference on Biomechanics. He was also Chairman of the Applied Mechanics Division in the American Society of Mechanical Engineers (ASME), U.S. National Committee for Biomechanics (now Honorary Chairman), and World Council for Biomechanics (now Honorary Chair). Dr. Fung has been a member of Microcirculatory Society Council, American Physiological Society Council of Circulation, American Heart Association Council on Basic Science, and World Congress of Biomechanics Steering Committee.

Awards and Honors

Dr. Fung's outstanding contributions to research and education in bioengineering are broadly recognized. He is a member of all three U.S. National Academies (National Academy of Sciences, National Academy of Engineering, and Institute of Medicine), as well as a Foreign Member of the Chinese Academy of Sciences and a member of Academia Sinica (Taiwan).

Dr. Fung has received all the major awards in his fields of endeavor, including the Eugene Landis Award from Microcirculatory Society (1975), the Theodore von Kármán Medal from American Society of Civil Engineers (1976), Lissner Award for Bioengineering from ASME (1978), Centennial Medal from ASME (1981), Worcester Reed Warner Medal from ASME (1984), Poiseuille Medal from International Society of Biorheology (1986), Excellence in Research Award from UCSD (1987), ALZA Award from BMES (1989), Timonshenko Medal from ASME (1991), Borelli Award from American Society of Biomechanics (1992), the Lifetime Achievement Award from the Association of Chinese Scientists and Engineers of California (1992), Distinguished Alumnus Award from California Institute of Technology (1994), Melville Medal from ASME (1994), Bioengineering Award from the Japan Society of Mechanical Engineering (1995). In the past decade, Dr. Fung has received three of the most prestigious awards: In 1998, he received the Founders Award, which is the highest honor given annually by the NAE to recognize "outstanding engineering accomplishments by an engineer over a long period of time and of benefit to the people of the United States". In 2001, he received from President

Clinton the United States National Medal of Science (Fig. 14), which is given to individuals "deserving of special recognition by reason of their outstanding contributions to knowledge in the physical, biological, mathematical or engineering sciences"; he is the first bioengineer to receive this prestigious honor. In 2007, he received the Russ Prize, which is awarded by the NAE "biennially to a researcher whose achievements are of critical importance in advancing science and engineering, ultimately improving the human condition".

Figure14. Dr. and Mrs. Fung with President Bill Clinton at the National Medal of Science ceremony.

Dr. Fung is a Fellow of the American Academy of Mechanics, American Institute of Aeronautics and Astronautics, ASME, American Institute of Medical and Biological Engineering, and Cardiovascular Section of the American Physiological Society. Dr. Fung is the recipient of Honorary Doctoral degrees and Honorary Professorships from many universities in China (including Hong Kong) and U.S.A. He has been invited to give many named lectures and plenary lectures, and he is an Honorary Member of several scientific organizations.

In honor of Dr. Fung, the ASME established the "Y.C. Fung Young Investigator Award" in 1986, the Chinese Association of Biorheology and Chinese Society of Biophysics, with funding from the International Society of Biorheology and the International Society of Clinical Hematology, established the Fung-Chien Young Investigators Award, and UCSD established the Endowed Chair "Y.C. Fung Professor of Bioengineering" (Fig. 15).

Figure 15. Professor and Mrs. Y.C. Fung and Shu and K.C. Chien at the ceremony celebrating the Y.C. Fung Chair Professorship at UCSD on September 15, 2006.

Dr. Y.C. Fung as a Renaissance Man

Dr. Fung has a wonderful marriage with his lovely wife Luna (Figs. 16 and 17) and they love their children Conrad and Brenda and their families. He enjoys being with long-time friends (Fig. 18) and with young people. Dr. Fung is not only a superb scientist and engineer, he is also a wonderful artist. He has excellent commands in calligraphy and poetry, and he has great talents in making Chinese chops (or seals, usually for people's names). Thus, Dr. Fung excels in Science, Engineering, and Art. He is a renaissance man, a peer of Leonardo DaVinci (Fig. 19).

Figure 16. Professor Y.C. Fung dances with Mrs. Fung.

Figure 17. Professor Y.C. Fung singing a poem he wrote and Mrs. Fung was beating the rhythm on a wooden drumlet at Shu Chien's 70th Birthday in 2001.

Figure 18. A good laugh with friends: Seated (from left): Mrs. Shirley Yih, Mrs. Luna Fung, Mrs. Ching-hua Wu, Ms. Hsien-Tzai Yu, and Mrs. Zhe-Ming Zheng. Standing (from left):, Drs. Zhe-Ming Zheng, Chia-Shun Yih, Theodore Wu, and Y.C. Fung.

Yuan-Cheng Fung

A Renaissance Man

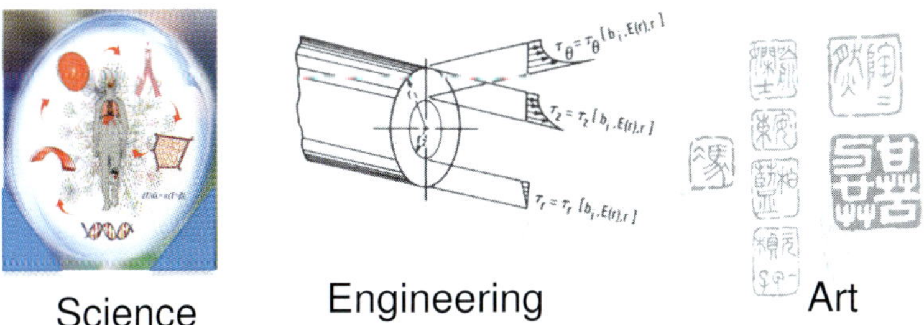

Figure 19. Professor Yuan-cheng Fung, the Renaissance Man who excels in science, engineering and art.

Congratulatory Note

Dear Bert, as we approach your 90th Birthday (Western Calendar), K.C. and I would like to express our warmest congratulations and best wishes for a Happy 90th Birthday, and Many, Many Returns for you and Luna! We look forward to the Centennial Celebration!

References

1. Y.C. Fung, *World Scientific*, Singapore. Part A (1084 pp); Part B (850 pp) (1994).
2. Y.C Fung, John *Wiley and Sons, New York*, (1955). Revised, Dover Publications, (1969).
3. Y.C. Fung, (editor), *Biomechanics. Proceedings of a Symposium at ASME* 1966, Winter Annual Meeting, Nov. 28 - Dec. 1, 1966
4. Y.C Fung. *Annals of Biomedical Engineering* **19 (3),** 237 (1991).
5. Y.C. Fung and S.Q. Liu, American *J. Physiology: Heart and Circulatory Phys.* **262,** H544 (1992).
6. J.H. Omens and Y.C Fung, *Circulation Research* **66 (1),** 37 (1990).
7. H.C. Han and Y.C. Fung, J. *Biomechanics* **24,** 307 (1991).
8. C.J. Chuong and Y.C. Fung, *J. Biomech. Engrg.* **105,** 268 (1983).
9. Y.C. Fung and S.S. Sobin, *J. Biomech. Engrg.* **103,** 121 (1981).
10. S.S. Sobin, Y.C. Fung, H.M. Tremer, J. *Appl. Physiol.* **64 (4),** 1659 (1988).
11. J.G. Pinto, J.M. Price, Y.C. Fung, and E.H. Mead, *J. Appl. Physiol.* **39,** 863 (1975).
12. Y. Lanir and Y.C.Fung, *J. Biomech.* **7,** 29-34 & 171 (1974).
13. F.C.P. Yin and Y.C. Fung, *Am. J. Physiol.* **221,** 1484 (1971).
14. C.Dong, E. Mead, R. Skalak, Y.C. Fung, J.C. Debes, R.D. Zapata-Sirvent, C. Andree, G. Greenleaf, M Cooper, and J.F. Hansbrough, *Annals of Biomedical Engineering* 21, 51 (1993).
15. H. Gregersen, G. Kassab, E. Pallencaoe, C. Lee, S. Chien, R. Skalak, and Y.C. Fung, *Am J Physiol.* **273,** G865 (1997).
16. Y.C. Fung and P. Tong, *Biophys. J.* **8,** 175 (1968).
17. E. Evans and Y.C. Fung, *Microvasc. Res.* **4,** 335 (1972).
18. P.C.Y. Chen and Y.C. Fung, *Microvasc. Res.* **6,** 32 (1973).
19. Y.C. Fung, *Microvasc. Res.* **5,** 34 (1973).
20. G.W. Schmid-Schoenbein, Y.C. Fung, and B.W. Zweifach, *Circ. Res.* **36,** 173 (1975).
21. J.S. Lee, and Y.C. Fung, *J. Appl. Mech.* **37,** 9 (1970).
22. S. Chien, *Ann. Biomed. Engineering* **36,** 554 (2008).
23. Y.C. Fung and S.S Sobin, *J. Appl. Physiol.* **26,** 472 (1969).
24. T.H. Rosenquist, S. Bernick, S.S. Sobin, and Y.C. Fung, *Microvasc. Res.* **5,** 199 (1973).
25. Y.C. Fung and S.S. Sobin, *Circ. Res.* **30,** 451 (1972).
26. Y.C. Fung and S.S. Sobin, *Circ. Res.* **30,** 470 (1972).
27. F.Y. Zhuang, Y.C. Fung, R.T. Yen, *J. Appl. Physiol.: Respirat. Environ. Exer. Physiol,* **55,** 1341 (1983).
28. G.S. Kassab, K. Imoto, F.C. White, C.A. Rider, Y.C. Fung, and C.M Bloor, *Am. J. Physiol., 265,* H366 (1993).
29. Y.C. Fung, Biomechanics: Motion, Flow, Stress and Growth. (Springer-Verlag, New York, 1990).

30. S.Q. Liu and Y.C. Fung, *J. Biomechanical Engineering* **111,** 325 (1989).
31. W. Huang, Y.P. Sher, K. Peck, K., and Y.C. Fung, *Proc. Nat. Acad. Sci.* **99,** 2603 (2002).

Chapter 23

TRIBUTE TO A FRIEND AND A MASTER

TED WU

California Institute of Technology, Pasadena, CA, U.S.A.

To Professor Yuan-Cheng Fung
At His Ninetieth Anniversary Celebration

*How gratifying in life
        to know a friend and master
                so virtuous, so kind, and so learned,
        as to lift us to climb ever higher,
                to exhaust a view on truth so comprehensive and profound;
You have been a marvelous mentor pointing out to us
        how best to revere those from whom we learn,
                to assist those with whom we till and toil together,
        to cast more light on other's course of learning
                with what comes from heart, touches heart;
The high spirit you have so long inspired upon us,
        I believe,
                has sown the seeds of all intellectual innovation,
        may its fruit be ever more fruitful
                when being passed from hand to hand in mood of creative notion;
In our homeland,
        Ninety-ninth Anniversary embraces an endless room
                for celebrating a colorful White Longevity,
        keep radiating brilliant guiding light for your earnest followers
                to inherit fully your splendid spirit in mood of creativity.*

Ted and Chinhua Wu
September 2008

Chapter 24

TRIBUTE TO PROFESSOR AND MRS. YUAN-CHENG FUNG ON PROFESSOR FUNG'S 90TH BIRTHDAY

PIN TONG

*Department of Bioengineering, University of California at San Diego
La Jolla, CA 92093-0412 U.S.A.*

In 2007 Professor Satya Atluri and I discussed how to honor Professor and Mrs. Fung on Professor Fung's 90th birthday. We decided to publish a special series in the Journal of Molecular & Cellular Biomechanics (MCB) in Professor Fung's honor and bind the papers of the special series into a book, entitled "Molecular & Cellular Biomechanics". This is a very natural thing for MCB to do as Professor Fung is its Honorary Editor. He helped found the Journal and guided its growth.

In addition, we would hold a symposium entitled "Genomic Biomechanics – Frontiers in Biomechanics of 21st Century" on September 14, 2009 in Professor Fung's honor and make the day a day of festivities paying tribute to the happy couple, Professor and Mrs. Fung, for their more than sixty years of togetherness.

We are heartened that many world leaders in science and engineering, many of whom are friends and colleagues of Professor Fung, lined up to contribute to this special series. There are seven issues with 35 papers covering a wide range of topics in biomechanics spanning from genes, proteins, tissues, organs to human. The first copy of the book, signed by all participants of the symposium and the festive activities, was presented to Professor Fung. Figures 2 and 3 are the covers of the first special issue and the book, respectively.

Thanks to Professors Satya Atluri, Geert Schmid-Schonbein, Shu Chien, Savio Woo, and Peter Chen for organizing the symposium and putting the festive activities together. Also many thanks to all participants of the symposium and the festivities.

Fig. 1. The Happy Couple – Professor and Mrs. Fung

Fig. 2. The first issue of the special series of MCB in honor of Professor Fung.

Fig. 3. The book "Molecular & Cellular Biomechanics" – a collection of the papers of the special series paying tribute to Professor Fung. The portrait of Professor Fung was drawn by an anonymous artist.

A Giant

Professor Fung has rightfully earned a place in history as the father of modern biomechanics. He has reached the pinnacle in science and engineering like this poem of the Tang Dynasty:

….At the tallest peak of the mountains, the surrounding hills look small.…

Professor Fung was born in China in 1919. He studied airplane design in college, during the war-ravaged China, with the goal of helping China fight for its survival. After the war, he entered California Institute of Technology (Caltech) in 1946, received his Ph.D. in aeronautics and mathematics in 1948 and stayed on as a faculty member until 1966. Then he left his illustrious career in aeronautics and moved to the University of California at San Diego (UCSD) in order to concentrate on biomechanics.

Professor Fung published his first book on soaring and gliding in clouds in 1944 when he was only 25. His work in aero elasticity led to a fundamental understanding of flutter, especially in supersonic flow. At the age of 36, he published his seminal text book, *An Introduction to the Theory of Aeroelasticity*, which was required reading for generations of aeronautical students and engineers concerned with the safe design of high speed flying-craft. In the years that followed, he solved many problems in structures, vibrations, elastic waves, stochastic processes and protection of structures against nuclear bombs. In 1966, he published *Foundations of Solid Mechanics*, which lays out a vast spectrum of solid mechanics concepts with great insight, clarity and authority, helping readers understand the unifying concepts of continuum mechanics. The book became the vehicle of basic education in solid mechanics for many students and engineers ever since. The book was subsequently revised as *Classical and computational solid mechanics* (coauthored with me) in 2001 with expanded discussion on theories of plasticity, large elastic deformation and computational mechanics.

The work which Professor Fung completed before 1965 was of sufficient magnitude to constitute a life-time career for most people. But it was not for Professor Fung. He was always fascinated with matters of nature. By 1965, he shifted his focus on various aspects of physiology. In 1957, out of concern of his mother's glaucoma, he translated newly published articles on glaucoma into Chinese and sent them to his mother and her surgeon. This got Professor Fung interested in medicine. His strong curiosity with how the human body works eventually drew him to the field of biomechanics. He said:

> "…force, motion, flow, stress, strength, and remodeling – pervade the living world; I resolved to dedicate myself to the development of biomechanics."

Professor Fung has made tremendous contributions in biomedical fields. Armed with a thorough understanding of the laws of physics and the great mastery of applied mathematics, he has applied the principles of mechanics to solving many important and difficult problems in biomechanics. His work has spanned from basic constitutive relations to applications in microcirculation, the heart, the lung, the tissue and blood vessel mechanics, growth, healing and remodeling. He started the field of 'tissue engineering'.

He has written several books and published over 300 technical papers in the field. His vision of biomechanics is "genomics biomechanics", the bridge linking gene expression, protein function, and cell behavior to tissue, organ and continuum behavior. Genomics biomechanics will be the most relevant and intricate field of research and study for years to come.

Through his work and leadership, biomechanics has been transformed from a relatively unknown discipline in the early 1960s to a major field of endeavor today. He was the founder or a cofounder of a number of major organizations, which have led and promoted research and development in biomechanics. He was a founder of the Bioengineering Department at UCSD, one of the nation's top such departments. No single individual has done more to shape the modern development of biomechanics and deserves more recognition than him.

Due to his many pioneering contributions to aerospace sciences and bioengineering and for his ability to tackle fundamental questions, Professor Fung was elected to be a member of the US National Academy of Sciences (NAS), a senior member of the US Institute of Medicine of NAS, a member of the US National Academy of Engineering, a member of Academic Sinica, and a foreign member of the Chinese Academy of Sciences. He has received numerous prestigious awards and medals, culminating in President's National Medal of Science, the highest scientific honor bestowed by the nation, awarded in person by President Clinton in 2000 (Figs. 4 and 5).

Fig. 4. Professor Fung and the National Medal of Science – year 2000

Fig. 5. Professor and Mrs. Fung with President Clinton – year 2000

The secret to Professor Fung's success is that he enjoys what he does. Professor Chia-Shun Yih, Fung's friend since childhood, said that Professor

Fung seemed to be singing while working, and quoted the poem by Wang Si-Cheng of Ching Dynasty to describe Professor Fung's enthusiasm:

> The winds of May have blown so much green into the willows,
>
> That the swallows cannot refrain from coming home

Professor Fung himself described the excitement of working and discovery:

> ....Waves of excitement went through me. Again and again the ideas...generated in my mind beautiful vistas of the future and immediate plans for new investigations to be made.

He used Wang Kuo-Wei's quotation of three stanzas from three famous Chinese poems to describe the moods of the stages of any significant accomplishment - the vision; the hard work and frustration; and the joy of sudden and unexpected success:

I (Yen Shu, 991-1055)

Last night the west wind ringing
Stripped leaves from the green tree.
I climbed alone to the top floor
of my house, and looked up the road
as far as the horizon.

II (Lie Yung, 990-1050)

For love of her, I grew pale;
I turned thin and haggard.
For love of her, the belt hung loose on my robe.
...Yet I regretted nothing.

III (Hsin Chi-his, 1140-1207)

I look for her in the crowd
a hundred, a thousand times.
Then suddenly, as I turned
my head, I glimpsed her
in the dim light, in half darkness.

Fig. 6. A very hard and happy worker. Professor Fung really loves to work. He was smiling even washing car.

A Genius

A genius sees significance from the obvious and the not-so-obvious. Professor Fung's extraordinarily high intelligence allows him to ask provocative questions and find the right answers. Few examples to illustrate his ingenuity are as follows:

Capillary is a small tube connecting arteriole to venule. In *in vivo* observation, capillaries behave like rigid tubes under hydraulic pressure from 20 and 200 mm Hg. The high rigidity is generally accepted because of the inaccessibility and difficulty to measure and characterize the capillaries. Professor Fung recognized that the capillaries are imbedded in connective tissues (obviously) and gave the explanation for the apparent rigidity as a tunnel (the tube) in gel (the connective tissue matrix) rather than a discrete tube with high elastic modulus.

Mammalian red blood cells appear as biconcave disks under microscope and swell into spheres in hypotonic solutions. Red cells deform severely in capillaries. The exact determination of the red-cell geometry was difficult in the 1960s. It was even harder to analyze the severe deformation in flow. Based on the principles of mechanics, Professor Fung concluded that the cell has a liquid interior. This conclusion implies a low pressure difference between the interior and the exterior of a red cell in static equilibrium, permits sharp trailing edge for deformed cells in flow, and allows a cell without complicated anisotropic properties of the cell membrane to deform to a sphere in hypotonic solution.

Before the early 1960s, the capillary blood vessel network in the lung was idealized as short circular cylindrical tubes arranged in a hexagonal pattern. Professor Fung and Dr. Sid Sobin saw that the pulmonary capillary vessels in alveolar walls form a dense two-dimensional network, called *capillary sheet* supported by tissue between the vessels. This unique network gives hemodynamic closer to nature and allows blood vessels to remain smooth in the lung as the blood pressure fluctuates. The sheet model establishes the relationship between the sheet thickness, transmural pressure, and tissue stress that agrees with model experiments. Better understanding of the hemodynamics has led to improved treatment of lung injuries such as injury caused edema.

Professor Fung identified the expressed genes in the cells of blood vessel wall under hypertension. As gene expression controls protein production, the correlation with mechanical stresses points the way to study the mechanics of cells and how mechanical stimulation promotes cell growth, remodeling and division.

Professor Fung has an extraordinary memory. He has written many books and papers with numerous equations and references. Once I asked him how he managed to assure the consistency of cross-referencing the equations and the references before the age of personal computers and word processors, as he had to make changes during writing. He said, 'I remember every citation of equations and references in the text.'

Professor Fung's Teaching

Professor Fung has been a teacher, mentor and role model to many. He is a source of strength and encouragement for biomechanics of the twentieth and twenty-first centuries. His teachings have provided clear and forward looking advice to investigators and researchers in the field:

Fig. 7. Pictures of past giants in mechanics

to a modern worker, it appears that much of the work has to begin from the very beginning. To an analytical mechanist, the most serious frustration lies in the dearth of information about the material properties, i.e., the stress-strain history laws of living tissue. Without the constitutive law, no analysis can be done. On the other hand, without the solution of boundary value problems, which serious analyses (usually quite difficult because of non-linearity) have to be done for hypothetical materials in the hope that experiments will yield the desired agreement. If no agreement is obtained, new analyses based on a different starting point would be necessary (1968).

Professor Fung always takes a unique approach to any problem and to his teaching. In one introduction to mechanics lecture, he showed a slide of past giants in mechanics (Fig. 7). He told the students that several of these giants were actually horse doctors. What does a doctor for horses have to do with mechanics? A mechanics problem is like a horse – big and tough. Think like a horse doctor, and you can handle any big and tough problems of mechanics. The students got the point.

Hundreds of years ago, it was people like horse doctor in biomedical field leading the development of continuum mechanics. Decades ago, Professor Fung led the mechanics back to biomedicine.

Professor Fung taught us to tackle all questions from molecules to man (Fig. 8). He raised the fundamental question in biomechanics of today, i.e. **"What Axioms of the Classical Continuum Mechanics Have to Be Changed for Biology?"** This is actually the central theme of our symposium on "Genomic Biomechanics."

A Kind, Happy and Gentle Man

Professor Fung is a gentleman and a scholar. He believes in harmony and gentility. He treats everyone collegially and does not compete with anyone. He taught us that by working together, sharing ideas, and making constructive suggestion to and graciously receiving input from others, we all benefit and the entire field advances.

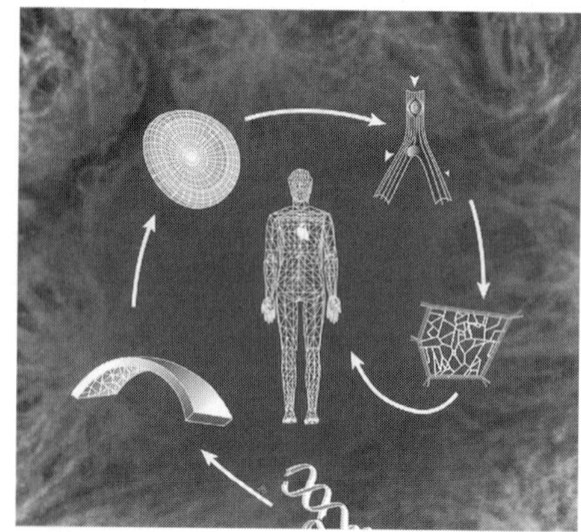

Fig. 8. From Molecules to Man - Professor Fung's Teaching (From Cover of Y.C. Fung: Introduction to Bioengineering, World Scientific Publishing Co., 2000, which was taken from the cover of Department of Bioengineering Student Handbook based on Fig. 1 in S. Chien: Microvasc. Res. 29:129-151, 1985 and the work of C. Galbraith, R. Skalak and S. Chien (background))

He is generous with his knowledge as well as his time. His caring and support have enriched the lives and launched the careers of many – his students, his post-doctoral scholars, his colleagues and people around him and learning from him.

Laughing loudly and being happy are the trademarks of Professor Fung. He appreciates people and things no matter how plain. Perhaps Mrs. Fung described him best at his 70th birthday:

> He laughs frequently and heartily. He loves deeply. He loves his family. He loves his friends. He lives simply. He does not demand anything. He does not compete with anyone, and never tries to keep up with the Joneses.

Professor Fung has benefited from his partnership of the past 60 years with and the love of Mrs. Fung. He could not have achieved what he has without her support. She is as much a member of the biomechanics community as he is. We all are in debt greatly to her.

Impact on Me

Helping people - even strangers - is part of Professor Fung's style. In 1962 shortly after I arrived in the United States from Taiwan, I was waiting to meet with a professor in a hallway at Caltech, when a tall stranger with a gentle smile asked me, "Can I help you?" I told him that I was there seeking financial aid. The stranger said, "Why don't you drop by my office afterwards?" The stranger turned out to be Professor Fung, and he asked me to work for him. I began to work the next day. This was the start of a journey of friendship and collaboration that has lasted many decades.

Professor and Mrs. Fung have always been kind and generous to me and my family. Our relationship has grown ever stronger over the years. Both Professor Fung's office and his home have always been open to me and my family. Their caring and support have helped make what I am today.

When he and I coauthored the book "Classical and Computational Solid Mechanics", I often wished I could write as clearly and precisely as he did. Even for this write-up, my daughter, Betsy, edited it many times before I got it to this shape.

I am a theoretician, but Professor Fung has taught me to appreciate experimentation. When I developed a micromechanics theory to illustrate the size effects of structures at micron scales, I wanted to just publish the theory. However, Professor Fung said "Show me the data." I experimented for two years and the resulting test data strengthened the theory.

After leaving Caltech, I ventured into different areas of mechanics and engineering. It took me more than 40 years to return to biomechanics. When I was retiring from the Hong Kong University of Science and Technology in 2003, Professor Fung suggested I work with him. I did. The journey that started almost 50 years ago comes back to its roots.

Here I am at UCSD (Fig. 9) pondering **"What Axioms of the Classical Continuum Mechanics Have to Be Changed for Biology?"** I too smile while working just like Professor Fung taught me!

In summary, Professor Fung has reached the pinnacle in science and engineering and rightfully earned a place in history as the father of modern biomechanics. I have had the privilege and great fortune of learning from, being associated with, and bestowed the friendship by Professor and Mrs. Fung. They have profoundly influenced and enriched my live. I thank them from the bottom of my heart for what they have done and continue to do for me and all in the community. It is indeed my high honor to pay tribute to Professor and Mrs. Fung.

Fig. 9. I too smile while working!

Chapter 25

TRIBUTE

ARNOST FRONEK

*Departments of Surgery and Bioengineering, University of California at San Diego
La Jolla, CA 92093, U.S.A.*

KITTY FRONEK

*Department of Bioengineering, University of California at San Diego
La Jolla, CA 92093, U.S.A.*

We would like to express our admiration and also thanks for your active contributions and collaboration in our cardiovascular physiology studies and help in our work.

Your combined interest and knowledge ranging from engineering to biologic sciences, expressed finally in your concept of bioengineering, helped us, working in the field of cardiovascular physiology, to apply more strict bioengineering principles in our daily work. We appreciate your readiness to answer our questions and advices you gave us in the past.

We do hope that you will be here at UCSD for many more years to the benefit of your family and all of us

Chapter 26

FOREVER GRATEFUL

PETER CHEN

*Department of Bioengineering, University of California at San Diego
La Jolla, CA 92093-0412, U.S.A.*

Dr. Fung is the Father of Biomechanics and is referred to as a Renaissance man by several authors in this tribute volume. He has set high standards for himself and others to follow and he has inspired, influenced and affected the lives of many who crossed his path.

I Came to UCSD in 1968 from Hong Kong. In deciding on a major I talked with Paul Saltman's lovely assistant Nancy at the Revelle provost office and she told me to contact Dr. Zweifach regarding a new bioengineering program. I went to the bioengineering office on the 5th floor of the Basic Science Building and met Perne Whaley. With a gentle smile Perne said Dr. Zweifach was out momentarily but Dr. Fung will see me. Dr. Fung was busy reading some manuscripts and while glancing at the papers frequently he explained briefly the requirements for this new major. I wasn't too sure what I may be getting into so I asked about job prospects. To this question, Dr. Fung put down the papers, looked at me and with his signature infectious laugh said, "If you are the best in your field there is no problem finding a job." Many of his students have achieved that status.

Knowing I was a foreign student Dr. Fung offered me a lab assistant position helping Tang Ling with computer data analysis. I worked in the office with her and Dr. Fung's secretary Barbara Johnson. Barbara's daughter will come in after school and sometimes I helped with her homework. It was like being with family. The first bioengineering class I took was continuum mechanics taught by Dr. Fung. One day Dr. Fung told us he will be away the following week and the TA, his graduate student, will take over. At the next class, shortly after the TA started lecturing and writing on the blackboard Dr. Fung quietly walked in and sat down in the front row, his meeting was cancelled. When the TA turned around and saw Dr. Fung he froze. After a full minute of silence he told Dr. Fung he cannot teach with his presence. Dr. Fung said he will move to the back and he did. After a few more minutes the TA said he really cannot continue so eventually Dr. Fung took the chalk and finished the class. The TA was doing a great prior to Dr. Fung walking in and it took me quite some time to realize how much we want to push ourselves to be like Dr. Fung.

At the encouragement of Dr. Fung I continued with the graduate program at UCSD. When I started, Frank Yin, Evan Evans, Hyland Chen, John Pinto, Mike Yen, Don Vawter, Larry Malcolm …. were well on their way in testing various tissues and deriving constitutive equations for their thesis. Most of their experiments required custom instrumentation and Eugene Mead was the engineer who was instrumental in designing and building the devices. The graduate students worked very hard. I remember John sleeping in a cot in the lab when he had to monitor papillary muscle over 24 hours and Hyland would come in at 4:00 a.m. to run his mesentery experiments. Every time we went to talk with Dr. Fung in his office we always came out telling ourselves we should be better prepared next time. In addition to being our thesis adviser Dr. Fung and Mrs. Fung also made sure all our other needs were taken care of. John Pinto called Dr. Fung 'Big Daddy' and I think that says it all.

During that time Yuji Matsuzaki was a visiting researcher in the lab working on air flow dynamics. When he was ready to return to Japan Dr. Fung was out of town and Conrad was the designated host for a farewell dinner. I remember gathering in Dr. Fung's house and when it was time Conrad called us together, drove to Lubach's in downtown, sat us down accordingly, and conducted that evening in a most professional and proficient manner. It was the first time I had the chance to interact with Conrad and I was impressed.

I did my postdoctoral training with Dr. Fung's colleague and good friend Dr. Sidney Sobin at the University of Southern California before returning to UCSD to work with Dr. Zweifach on microcirculation problems. I am fortunate to be close to Dr. Fung for the past 40 years and I really appreciate his advice, guidance, encouragement, care and kindness. In 1996 Dr. Fung started a new class BENG1 'Introduction to Bioengineering' for incoming bioengineering majors. There were about 30 students in that class. At the encouragement of Dr. Marcos Intaglietta I started teaching in 2000 and last few years Prof. Shu Chien and I have been co-teaching BENG1 because that class has grown to 300 students and they needed more than one instructor. We still invite Dr. Fung to speak to the students. I was with Dr. Fung last year when he prepared his lecture and he wanted to share with the students his birthplace and his journey to this country. Following is a photo of Yuhong, Dr. Fung's birthplace in China. I am sure Dr. Fung wants to share it with us also.

Yuhong, Wutzin, Kiangsu. Dr. Fung's birthplace

Gathering in John Pinto's house with former students, colleagues and friends during the First World Congress in Biomechanics in 1990. That was the last time we saw Dr. Zweifach, Dr. Sobin, Gene Mead, John Pinto and Perne Whaley together.

We are grateful to Dr. Fung for shaping our characters and urging us to aim high. I believe he will be most satisfied when one day he sees some students surpassing his achievements. That will be the greatest birthday gift!

Chapter 27

TRIBUTE TO Y. C. FUNG, WITH FONDNESS, ADMIRATION AND APPRECIATION

SHELDON WEINBAUM

Department of Mechanical Engineering, City College of New York, NY, U.S.A.

These are the opening comments of the talk that I gave on the occasion of Bert's 90th birthday celebration at UCSD. The talk itself entitled "An airborne jet train that flies on a soft porous track at 700 km/hr" has been submitted to Phys. Rev. Lett.

I am just one among many whose lives have changed dramatically through the generosity and wisdom of YC Fung. It is a story that many of you sort of know but not in its details. It is the details that make it special. In 1969 the Grove School of Engineering at CCNY celebrated it's 50th anniversary. Each of the then four engineering departments was given the opportunity to invite one special speaker to come to the college for a week to give a series of lectures that would be shown throughout the university on CUNY TV. As a member of the ME Department I had submitted a proposal to invite Bert. I believed my proposal would be a long shot, both because I was seen as a trouble maker and I hardly knew Bert except as a pioneering figure who had gone from Cal Tech to UCSD and had changed one field, aeroelasticity, and was about to start another. The year before I had been censored by own faculty for the role I had played in the seizure and occupation of the CCNY campus by its Black and Latino students in the CUNY Open Admissions struggle. I had met Bert only in passing and it seemed far-fetched that someone would come and lecture, not for a day but an entire week, at the bequest of a young faculty member who was just getting his toes wet in BME. I sometimes wonder if I would have done the same.

Bert not only came but took it upon himself to introduce me to other investigators in NYC who he knew had similar interests. He said 'there are two special people at Columbia I think you should meet'. One was Dick Skalak and the other Shu Chien. As I think many are aware, I was sort of adopted by both. Dick eventually served on the examining committee of nearly every PhD student that I had between then and the time he left for UCSD in 1987 and Shu invited me and Bob Pfeffer to be guests in the physiology course he supervised at

Columbia P&S. This was the beginning of a collaboration that has lasted till this day.

It is striking that just 30 years later in 1999 I made another request of Bert. CUNY had not started a new PhD program since 1982 when it started its now well known doctoral program in music. There had not been a new doctoral program in engineering since 1964.

As a public university the start of a new doctoral program is a major event involving a lengthy proposal and extensive evaluation by outside examiners for the NYS Board of Higher Education. Our two outside examiners were Bert Fung and Roger Kamm. We are forever grateful for their extensive and glowing report. The President of the Graduate School was well aware of Bert's eminence in the field and his words made an indelible impression. It was if we had been blessed.

Last year the college and the BME Department celebrated my 70th birthday. One of the very special events was a video from Bert and Shu imparting a special message that I often listen too when the going gets rough. It is a great pleasure for me to contribute to this special volume to toast the many contributions that Bert has made to all of us. You are a very special person who will be remembered through the ages and a symbol for the rest of us who follow in your footsteps.

With fondness, admiration and appreciation for all that you have done.

Chapter 28

DR. Y. C. FUNG: MY RESPECTED MENTOR AND CHERISHED FRIEND

SAVIO L.-Y. WOO

Musculoskeletal Resarch Center, Department of Bioengineering
Swanson School of Engineering, University of Pittsburgh, PA 15228, U.S.A.

Dear Dr. Fung and Luna, Conrad, Brenda, Tony and Michael, Colleagues, Friends and Guests:

Good evening.

- Pattie and I, together with Kirstin, Jonathan, Adam and our baby granddaughter (Zadie arrived on February 3, 2009) wish you, **Dr. Fung**, a Happy 90th Birthday! May your happiness extend beyond the Eastern Seas (of China) and your longevity, taller than the Southern Mountains.

- It has been a pleasure to work with Geert, Shu, Peter and Pin (pictured below) in organizing this beautiful event. I wish to sincerely thank them for giving me the privilege to be tonight's banquet speaker.

- I have titled my talk as "Dr. Fung: My Respected Mentor and Cherished Friend". Many of you have already had this experience. But, it would really make me happy if you would let me indulge with you in the next 20 or so minutes to assert that – Yes! He is indeed "Respected" and "Cherished".

- For over 60 years, Dr. Fung has written treatises on aeroelasticity, applied mechanics, and biomechanics. We all know about his first book "Introduction to Theory of Aeroelasticity", a true classic in Dover Publication. In applied mechanics, we have all used his "Foundation of Solid Mechanics" published in 1965 and used it to pass our doctoral exams. Then came his "1st Course in Continuum Mechanics" in 1969 to be followed by a series of books on Biomechanics and Biodynamics – "The Mechanical Properties of Living Tissues" published in 1981 (and the second edition in 1993) and "Circulation" in 1984 and "Motion, Flow, Stress and Growth" in 1990. When I studied these books, I can easily visualized Dr. Fung's fascination with matters of nature. Through his thorough understanding of laws of physics and armed with mastery of applied mathematics, he has solved many important and difficult problems in all these fields. Using his distinct and clear writing skills, Dr. Fung taught me (and us) in a manner that was easy to understand. And I (we) have learned.

Listening to his lectures, Dr. Fung has explained to me (us) the reasons for the sphering of red cells, the need of sheet flow phenomena for pulmonary alveolar microcirculation, the importance of strain energy density function to describe the pseudoelasticity of living soft tissues, the appropriateness of the quasi-linear viscoelastic theory for soft tissues, and the scientific basis for stress-dependent homeostasis, as well as the existence of residual stress, the importance of wave propagation theory in reducing the trauma caused by impact on the chest wall and so much more.

Let me also refer you to one of the best summary on Dr. Fung's collected works written by Ghassan Kassab, his most recent PhD student. The title is "Y.C. "Bert" Fung: The Father of Modern Biomechanics", published as the leading article in the first issue of <u>Molecular and Cellular Biomechanics</u> in 2004.

- Now that I have covered the technical aspects, may I turn and share with you a few personal short stories.

- In 1964, I had my first sighting of Dr. Fung. I attended the Space Technology Summer Institute directed by Dr. Ernest Sechler (Dr. Fung's mentor) at CalTech. One afternoon, several of us walked quietly by Dr. Fung's office and had a glimpse of him – as you know, his office door is always open. We wispered to each other, admiringly, "Oh look, that is Dr. Y.C. Fung".

- In 1968 (or was it 1969?), Dr. Fung came up to Boeing, in Seattle, to give a lecture on Biomechanics. My professor, Dr. Albert Kobayashi, and I went to his lecture. After he had finished and answered all the questions, Albert and I were inching our way towards him. It seemed that no one had been assigned to give him a ride back to the hotel near the university. All the time, we thought that this might be our chance and we could introduce ourselves and present our biomechanics research to him as well as to pick his brain. As we get within about 5 or 10 feet, the host, Dr. Jonathan Turner, said "Hey Bert, let me drive you back to your hotel". We were so thoroughly disappointed!

- In the Spring of 1970, I was finishing my PhD studies and was fortunate enough to have received job offers – one to remain in Seattle, one from Chicago and one from UCSD. Well, Dr. Fung was at UCSD and so my choice was obvious. In May, I traveled to La Jolla to have my interview.

1988 – Graduating our doctoral students at UCSD

Of course, I visited Dr. Fung. He invited me to sit across the table from him in his office on the 5th floor of the Basic Science Building (BSB). Dr. Fung made me feel at ease and taught me what I needed to know about the academic system at UC and advised me what I needed to do and how to proceed. I was like "WOW! What a remarkable experience for me to talk with this **"giant"** for more than one hour!" He was extremely generous with his knowledge as well as his time. Those of you who have had the good fortune of knowing Dr. Fung recognize that he is indeed this way with everyone.

Then, in August, I joined the Departments of Surgery and AMES and became his junior colleague. It was indeed my great fortune that my office was in the same BSB Building so that I could learn from Dr. Fung from close range for the next 20 years. On countless occasions, he not only taught me biomechanics, but also counseled me on difficult matters involving colleagues and students. Our friendship grew!

Meanwhile, our families are getting closer together as Dr. Fung and Luna watched our children grew and sometimes they are like grandparents to our children, Kirstin and Jonathan.

We frequently had lunch together at a small cafeteria in the building. There, Dr. Fung and I enjoyed many cheeseburgers with French fries and coke! Later, our lunch menu had expanded, thanks to our Japanese/American chef, Susie. She offered a tasty special of teriyaki steak over California rice with salad for $1.35! I also learned from Mike Yen that soy sauce is a great salad dressing!

Dr. Fung moved over to the Engineering Building and soon after that, the UCSD Faculty Club opened. We could splurge there and really dine! We regularly met for lunch on Fridays. For several years, on every Monday morning, I would begin to look forward to our private luncheon on Friday. **<u>Yes, he is my cherished friend.</u>**

In 1991, I was lucky to be inducted into the Institute of Medicine. But more importantly, Dr. Fung was also inducted as one of the only 5 senior members (a very difficult achievement!) in the same class. It was really a humbling experience that I could sit alongside with him at the induction ceremony. Definitely, that was a highlight of my career.

- **I have learned** a great deal from Dr. Fung.

  - **I have learned** how to do research properly. Dr. Fung has outlined for us, the following 8 steps on solving biomechanics problems.

1990 – With my students at the First World
Congress of Biomechanics in La Jolla,
California

1.   Learn anatomy/morphology
2.   Obtain mechanical properties of materials
3.   Apply fundamental laws of physics
4.   Use meaningful boundary conditions
5.   Solve boundary-value problems
6.   Perform physiological experiments to test solutions
7.   Do validation, and
8.   Employ theory to predict the outcome of other conditions

I have religiously followed these steps in my work and have passed these on to all my students.

- **I have learned** that it is important to include professional services to our field as part of our academic duties. Dr. Fung led us in many organizational activities and every one of the societies that he created has enjoyed paramount success.

With President Yang and
Professor Kang

1995 – 4th China/Japan/USA/Singapore Conf. on
Biomechanics in Taiyuan, Shanxi, China

1. In 1972, he wanted to gather all of us together to discuss Biomechanics. Using the American Society of Mechanical Engineers as his platform, he established the biannual Biomechanics Symposium and had our inaugural meeting at Georgia Tech the following summer. It was my lucky day when he chose me as the secretary of the organizing committee – what a huge jumpstart for me to serve my profession! You all know how successful the Biomechanics Symposium has become and now, our younger folks have further upgraded it to our Annual Summer Bioengineering Conference and is without a doubt, the most "look forward to attend meeting" by all of us!

2. Only a few years later, he founded the Journal of Biomechanical Engineering and then orchestrated the United States National Committee on Biomechanics.

3. We, also, worked together on the China/Japan/US (later Singapore) Conference on Biomechanics. The first meeting was in Wuhan, China in 1983, and then Osaka, Japan in 1987, Atlanta, GA in 1991, and Taiyuan, China, in 1995 and so on. At the Taiyuan meeting, Dr. Fung wrote the following:

   "We see how topics change in a span of a dozen years. How some concepts barely mentioned at one time took center stage at a later time. How some other topics became mature and less discussed".

   What he said is exactly what has happened in the ensuing years. Indeed, many changes in Biomechanics have occurred, and we have gone from tissue to cellular to molecular to where Dr Fung is pointing for us to go - **"genomic biomechanics"**.

4. In 1983, we climbed beautiful Yellow Mountain (indeed, a most memorable experience!). Dr. Fung, at that time, had some knee pain. But, after climbing the 7,000 steps up +10,000 steps down, he said the knee pain had magically disappeared. Now, I am supposed to be the expert on knee biomechanics and I can confess that I was totally clueless! In Taiyuan, we climbed the historical Wutai Mountain, which was another fantastic experience.

1983 – Scaling the beautiful Yellow Mountain

5.   Dr. Fung also spearheaded the effort for the World Congress of Biomechanics. The first meeting was held in La Jolla, of course and every four years hence, in Amsterdam, Hokkaido, Calgary and Munich. All have been big hits as the attendances and enthusiasm continued to rise to newer heights Now we are looking forward to the next congress in Singapore 2010 with great anticipation. He also founded the World Committee for Biomechanics. It is because of Dr. Fung that Biomechanics is now a very visible discipline.

1994 – 2nd World Congress of
Biom. in Amsterdam, Netherland

1998 – 3rd World Congress of
Biom. in HoKKaido, Japan

- **I have learned** how to become a better teacher and person. Some of Dr. Fung's extraordinary attributes must have rubbed off on me, because of our close proximity.

In the 70's, I had job offers from Chicago, Cleveland, Iowa, Seattle, and Boston. Like everyone, I would first seek Dr. Fung's advice on each occasion. He would always advise me to stay at UCSD and gave me his cogent analysis. He made me look ahead, say 10 years down the line, taught me not to just look at those (perceived) problems or road blocks placed in front of me, urged me to be calm and think things through and only to change the work place for the right reasons. I listened to him and decided to stay at UCSD each and every time. I have adopted Dr. Fung's methods in dispensing my advice to all my students, fellows and younger colleagues.

Incidentally, when it comes to the discussion that my offers include substantial increases in salary, he would simply say, "Well, La Jolla is worth $5,000", in the early 70's or "Well, La Jolla is worth $10,000" in the late 70's – you see, there was actually inflation even in the 70's! By the way, $5,000 was a sizeable amount in those days – like about 40% of an Assistant Professor's salary. To put the issue of salary in perspective, the starting junior faculty salary at UCSD in the early 1970's was lower than the average Caucasian's salary in Orange County.

- **I have learned** how to be generous in singing praises to my students, friends and colleagues when called upon.

1. In early 1990, I was contemplating leaving UCSD for the University of Pittsburgh. Of course, I consulted Dr. Fung – both at his home and in his office. At first, he told me that he did not think that I should go to Pittsburgh. But later, and after some thought, he decided it was indeed a good opportunity for me to move and supported and guided me through the negotiation process. He lovingly confessed that he just did not want to see me leave La Jolla (that really warmed my heart!!).

   On August 5th, Dr. Fung, Luna together with Shu and many of my friends in Bioengineering at UCSD organized a beautiful luncheon for my family at the La Jolla Village Inn (now the Sheraton). There, he gave me his new book with the following inscription:

"Dear Savio. . . . I know your first paper: Your thesis. It was on cornea and sclera. It was the first paper in Biomechanics which treated pressure in an incompressible material correctly. . . .

He also gave me some advice:

". . . .As an older brother, I like to give you three parting advice's: Slow down to enjoy, detach to preserve perspectives, and take care of your health always. Love, Y.C. Fung."

I have treasured these lines for all of the last eighteen autumns and every time I read these cogent works of wisdom, my eyes well up.

1990 – The Fungs, Woos and Cheongs at our farewell
party in La Jolla, California

2. When asked to be an external reviewer for the selection committee for the Olympic Gold Medal and Prize. There, he put me (albeit, undeservedly) in the same league "with F. Pauwels and Julius Wolfe" and called me "a brilliant educator and congenial mentor for those who

are fortunate enough to work with him". Clearly, I have not earned these accolades. Nevertheless, I have adopted his generous approach when asked to write recommendation letters. Maybe the words are out and that is why I have so many requests!

- **Celebrating Dr. Fung's special milestone birthdays** have been one of the most favorite things for me (and many of us) to do!

  - In July of 1984, I, together with Geert, Peter, Michael and several others, organized a 3-day long symposium entitled "Frontiers in Applied Mechanics and Biomechanics" in La Jolla in honor of Dr. Fung's 65th Birthday. Many of his students, friends and colleagues came for this event and all had superlative words to say about him. Everyone thanked him for guidance as well as his friendship and advice and felt fortunate to have been associated with him.

  There were many laudations from Dr. Fung's dearest friends, Chia-shun Yeh; Sid Sobin; Jen Ming Woo, Ben Zweifach; Erik Reissner and Chi-Ming Chen as well as 2 poems presented by my father-in-law, Mr. In-Cheong Cheong.

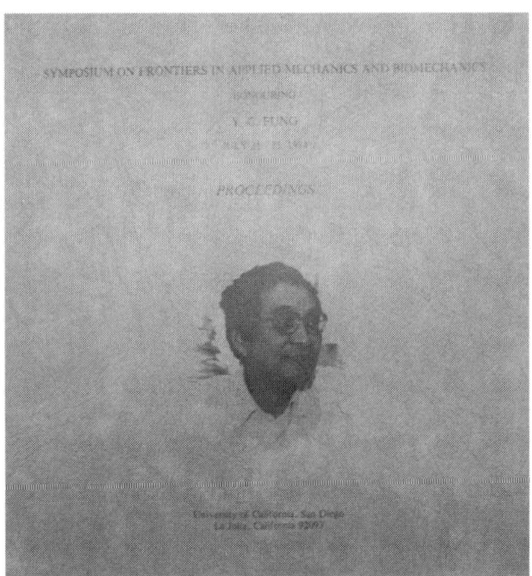

1984 – Proceeding of Symposium celebrating
Dr. Fung's 65th Birthday

With your indulgence, I would like to read part of what Sid had so skillfully and eloquently written

> "How can one capture on a page the essence of an individual – his effervescence, vitality, laugh and spontaneity, that counterpoise with intellectuality, introspection, intuition, creativity and so much more; yet cloaked with nobility, humanness, humility, generosity and integrity, that he provides inspiration to so many, and even awe at the accomplishments . . . . "

It is worth noting that the excess funds donated for the symposium was parlayed to set up the Y.C. Fung Young Investigator Award in the American society of Mechanical Engineers – first as a Bioengineering Division award and later, and with the efforts of many and funding from the Division, the Fung Award turned into a societal award with a medal.

- In 1989, Dr. Fung celebrated his 70th birthday with us. After dinner, Luna had this to say about the love of her life.

1989 – 70th Birthday Party in La Jolla (Jonathan at the podium)

"He laughs frequently and heartily. He loves deeply. He loves his family. He loves his friends. He lives simply. He does not demand anything. He does not compete with anyone and never tries to keep up with the Jones."

I might add: Of course, when one is head and shoulders above everyone, there is no competition and certainly no one to keep up with.

- In 1994, we celebrated Dr. Fung's 80th birthday at the Summer Bioengineering Conference in Big Sky, Montana. I, together with Van Mow, Shu, and Rik Huiskes organized a Biomechanics Symposium in his honor. We covered heart, soft tissues, bone, modeling, large vessels, small vessels as well as cells. At the banquet, we presented him with a huge birthday cake and over 400 of his admirers sang the "Happy Birthday" song to him.

1999 – A Very happy Dr. Fung with Luna in
Bigsky, Montana

We were fortunate to have another celebration of his 80th Birthday simultaneously with our 30th Wedding Anniversary in Laguna, California.

1999 – In Laguna, California

-     Now, here we are, celebrating his 90th birthday. We want to be here because Dr. Fung is a beautiful person who appreciates people and things no matter how plain. We admire his many talents including calligraphy, painting and carving as well as having a "green thumb". Dr. Fung laughs heartily and a lot. Those of you who know him can always hear his laugh even in his absence. **<u>Yes, he is our cherished friend.</u>**

2009 – At the symposium celebrating Dr. Fung's 90th
Birthday

- As you all know, Dr. Fung has received many, prestigious honors and highest awards. With these, he has broken many glass ceilings for biomedical engineers. Last year, he received the Fritz and Delores Russ Prize given by the National Academy of Engineering. His entire family, plus several of us, was there to help him celebrate this joyous occasion. The Russ Prize is given to Dr. Fung for his significant contributions to humankind rather than his lifetime achievement. As the Chair of the Selection Committee, I had the pleasure of hearing my committee members elaborate on the novelty of his contributions, the large number of people his work has benefited, and how he has removed barriers. He was cited for how his discoveries have enhanced vehicular safety that saves many, many lives and how he has enlightened the public to better understand the importance of engineers. To conclude, they gave him the revered title "Father of Biomechanics and Tissue Engineering". **<u>Yes, he is our respected mentor.</u>**

2008 – At Russ Prize Celebration and Banquet

It is indeed an honor to have this opportunity to tell you a little bit about my "respected" mentor and "cherished" friend, **Dr. Y.C. Fung**. But, in spite of my best efforts and intensions, I feel that I have not at all adequately conveyed the excellence of this gentle man. I ask that you give me 5 more years to work on my story. When we all come back to celebrate his 95th Birthday, I hope to do a much better job! Should I fail; Heck! There is always his 100th birthday.

2009 – With family at the Banquet for Dr. Fung's
90th Birthday

Thank you so very much for being so patient with me.

Savio L-Y. Woo

September 14, 2008

San Diego, California

Photographs courtesy of Mrs. Pattie Woo

Chapter 29

A TRIBUTE TO PROFESSOR YUAN-CHENG FUNG ON HIS 90TH BIRTHDAY

G. W. SCHMID-SCHÖNBEIN

Department of Bioengineering, Institute of Engineering in Medicine
University of California at San Diego, La Jolla, CA 92093-0412, U.S.A.

We celebrate the 90th Birthday of one of the giants of our time. The grand vision of Professor Y.C. Fung, often called the "Father of Biomechanics", has become a reality. Mechanics and Biology have started to fully embrace and a new generation of bioengineers is starting to emerge that has never known engineering other than being based on a merger of four basic sciences: biology, chemistry, physics and mathematics. We owe a great dept to Professor Fung, congratulate him for such a vision, and thank him for decades of inspiration and warm friendship.

The celebration of Professor Yuan Cheng Fung is also a celebration of engineering and science. YC Fung is a unique individual in our times. Among many who have contributed to the modern development of engineering, few have reached his level of impact. He initiated the development of a new academic engineering discipline: bioengineering.

1. Professor Fung and the Beginning of Bioengineering

YC Fung's leadership and contributions have been celebrated repeatedly (3). A summary of his life and his impact and his many recognitions was written by Professors Pin Tong and Satya Atluri in the Introduction to the Festschrift to his Chinese 90th Birthday (4). In 2008 we celebrated YC Fung's 90th birthday according to the Chinese calendar, in which a person's day of birth is counted as the 1st birthday, and so it seems appropriate for somebody who bridges Chinese and Western Culture to celebrate his Western 90th Birthday in 2009 as well.

Starting in aerospace engineering, he produced a long string of contributions that would define Biomechanics in a form and scope unseen before. He showed the way to apply systematic engineering thinking to a wide variety of medical problems. In the process he also developed new thinking and formulated new problems in mechanics. In his writings he turns medical phenomena into rational mechanics problems with crisp mathematical analysis. Reading YC Fung's manuscripts is like listening to a Mozart symphony - pieces of art in clarity, originality and freshness. He has given the engineering community a taste of the richness of biology and medicine.

Having shown the way for generations of biomechanics students how to analyze living tissues with rigorous engineering analysis, his greatest, although less tangible, contribution is the fact that he spearheaded the establishment of a modern-day bridge between medicine and the physical sciences. This effort has served to reverse a longstanding separation between engineering and medicine.

2. Bridging a Historical Divide Between Engineering and Medicine

To understand this contribution of YC Fung, one needs to look into the history of science. After World War II when bioengineering was started as an academic discipline engineering and medicine were separated and far apart in terminology and focus. Even at the times of Leonardo da Vinci physiology, medicine and engineering were close to each other and one field. One of the reasons for the separation was that in the 19th century medicine started to systematically catalog diseases and for example introduce Latin terms to describe diseased tissues, while engineers started to introduce rigorous mathematical tools, like tensors. Medicine in those days had hardly any use for tensors, and although there were engineers who worked on medical problems the thinking in the two disciplines was widely separated from each other. The differences between engineering and medicine were visible in their educational expectations, their thinking and in their language.

Even though many individuals over the past centuries made contributions to Bioengineering (see for example Chapter 1 in the book entitled *Biomechanics: Mechanical Properties of Living Tissues*, (1)), YC Fung had the vision to make Bioengineering and specifically Biomechanics a formal academic discipline. At UCSD, he and his cofounders of Bioengineering, Benjamin W. Zweifach and Marcos Intaglietta, introduced in 1967 a formal University curriculum for bioengineering undergraduate and graduate students with a full set of courses and degrees. A small new breed of engineers was being educated and working on living tissues using the language of engineering and medicine.

The very fact that in 1971, as an undergraduate student in physics and mathematics, it was considered to be quite special to study problems on living tissues, is an indicator for the gap between the two fields at that time. So with some uncertainty in mind, I decided to interrupt my physics studies at the University of Giessen, Germany, and explore "for a year" this new field of Bioengineering as a formal course degree spearheaded by a small group of pioneers at the University of California in San Diego. This is how I met Professor YC Fung. His personal charm and his determination to work on biological problems had captivated me. He had grasped a unique opportunity and decided to leave behind the security of the California Institute of Technology in Pasadena, and join a new University with a brand new Medical School at UCSD. Professor YC Fung is in many respects also an adventurer. In my case, out of the one year grew a lifetime with YC Fung for which I am ever grateful.

3. A New Thinking in Engineering and Medicine

The consequence of his type of initiative is that today the gap between engineering and medicine has largely disappeared. A new thinking has taken root. Today, medicine is *expecting* contributions from engineering, and new institutions educating Bioengineers have sprung up in many Universities. The old concept, that the Engineering School can be in one part of town and the Medical School separately in another part of town, is replaced by long-term plans to integrate the two disciplines, just like in da Vinci's times. Today the thinking is directed to make optimal use of engineering in medicine. Engineering is expected to help in many medical disciplines and with increasing demands, not only to build new devices but also to analyze, understand and even synthesize living tissues in health and disease. The engineering analysis of living tissues has become a sine-qua-non without which we will not be able to understand and prevent disease processes. Engineering in turn has recognized the many and intriguing opportunities that medicine and biology has to offer with countless disease processes to be worked on. Today hardly any biological or medical processes are not subject to a more and more rigorous bioengineering analysis.

4. Bioengineering in the New Century

New opportunities arise now all the time, from tissue engineering to detailed multiscale modeling of organs.

Engineering analysis of disease has become a major focus. There are already many diseases subject to detailed engineering analysis, several of them were started by YC Fung. But the list of important diseases is long and there are many hardly touched by engineering. Exciting and potentially important discoveries about the causes of diseases are possible. Any disease, including the most severe, benefits from some systematic engineering analysis. Engineering thinking can generate new understanding and innovative ideas for early intervention and prevention (2). Biomechanics is the tool of choice as the foundation of any living tissue analysis.

But there are other important opportunities. Biomechanics applied to genetic problems – a field we designated with Pin Tong as *Genomic Biomechanics* during the YC Fung 2008 Symposium – will one day serve to explain the shape of a living organism from its genome. Such an analysis requires mechanics applied to the DNA and protein synthesis, cell division, tissue growth and it will allow application of mechanics to the entire living world. It requires new paradigms that are specific for DNA and the enormous number of biological molecules that make up life. Genomic Biomechanics will one day predict the shape of a bacteria, a flower, a tissue or even a limb from its genome. May be it will help to predict one day the shape of living objects outside earth.

Future generations of engineers will look back and wonder how engineering could not always have been part of biology and medicine. YC Fung asked the same question and had this vision to assure engineering and medicine become as much integrated as biochemistry or statistics and medicine are.

5. A Happy Birthday

So we celebrate an individual who is a pioneer and giant in our times, selfless and generous, ingenious, precise and endlessly resourceful. Yuan Cheng Fung inspires the best among his friends and followers and I am most grateful to be given the opportunity to be one of them.

Congratulations, dear Bert, our warmest greetings to Luna, Brenda and Conrad and all of your close and distant family and our best wishes for the next decade. We love you, honor you, and follow in your footsteps.

Van Mow, Yuan Cheng Fung, Savio L.-Y. Woo and Geert W. Schmid-Schönbein (l. to r.) taken at The Future of Bioengineering and Biomedical Engineering Workshop in December 2008.

Left panel: Yuan Cheng Fung and two of his grandchildren Tony (l.) and Michael (r.) on the occasion of the Award Ceremony of the Russ Prize by the National Academy of Engineering in Washington DC on February 20, 2007. Right panel: Luna and Yuan Cheng Bert Fung on January 27, 2007, celebrating Luna's Birthday and Bert's Russ Prize in San Diego.

References

1. Y.C. Fung, *Biomechanics: Mechanical Properties of Living Tissues. 2nd Ed.* (Springer Verlag. New York, 1993).
2. G.W. Schmid-Schönbein, *Mol Cell Biomech* **5**, 83 (2008).
3. G.W. Schmid-Schönbein, S.L.-Y. Woo and B.W. Zweifach, eds., *Frontiers in Biomechanics* (Springer-Verlag. New York, 1986).
4. P. Tong P and S. N. Atluri, eds., *Molecular and Cellular Biomechanics. In Honor of the 90th Birthday of Professor Yuan Cheng Fung.* (Tech Science Press. Norcross, VA, 2008).

Chapter 30

TRIBUTE TO A WONDERFUL MAN

PETER HUNTER

Auckland Bioengineering Institute, University of Auckland, New Zealand

I first met Dr. Fung when he came to Oxford in 1975 to examine my D.Phil. (Ph.D.) thesis on the mechanics of the heart. I had been told by my supervisor, Derek Bergel, six months earlier that Dr. Fung was coming on this particular day and that I needed to have my thesis finished and submitted in time. In the UK the time allowed for a Ph.D. thesis is only three years so the pressure was on! I had gone to Oxford on a Commonwealth scholarship to undertake doctoral work on the heart with Derek Bergel in the University Laboratory of Physiology. We had begun our analysis of pressure-flow relations in the left ventricles of dog hearts using models based on electrical circuit analogs. I quickly came to realize that we were not going to learn very much about cardiac mechanics from these source impedance models and I decided that continuum mechanics modeling together with finite element solution methods provided a more promising approach. I discovered a couple of wonderful books to help me, one by Tinsley Oden called 'Finite Elements of Continua' and another by YC Fung called 'Foundations of Solid Mechanics'. That book was my introduction to the astonishing breadth and scholarship of this great man.

Two years later in 1977, while I was on a postdoctoral fellowship at Oxford, I visited Dr. Fung at UCSD to give a seminar in the newly formed Bioengineering Department. I have a number of wonderful memories from this visit. He showed me proudly around the UCSD campus, which was much smaller than it is today and I remember the astonishing number of people who greeted him so warmly as we walked around. He was clearly a celebrity on his own campus but had a characteristic humility and graciousness that made him many friends. He invited me to his house in La Jolla for dinner and then insisted on driving me to the downtown bus depot where I was due to catch a Greyhound bus to Los Angeles. I remember that he was not prepared to leave me until I was seated on the bus – concerned I think at the risk of a late night mugging in a less salubrious part of San Diego. It was astonishing to me to have a person of Dr. Fung's eminence prepared to give up a whole day to a humble kiwi student.

I returned to New Zealand at the end of 1978 to a lectureship in the Department of Theoretical and Applied Mechanics (now Engineering Science) and began a collaboration with the physiologist Bruce Smaill that still continues 30 years later. However, it was many years before I was able to travel out of New Zealand again (a consequence of heavy teaching duties and lack of travel funding) and I did not visit UCSD again until I went on sabbatical leave in 1986. I reestablished contact with Dr. Fung during this period and one of the outcomes was the offer of a short term teaching appointment (to teach courses that Dr. Fung had established) to a very talented and recently graduated young New Zealand student called Andrew McCulloch. Andrew had completed his Ph.D. on cardiac mechanics with Bruce Smaill and me in 1986 and he accepted Dr. Fung's offer to become an Assistant Professor of Bioengineering at UCSD in 1987. That soon became a long term career move for Andrew, following an NIH grant in 1988 and an NSF Presidential Young Investigator award in 1991, and he went on, under the mentorship of Dr. Fung and Dr. Shu Chien, to build one of the most productive cardiac mechanics research groups in the world - and to become Chairman of the UCSD Bioengineering Department from 2003 to 2008.

My major focus over the past 10 years has been helping to lead the Physiome Project, established by the International Union of Physiological Sciences (IUPS) in 1997 to bring bioengineering approaches and computational techniques to the understanding of multi-scale, structure-function relations in physiology. This attempt to put physiology onto a robust, quantitative, bioengineering foundation has benefitted from the vision and passion of highly quantitative physiologists like Denis Noble and Jim Bassingthwaighte, but it has also benefitted enormously from the bioengineering foundations in continuum mechanics established above all by Dr. Fung. I am very grateful to have had the privilege and benefit of a small involvement with this wonderful man – and may it continue for many years to come!

The author with Dr Fung at the workshop on graduate curricula for BME programmes organized by Dr Shankar Subramaniam at UCSD in December 2008.

Chapter 31

A RENAISSANCE MAN: DR. Y. C. FUNG

LILLY LI-RONG CHENG

School of Speech, Language, and Hearing Sciences, San Diego State University
San Diego, CA 92181-6060, U.S.A.

Professor Yuan-Cheng Fung is a leading scientist in the field of bioengineering. His devoted research in this field covers the span of over four decades. His vision is to understand the world and life; his mission is to make contributions in the advancement of science; his passion is the ongoing quest of knowledge and the discovery of truth. Norman Cousins once described Dr. Jonas Salk, the famous developer of the polio vaccine, "You represent the perfect marriage of science and art as it is and as it should be." Professor Fung is a member of the National Academy of Sciences, National Academy of Engineering and Institute of Medicine. He received the National Medal of Science in 2001.

1. Would you please tell us about your personal journey to this country?

My journey to this country is actually quite an interesting story. During World War II, a group of American professors went to China and met with Mr. Li-Fu Chen, the Minister of Education at the time, in Szechuan Province. They asked him why there were no Chinese students going to study in the U.S. Mr. Chen informed them that the country was at war, and there was economic hardship. Mr. Chen was delighted to facilitate the study of students abroad if financial assistance could be secured.

After this group returned to the US, they secured 40 scholarships and offered them to Chinese students. When the good news arrived, the government held a nationwide examination to select the best students to go abroad to study. I was among this lucky group.

I arrived at California Institute of Technology in 1946. Upon arrival I learned that this specific scholarship was offered two years before and was not available. Fortunately, Professor Sechler, my mentor, informed me that although I missed the first quarter I could audit the classes and take the examinations as well. He also offered me a job to working in his laboratory. The best part of this offer was that the results could be counted toward my degree requirements if I passed the tests. At the end of the first year, I passed most tests without having even registered as a formal student. I was truly one of the fortunate ones! I also had the opportunity to meet other Chinese students on campus including Dr. Hue Sheng Tsien.

2. When you first came to the US, you studied aerospace engineering. How did you end up with bioengineering? What motivated you to switch your field?

One of the incidental reasons leading to my switch from aeronautics to bioengineering was my mother's eye disease. In 1957, she had glaucoma. I was so worried but I was not able to go back to China to take care of her. I was not even allowed to send money back to her to visit doctors. Finally, I came up with the idea of clipping all medical briefs I could find about her disease and translating them into Chinese. She took them to a local doctor at my hometown. Fortunately, my mother's eye operation was successful. Many years later, when I went back to my hometown, this doctor even showed me his collection of material I sent back to him. From then on, I started to study biology and physiology myself. I started to realize the potential contribution of engineering to biology. That eventually became my major interest and set the agenda for my research.

3. Can you share the early days of biology research with us?

The most important person in biology in the western word was Aristotle. He did a great study of human anatomy. Since he was the teacher of Alexander the Great, he had the opportunities to go to wars with the King and studied the bodies of the wounded or dead soldiers. He studied the blood and heart. He produced detailed pictures about human anatomy. However, he never realized the connection between the heart and lungs.

Not until about a thousand and five hundred years later, did Harvey make the connection between the heart and the lungs. Indeed, Harvey realized that two ounces of blood being pumped out of the heart during every heartbeat had to go somewhere and concluded that the blood went to the lungs from the heart.

4. How about the early history if biology research in China?

Chinese had a much earlier and much better understanding of blood circulation than other countries in the early days. Ancient Chinese conducted clinical trials to understand how herbal medicine worked in the human body thousands of years ago. The legendary Chinese leader, Shen-Long-Shi, and his cabinet members tested thousands of herbs to investigate the efficacy of the Chinese medicine. Even though the documentation was extensive, the classical Chinese medical documentations rarely had good graphical description of these findings.

As early as Yellow-Emperor (ca 2697 B.C.), the Chinese Book-of-Internal-Medicine had already mentioned the interaction between the heart and lungs. That discovery was made several thousand years before the western world discovery.

During the Ming dynasty, Li Shizhen edited the classic, "Pen Ts'ao Kang Mu" (The Great Herbal Catalog) which had monumental impact on the development and research in Chinese herbal medicine. Before his days, medical documents were scattered with no systematic classifications and often inaccuracies. Li spent about forty years on his book that included 1,891 recipes of herbal medicines. This book was finished in 1593 and had been translated into several languages in Japan, the Netherlands, France, Germany, and Great Britain. Amazingly, the information he provided then was essentially all accurate.

5. In the early 1980's, you were trying to study the human pulse that is used in traditional Chinese medicine (ba mai). What did you learn?

Human pulse is a very complicated phenomenon. In China, the medical practitioners put three fingers around the wrist and try to diagnose the body's condition by feeling the pulse with the fingers. They believe that a matrix of six points around the wrist can provide a synopsis of the health conditions of individuals. People with different health conditions will produce different types of pulse movements. We tried to build a mechanical device to mimic the functions of the human fingers as a diagnostic tool and to collect data for analysis. However, we found that the device was not as flexible as human fingers. For example, human fingers can adjust to the proper positions when a patient's hand moves; however, it is not easy for the mechanical device to adjust to that kind of movement. The device requires patients to stay still for the entire duration of applications.

However, the most difficult part is that the pulse signal-noise ratio is too small to make any concrete conclusions in this kind of study. As in any clinical, patient variations are so great, even a sophisticated statistical model cannot filter out a significant pattern of signal to overcome the noise. That really makes the detailed study more difficult.

6. How about today?

There are a good number of Chinese researches in Asia working on Chinese medicine, including the method of diagnosis. Actually, there was a recent conference in Asian discussing the development of Chinese medicine and method of diagnosis. They hope to create some synergy by using the classical and more modern knowledge of Chinese medicine. Interestingly enough, the experts concluded that the book edited by Li Shizhen is amazingly useful and accurate. It really is a great treasure in Chinese medical research. However, because of the competition of funding and resources, the effort to sustain this line of research in America is somewhat difficult.

7. Looking back 10 or 20 years, could you share with us some of the great milestones in bioengineering that greatly enhanced human lives?

There are a lot of great developments through the effort of bioengineering to benefit human beings. Almost all the inventions of medical devices are through the joint effort of bioengineering and medicine. Examples include the heart-lung machine which enables patients to maintain the proper function of the body while they are undergoing major medical procedures, the artificial heart which enables patients with heart problems to continue their lives, the equipment and material to perform orthopedic surgeries, etc. There are really quite a lot if we think about it carefully.

By the understanding of human anatomy and the special expertise of the engineering thinking, we can create something useful to enhance the quality of human life. For example, by studying the human blood cells, we discovered that the red blood cells were shaped like donuts. With this shape, the blood cells has zero inner pressure that makes it very easy and very flexible to travel anywhere in the body and, therefore, optimize its functions. This finding can potentially be used to develop other devices for other devices for other purposes which are not necessarily limited to medicine or biology. In addition, we also discovered that the shape of the white blood cells changed when the animal died. Up to today, this shape change still serves as the most clear cut boundary of life and death. If one can come up with a procedure to slow down or reverse this change, human life can potentially be prolonged. However, that may also delay someone from going to the heaven (laughs).

8.   How about the future research in bioengineering?

It has been fifty years since people discovered the structure of the DNA molecule. The more recent biological research has very much been directed toward the function of DNA. Genomics research is the hot topic and is also the primary direction to go in these days. The success in this area will someday enable people to design individualized medicine with better treatment efficacy. At the University of California, San Diego, we have very strong research programs in this area. In addition, we also have a booming biotechnology industry in Southern California that will only make the research in this area moving even faster and more productive.

However, the competition of research is extremely fierce. The sharing of results and the openness of discussions have recently been greatly reduced. Part of the reason is the government policy. Ever since the government permitted researchers to patent their findings in this area of science, researchers started to safeguard their research results with increased secrecy in hopes to reap the profits of their work someday. This is not helpful to the advancement of scientific research.

There are still lots of interesting research opportunities in this area. I strongly encourage people to pour more effort in the research of this area though cross-disciplinary collaborations. When different people from different training backgrounds look at the same problem from different perspectives, the combination of their diverse expertise is the best way to move research ahead and make the most advances.

9.   You are truly a pioneer, gentlemen, and scholar. We are so privileged to talk to you. Could you share your personal philosophy with us?

In the classical Chinese book of Zhuang-Tse, there was a story about a butcher and how he dissected the cows. He was so proficient at his job and so skillful of his performance; whatever he did became an enjoyment for him. This is the Chinese philosophy of living a long life. I believe in it. In everything we do, we should go with the natural flow and not to force the issue. We need to be passionate about what we do and what we will learn the enjoyment of our craft.

Chapter 32

TRIBUTE TO A MOST RESPECTED TEACHER

RUIJUAN XIU

Institute of Microcirculation, Chinese Academy of Medical Sciences
Peking Union Medical College, Beijing, P.R. China

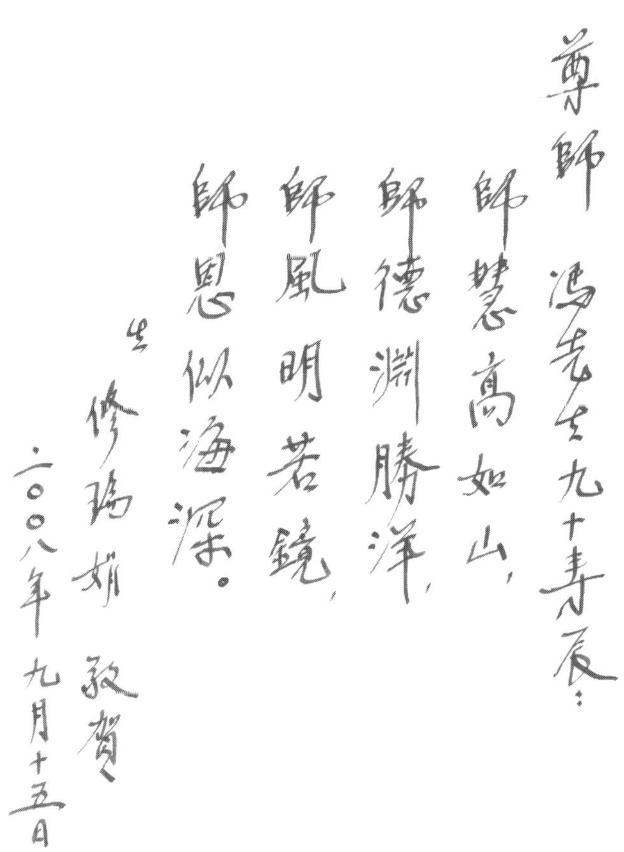

Opening Ceremony - Fung Auditorium at the Institute of Microcirculation in Beijing, 2002

Chapter 33

Y. C. "BERT" FUNG: A MASTER

GHASSAN S. KASSAB

Department of Biomedical Engineering, Surgery, and Cellular and Integrative Physiology, IUPUI, Indianapolis, IN, U.S.A.

It is with great pleasure that I reflect on Prof. YC Fung's 90th Birthday. I was very fortunate to be one of his last students. I recall the journey vividly and especially my first encounter. I was a junior undergraduate majoring in Chemical Engineering when I noticed a possible technical elective called "Biomechanics". I knew little of the subject but only that "Bio" was life and "Mechanics" was a branch of physics that dealt with motion and forces that caused them. This course was taught by YC Fung, part of a curriculum he had established at UCSD. I recall he came to class empty handed but before the lecture was over, the board was filled with well proportioned conceptual drawings (he is an artist) of function of organs (he is a physiologist) with equations that described the concepts (he is a mathematician) and many deep insights that revealed the essence of the problem (he is a poet). But actually he is a bioengineer; one of the many terms he coined in the field he seeded and nurtured (Kassab, 2004).

As the lectures went on, I was struck by his insights and perspective. He painted a story with large brush strokes, not pausing for details but rather focused on visions and paths, on essence and realism, on form and function. I knew then that he was a Master that I would follow. The next year and a half was painful as I had to finish a major I had committed to before attending graduate school in bioengineering to train with Prof. Fung. This was a life changing decision.

The journey with a Master was extraordinary. I watched YC create new theories, envision new experiments, conquer hurdles and all with grace and humility. It was a joy watching him think. He approached every discussion as an opportunity for discovery. As we worked on new and difficult projects, he was full of enthusiasm and optimism and could typically predict the outcome. He always held the course despite obstacles and we all benefited from the fruits of his convictions. He taught me that many small successes add up to a big success and to remain well rooted. He taught me to focus and to set high standards. Most of all, he taught me to love my trade.

The term "Master" has special significance to me as I have trained in Eastern martial arts for many years. I trained with a great Sensei and I have learned from many others. My Sensei always reminded me that "things cannot be taught, they must be caught". A true Master helps you "catch" things (physical or intellectual concepts, visions, morals, lessons, ethics and so on). YC Fung is brilliant at energizing others to catch and to learn.

A dear friend recently shared some thoughts from CS Lewis's Mere Christianity which are particularly relevant:

"Already the 'new men' are dotted here and there all over the earth. Some, as I have admitted, are still hardly recognizable: but others can be recognized. Every now and then one meets them. Their voices and faces are different from ours: stronger, quieter, happier, more radiant. They begin where most of us leave off. They are, I say, recognizable, but you must know what to look for. They will not be very like the idea of 'religious people' which you have formed from your general reading. They do not draw attention to themselves. You tend to think you are being kind to them when they are really being kind to you. They love you more than other men do, but they need you less. (We must get over wanting to be needed: in some goodish people, specially women, that is the hardest of all temptations to resist.). They will usually seem to have a lot of time: you will wonder where it comes from. When you have recognized one of them, you will recognize the next one much more easily. And I strongly suspect (but how should I know?) that they recognize one another immediately and infallibly, across every barrier of color, sex, class, age, and even creeds...."

This description embodies YC Fung so well and I must say that it was easy to recognize him as such. His enormous energy, his loud and wholehearted laughter, his insightful thoughts, his wise words and his joyful presence are very hard to miss.

On this very special occasion, I wish Prof. Fung a wonderful 90th birthday and to thank him for the remarkable legacy.

References

1. G. S. Kassab, *Mechanics and Chemistry of Biosystems* **1(1)**, 5 (2004).

Chapter 34

TRIBUTE TO A FRIEND AND COLLEAGUE

JOHN WATSON

Department of Bioengineering, University of California at San Diego
La Jolla, CA 92093-0412, U.S.A.

Dear Bert,

Congratulations on your ninetieth birthday. This special occasion is a propitious opportunity to share with you the effect your career had on the National Institute of Health's (NIH) view of the value of engineers in biomedical research and how your achievements have advanced the field of bioengineering nationally.

When I arrived at the NIH in 1977, bioengineering was not well understood. It was difficult for NIH Institute Directors to accept that bioengineers contributed to basic understandings of medical and biology. NIH Directors categorized bioengineers as producing widgets and supporting the research of biologists and physicians. Years of my visits with NIH Directors, presenting the basic research results of bioengineers around the world, lead to a very interesting common comment. "Yes, this is a very exciting basic mechanism or basic research result but bioengineers perform applied research and development."

Your selection for the National Medal of Science in 2000 was a turning point for Bioengineering. After your award, I later visited more than a dozen NIH Directors. The recognition associated with the Medal of Science award convinced them that bioengineers were capable of equal contributions to basic medical and biological research as any other investigators. This was a critical turning point in the growth of bioengineering in the NIH extramural programs. This was the same year that NIH established the National Institute of Biomedical Imaging and Bioengineering (NIBIB). The Congressional action and the NIH response to form the NIBIB were decidedly based on the accomplishments of bioengineering rather than biomedical imaging. Although together, they formed a more compelling research capacity for the Congress to consider passing and for the President to sign the NIBIB act.

So enjoy your ninetieth year and the knowledge that your brilliance in using engineering principles in biological research has led to a new era of opportunities for young men and women bioengineers to pursue careers in medical and biological research.

All the best from your friend and colleague,

John Watson

Chapter 35

A TRIBUTE TO DR. YUAN-CHENG B. FUNG

JASON X.-J. YUAN
AYAKO MAKINO

Department of Medicine, University of California at San Diego
La Jolla, CA 92093-0725, U.S.A.

Jason Yuan, MD, PhD, Professor and Vice Chair for Research, and Ayako Makino, PhD, Assistant Professor, are currently in the Department of Medicine, University of California, San Diego (UCSD). Dr. Makino worked in the Department of Bioengineering at UCSD in 2002-2006. Dr. Yuan's research centers on pulmonary vascular pathobiology, one of the research interests of Dr. Fung's during the last 40 years. Dr. Makino studies mechanisms of coronary vascular dysfunction in diabetes.

When I finally decided to move from the University of Maryland at Baltimore to the University of California, San Diego (UCSD) in 1998, I wrote a letter to Dr. Fung, a pioneer in studies on pressure, flow, stress and remodeling of the pulmonary vasculature. I did not expect to receive any response, but got a warm welcome letter from Dr. Fung within a week. Right after my arrival at UCSD in 1999, I had an opportunity to have lunch with Dr. Fung in the Faculty Club during which we talked about science, life, and the future direction in the research field of pulmonary vascular pathophysiology. Although I have never worked directly with Dr. Fung, I feel, since that lunch, I have learned a lot from him, probably even more than many of his graduate students and postdoctoral fellows. His smile, his patience, his, sometimes, loud laughter, his positive attitude about life, his passion about science and research, his knowledge about history and culture, his generosity, his willingness to spend time with those who need his guidance …My wife and I have asked ourselves many times, how come an eminent scientist and internationally recognized pioneer can still be so nice, approachable, and sensible. Dr. Fung is not only a role model for us in research and science, but also a guide for us to have an enjoyable and meaningful life.

I met my wife, Ayako Makino, in an international meeting on smooth muscle cells in Nagoya, Japan. Our relationship rose to a different level when she was a postdoctoral fellow in the Medical College of Wisconsin at Milwaukee. Because of me, she moved to work with Dr. Geert Schmid-Schonbein in the Department of Bioengineering at UCSD. So I had more opportunities to interact and meet with Dr. Fung and other professors (e.g., Drs. Shu Chien, Geert Schmid-Schonbein and Wei Huang). Dr. Fung communicated a paper in *PNAS* for us in which we showed that upregulation of TRPC6, a calcium channel, in pulmonary artery smooth muscle cells may contribute to the development of idiopathic pulmonary arterial hypertension [1]. Dr. Fung and I also wrote a review article with our research fellows for *Microvascular Research* [2] in which we tried to generate an overview on pathogenic mechanisms of pulmonary vascular disease based on different perspectives (morphometry and structure, mechanics and function, rheology and physiology, molecules and modeling...). The review turned out to be well received by the peers in the field. Without Dr. Fung's help, this paper would have just been an ordinary article buried in the long list of millions...that nobody would read. After the review article, we had so many plans to study, e.g., pulmonary venous abnormalities in patients with pulmonary vascular disease, we drafted several grant proposals to study vascular stress in patients with idiopathic and thromboembolic pulmonary hypertension, and we designed a series of experiments to study pathogenic role of stem cells in pulmonary hypertension... I really miss the time when I went to Dr. Fung's office talking with him about the plans, the future, the new concept (e.g. genomic biomechanics in pulmonary circulation), and what we should do next day. I want to go back to Dr. Fung's office, like those unforgettable days and nights, to talk with him about the projects that link (bio)engineering and mechanics to medicine and pathophysiology; I want Dr. Fung to show me where in his books I can find answers that guide my experimental designs; I want Dr. Fung to tell us stories about aeroelasticity (or the dynamic instability of airplanes and birds) and teach me the formula used to study dynamic remodeling of pulmonary blood vessels...and I want to learn more from him!

Ayako and I will never forget the day of our wedding ceremony; not only because it was the day we got married, but also because Dr. and Mrs. Fung joined us celebrating the most important day of ours. We sincerely appreciate

Dr. Yuan-Cheng Fung (right) and his wife, Luna, at his 90th birthday celebration party.

Dr. Fung's support and encouragement for our career, we are among many whose lives and career have changed dramatically through his generosity. We enjoy every moment being with him and Luna, and we wish we will have more parties to celebrate his scientific achievements and his life.

Dr. Fung, you are not only a great scientist and engineer, but also an admirable mentor, a nice father, a role model, and a close friend for us.

References

1. Y. Yu, I. Fantozzi, C.V. Remillard, J.W. Landsberg, N. Kunichika, O. Platoshyn, D.D. Tigno, P.A. Thistlethwaite, L.J. Rubin, and J.X.-J. Yuan, *Proc. Natl. Acad. Sci. USA* **101**, 13861-13866 (2004).
2. M. Mandegar, Y.-C.B. Fung, W. Huang, C.V. Remillard, L.J. Rubin, and J.X.-J. Yuan, *Microvasc. Res.* **68**, 75-103 (2004).

     J. X.-J. Yuan & A. Makino

Drs. Jason Yuan and Ayako Makino with Dr. & Mrs. Yuan-Cheng and Luna Fung and Drs. Shu and K.C. Chien at Dr. Fung's 90th birthday celebration party on September 14, 2008.

Chapter 36

THREE DEGREES OF SEPARATION

DARRYL D'LIMA

Shiley Center for Orthopaedic Research and Education at Scripps Clinic
La Jolla, CA 92037, U.S.A.

John Guare's play, "Six Degrees of Separation" refers to the connectivity of the human web on planet Earth. The theory states that if our opinions and influences reach every individual we know, then reach every individual they know, and so on, six times, our influence would reach everyone on the planet. However, in the biomechanical world, thousands of students, professors, engineers, and scientists around the globe are separated by a mere three degrees or less from the pioneering work of Prof. YC Fung. The three degrees can simply be stated as: those who used Prof. Fung's research and findings in their work, those who have been trained by Dr. Fung's students, and those who have had the privilege to study directly under him: at the center of biomechanics stands Prof. Fung himself. I am among the lucky few who can claim to have touched the center of the biomechanical world and it has left a lasting impression on my work and my life.

I was trained as an orthopaedic surgeon in India and in 1994 completed a research fellowship in hip and knee joint arthroplasty at Scripps Clinic under Dr. Clifford Colwell. Dr. Colwell, who inspired me with his passion for research and education, recruited me to manage research at his Joint Mechanics Lab, which changed my career path forever. My work in the lab was my introduction to what follows.

Around 10:00 one morning in 1995, I received a call from Peter Chen. A specimen that we were testing in long-term fatigue had failed and he wanted to start testing the next specimen. We were testing the shear strength of a bone–cement–implant interface. I was the humble "cement-mixer" who vacuum-mixed the bone cement (as a surgeon would). I remember this clearly, because it was Sunday morning and I was getting ready for a date, which was a rare event in the life of a budding research scientist. Peter insisted I have to fulfill my commitment.

This was my first connection to Prof. Fung's work, through three of his associates. Peter Chen and John Pinto were Prof. Fung's former students and Gene Mead was his developmental engineer. In their spare time, they collaborated with Dr. Colwell on biomechanical orthopedic projects. John was the theoretician; Peter, the experimentalist; and Gene Mead, the engineer and the "unofficial" graduate. Gene Mead built the equipment that allowed Peter to test John's theories. Together, the three could conquer almost any biomechanical challenge. I was completely impressed with their creativity and ingenuity. As I worked alongside Peter, John, and Gene, I learned how Prof. Fung directly influenced their expertise, dedication, and work ethic.

I learned most of my laboratory skills from Peter. Some of his fundamental habits have become deeply ingrained in my practice. Whenever I walk into a laboratory, I subconsciously look around for the barometer, the temperature and the humidity gauges. I got an education just watching Peter plan and run experiments. Peter never made me feel ignorant for asking "dumb" questions.

Prof. Fung had a special relationship with Gene despite his lack of formal training. I remember watching Gene build a device for biaxial testing of lung tissue. At that time such a machine did not exist. I remember trying to analyze his problem-solving skills, to figure out what led to his flashes of insight. Gene's lack of formal training undoubtedly enabled him to solve problems in a manner similar to what Edward de Bono calls lateral thinking.

John Pinto was a gifted theoretician and teacher. An innocent question would generate an entire lecture on the topic in question. John was never satisfied until he felt I truly understood the problem, all the possible solutions to it, and the reasons why his solution was superior. Together, these three men, all students of Prof. Fung, instilled in me a deep passion for biomechanics. But, my progress in biomechanical research was indirect. I initially learned experimental biomechanics and then tried to understand the theory. I ultimately received my formal PhD under Professor Shu Chien's direction. Prof. Chien exposed me to mechanobiology: How do living cells and tissue respond to mechanical forces? Professor Pin Tong, also a student of Prof. Fung, was one of my advisors and supported me in my biomechanical research. I now begin to see the fascinating interplay between the two disciplines.

My research interests have since broadened to include cell biology, molecular biology, and genetics. However, biomechanics remains closest to my heart. I use biomechanics in many scales: to analyze the contact mechanics of hip, knee, and shoulder artificial joints, to predict injury and repair in cartilage tissue, to study cellular response to mechanical stimuli, and to unravel the behavior of cytoskeletal proteins. I am constantly surprised by the fractal nature of the biomechanical fundamentals that change very little; a tribute to the solid foundation laid by Prof. Fung's pioneering work.

Prof. Fung continues to generate groundbreaking theories. My first experiments with Peter took place in 1995. During this time, I met Prof. Fung at several social events and it was years before I formally interacted with Prof. Fung on a biomechanical problem. Nearing 90 years of age, he is still active in research and is the principal investigator of a National Science Foundation grant to study the tensorial representations of intracellular cytoskeletal proteins to explain the biomechanical remodeling in vascular tissue. Peter Chen and I are now working on applying this novel approach to explain cytoskeletal remodeling in chondrocytes. With Prof. Fung's guidance, we propose to derive the constitutive equation to describe this process.

In chondrocytes, the native structure of the cytoskeleton is 3-dimensional (3D) and is exquisitely sensitive to mechanical forces. Chondrocytes cultured in monolayer display characteristic actin stress-fibers (Fig 1). On the other hand, chondrocytes in situ and in 3D culture (e.g. in hydrogels) do not display any stress-fibers. Actin forms a network below the cell membrane with abundant dynamic filamentous projections. We know less about the vimentin cytoskeleton in three dimensions. Vimentin in chondrocytes in 3D culture forms a fairly dense network that is very resistant to cell compression (Fig 2). Application of Prof. Fung's novel constitutive model of intracellular protein tensors will shed valuable insight into the mechanical behavior of cytoskeletal polymer proteins: in situ and in response to biomechanical and biological stimuli.

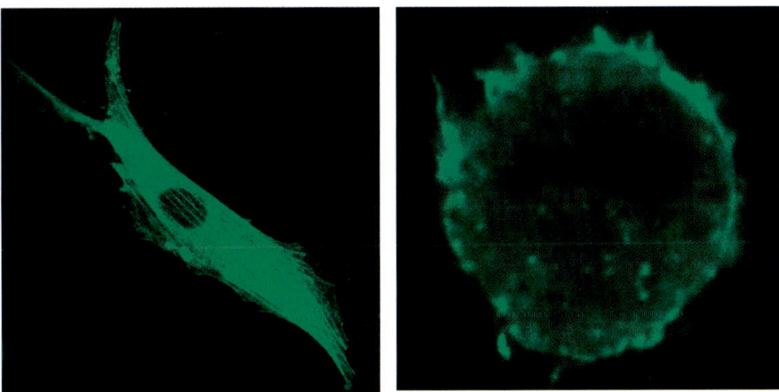

Figure 1. Live cell confocal laser microscopy image of a chondrocyte in monolayer culture (left) displaying prominent stress-fibers of GFP-tagged actin. Live cell confocal laser microscopy image of a chondrocyte in 3-dimensional alginate culture (right) displaying the typical cortical network and diffuse cytoplasmic distribution without any stress fibers.

　　　　　　　　　　　D. D'Lima

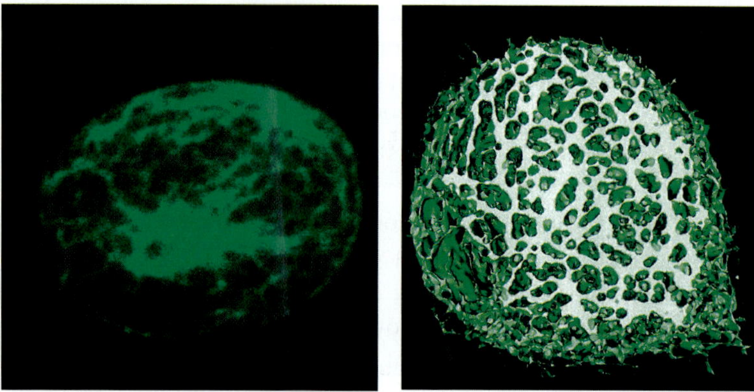

Figure 2. Live cell confocal laser microscopy image section of a chondrocyte in 3-dimensional alginate culture (left) displaying GFP-tagged vimentin network. 3-dimensional reconstruction of a stack of confocal images (right) displaying the fairly dense interconnected network of vimentin.

Prof. Fung's constitutive models for soft tissues have been implemented in mainstream commercial software. Although this would hardly impress him, at least one finite element software package includes a Fung soft-tissue material (Abaqus, Simulia, Providence, RI). The entire scientific community now has ready access to some of Prof. Fung's pioneering work. The major clinical relevance is that Prof. Fung's groundbreaking models are now used to design, develop, test, and optimize a wide spectrum of biomedical devices.

As I reflect on the accomplishments of Prof. Fung and think about John Guare's play, "Six degrees of Separation", I'm amazed at the influence one man has had on so many. To think of the thousands of students, professors, engineers, and scientists around the world involved in biomechanics and how they have been influenced by Prof Fung, all connected by only 'three degrees of separation', from his pioneering work. How one man has reached so many is truly remarkable. I consider myself among the lucky few who can claim to have touched the center of it all: Prof YC Fung.

Chapter 37

AN UNINTERRUPTED DIARY

CONRAD FUNG

Brookfield, Wisconsin, U.S.A.

BRENDA FUNG

Belmont, Massachusetts, U.S.A.

To understand our father's inner harmony and how he could achieve so much with such apparent ease, we cite one of his favorite sayings, "Easy to Do, Hard to Know," as a unifying theme for his life as a model of Choice coupled with Commitment.

1. An Uninterrupted Diary

There is no database more comprehensive or more effective than our father's diary, a personal record begun in school three quarters of a century ago and updated daily ever since, on consecutively numbered pages. The intervening days number over twenty-seven thousand. "What dedication," we once wondered aloud in his presence, to which he replied, "I decided to do it, and I did it."

Still in China, the diary came to be written in English. The early pages chronicled not only daily events but also plans for future learning, including books to be read – all in fastidiously neat penmanship.

Through a long career, feelings were recorded, as were guest lists of innumerable parties. Recently at home, our parents came across a photo of a forgotten event, on which the only objective data was the time stamp from the camera on which the photo was taken. A quick consultation of the diary for that date brought back the event, decades prior, as well as the participants and the topic of conversation.

An uninterrupted diary is a metaphor for commitment, a symbol of self-awareness of a life worth remembering. To keep the diary was one among many choices made long ago, and its continuation is just one example of our father's unwavering dedication in many aspects of his life.

2. To Understand Harmony

In contemplating how we could possibly add to a volume already filled with tributes from a world of friends and colleagues, we were humbled to realize how much everyone else already knows about our father. But that's the point—his has been a lifetime of generous sharing: of time and knowledge, of friendship and love. Clearly, giving his time freely to others did not slow his productivity; nor did his fierce pace of work and high expectations undermine his good humor and evident calm. His trademark laugh has been part of every interaction.

A dinnertime conversation from our younger years sheds light on how these apparently divergent behaviors can be in harmony in one person. To our moans about our travails of the time, he responded with his version of an old Chinese saying: "Easy to Do, Hard to Know" … that once a fundamental choice has been made, everything else follows. "Easy" turns out to be relative, of course, but the point was that the choice must come first.

One example of a fundamental choice was his decision to stop consulting when we moved to La Jolla in 1966. Replying to requests for proprietary work became easy: the answer was no; and much time was saved for work that could be given away. We overheard a phone conversation in which he declined to take on a private project, saying "Everything I know is already public."

From career to career, he has moved from strength to strength. And together with cumulative achievement has come ever-increasing generosity and gentleness. In the language of statistics, these are declining hazard rates; success predicts more success, as long life predicts longer life.

An inseparable part of his being has been his partner since youth, our mother, Luna Fung, the gentlest soul one could ever meet, and a person of few, but essential, words. In every birthday toast someone has said that when she speaks, you'd better listen. We like the image of our father traveling by ship, westbound from Calcutta, and of our mother also traveling by ship, eastbound from Shanghai, finally to meet again in the U.S. in a hug around the world.

Upon moving to La Jolla she gave up her career of teaching college math, but found new rewards in nurturing the International Center at UCSD, tutoring international visitors in English, and hosting innumerable parties – a fundamental role in both our parents' social happiness in the academic world.

Y.C. and Luna Fung in Pasadena, California, in 1949.

Season's Greetings from the Fungs, 1954.

We add this humble tribute to those of many others and will not repeat our father's long catalog of accomplishments. We hope we have recalled to the reader our father's infectious laugh, our mother's gentle support, their eternal optimism, and how they both have unjealously treasured their teachers, friends, and colleagues.

We also hope we have brought currency to that old saying, "Easy to Do, Hard to Know," by which one can understand how big choices (once made) can clear the mind. Our father's uninterrupted diary is just one example of Choice coupled with Commitment in his long, harmonious life.

Happy Birthday, Daddy!

Photo taken at the International Symposium on Genomic Biomechanics: Frontier of the 21st Century, September 14, 2008